Selected Popular Writings of E.U. Condon

Edward Uhler Condon (1902–1974)

Asim O. Barut   Halis Odabasi
Alwyn van der Merwe
Editors

Selected Popular Writings of E.U. Condon

Springer-Verlag
New York   Berlin   Heidelberg   London
Paris   Tokyo   Hong Kong   Barcelona

Asim O. Barut
Department of Physics
University of Colorado
Boulder, CO 80309-0390
USA

Halis Odabasi
Department of Physics
University of Colorado
Boulder, CO 80309-0390
USA

Alwyn van der Merwe
Department of Physics
University of Denver
Denver, CO 80208
USA

Library of Congress Catalogue-in-Publication Data
Condon, Edward Uhler, 1902–1974
    [Selections. 1991]
    E.U. Condon: selected popular writings/Asim O. Barut, Halis
Odabasi, Alwyn van der Merwe, editors.
        p. cm.
    ISBN-13:978-1-4612-7783-5

    1. Physics. I.Barut, A.O. (Asim Orhan), 1926–
II. Odabaso, Halis. III. Van der Merwe, Alwyn. IV. Title.
QC71.C652 1991   90-45176
530—dc20   CIP

Printed on acid-free paper.

© 1991 Springer-Verlag New York Inc.
Softcover reprint of the hardcover 1st edition 1991

Typeset by Best-Set, Chai Wan, Hong Kong.

9 8 7 6 5 4 3 2 1

ISBN-13:978-1-4612-7783-5   e-ISBN-13:978-1-4612-3066-3
DOI: 10.1007/978-1-4612-3066-3

Preface

"The middle third of the twentieth century was the era of hegemony of physics in American science. During that whole period Edward Uhler Condon was a leader in physics, in research of his own, in stimulating research in others, in applying physics, and in calling attention to the effect on all of us of its indiscriminate and irrational application. When he made his first contribution to theoretical physics in 1926, the word physics was not in the vocabulary of most Americans, and the revolutionary concepts of quantum mechanics and relativity were just being worked out in Europe; by 1960 the applications of electronics and solid state physics had begun to change our lives irreversibly, and the implications of nuclear physics were manifest to everyone. Ed Condon contributed to each part of this explosive evolution." — Philip M. Morse.[1]

We are happy to present the Selected Papers of Edward Uhler Condon, one of the most influential American physicists of this century, whose life has touched many fields and many people including our own. This volume presents selected articles he wrote not reporting on his current research. We have included several obituaries and other essays that report in detail on Condon's contributions and importance to physics. We also include "Reminiscences of a Life in and out of Quantum Mechanics," Condon's charming autobiographical essay. He presented it at the International Symposium on Applications of Quantum Mechanics to Atomic, Molecular, and Solid State Theory, and Quantum Biology, which was held in his honor on Sanibel Island, Florida, during the week of January 22–27, 1973.

Since his death, several obituaries of Condon appeared,[1–3] and the Fourth International Conference on Atomic Physics was dedicated to his memory. The Proceedings of this conference[4] contains contributions by L.M. Branscomb, I.I. Rabi, and B.R. Judd about Condon's accomplishments as a scientist and as a deeply concerned humanist. Another book,[5] edited as a tribute to Condon on the occasion of his becoming Professor Emeritus in the Department of Physics and Astrophysics in the University of Colorado in the summer of 1970, also has pertinent and detailed information in its preface and in its foreword written by Frederick Seitz.

Therefore, rather than repeating the details, we refer the interested reader to these sources, some of which are included in the present volume.

We would like to take this opportunity to acknowledge a grant from the Corning Glass Foundation towards the publication of this book. Thanks are due to Mr. Stephen Catlett, manuscript librarian at the American Philosophical Society Library, where the papers of Professor E.U. Condon are preserved, for his friendly help. We would also like to thank the Condon family and others for permission to edit the papers included in this book.

References

P.M. Morse, *Rev. Mod. Phys.* **47**, 1 (1975).

L.M. Branscomb, *Phys. Today* **27**, 68 (1974).

C. Eisenhart, *Dimensions* (the Technical News Bulletin of the National Bureau of Standards) **58**, 151 (1974).

Atomic Physics, Vol. 4, G.Z. Putlitz, E.W. Weber, and A. Winnacker, eds. (Plenum, New York, 1975).

Topics in Modern Physics — A Tribute to Edward U. Condon, W.E. Brittin and H. Odabasi, eds. (Colorado Associated University Press, Boulder, 1971).

A.O. Barut
Halis Odabasi
Alwyn van der Merwe

Contents

Contents

Edward Uhler Condon, A Personal Recollection

LEWIS M. BRANSCOMB

IBM Corporation
Armonk, New York

Edward Uhler Condon was one of the young American physicists who, in 1926, enjoyed a Fellowship at Göttingen and Munich, in time to participate in the revolution of theoretical physics arising from the new quantum mechanics. Son of a railroad engineer in the western United States, sometimes newspaper man in San Francisco, Condon's scientific career combined the abstract with the practical, theory and observation in a unique way. The most fruitful period of his work in theoretical physics occurred between 1926 and 1936, encompassing collaborations with many of the most distinguished scientists of that period. Others in today's symposium will discuss these contributions in more detail. I would like to recall his impact on science after he began to combine the academic and administrative life, including major positions in industry (Westinghouse and Corning Glass) as well as in government (Directorship of the National Bureau of Standards).

When Condon went to Pittsburgh as Associate Director of Research for the Westinghouse Electric Corporation in 1937, he set about strengthening the basic research and preparing Westinghouse for the capability to move into the nuclear field. His success in attracting highly competent young scientists to an industrial laboratory, while focusing the total research effort in a direction of enormous future economic importance, set a precedent that has since been followed by industrial research managers in many enlightened companies. It was not long, of course, when World War II caused his effort to focus in microwave technology in support of the radar work at the M.I.T. Radiation Laboratory. He also became involved in the nuclear weapons program, serving as Associate Director of the Los Alamos Scientific Laboratory, and later headed the Theoretical Physics Division of the Berkeley Radiation Laboratory.

In this connection, Condon became a close associate of the substantial British group, including Sir Mark Oliphant and Sir Harrie Massie. The research they did in electromagnetic separation of isotopes resulted in major advances in the science of mass spectroscopy. This was later reflected not only in Westinghouse products but also in the major contributions to atomic physics that resulted from mass spectrometry after the war.

Of all the American scientists whose concern about the use of atomic weapons at Hiroshima and Nagasaki led them to devote themselves to the cause of peace, none was more dedicated nor more effective than Condon. His efforts as Science Adviser to the Select Committee on Atomic Energy of the United States Senate after the war had much to do with the decision by the Congress to place government responsibility for atomic energy in a civilian rather than a military agency. The price he paid for his relentless pursuit of this goal is another story, taking us away from his scientific contributions.

His responsibilities as adviser to the Senate were concurrent with his tenure as Director of the National Bureau of Standards. Just as he had done at Westinghouse, he brought the Bureau of Standards in a very few years into the world of modern physics and chemistry. All of the Bureau's contributions in atomic collision physics, quantum chemistry and low temperature physics stem from his enthusiasm and the people he attracted to NBS. The program in atomic spectra, which predated his arrival, received the kind of theoretical underpinnings it required. But Condon did not restrict his interests at NBS to the areas of modern physics he knew best.

Probably his most important contribution was his recognition in the late 1940's that the field of building technology was probably the most important neglected area of technology in the country. Condon created the Building Research Division and set it on the path that now enables NBS to claim for its Center on Building Technology a leading position among such laboratories in the world.

Condon also anticipated by about 20 years the need for government to put research to work in behalf of the consumer. Indeed, he moved so effectively in this direction that his successor reaped a whirlwind over battery additive ADX2 after the elections of 1953.

Condon also created the Central Radio Propagation Laboratory of the NBS, locating it in Boulder, Colorado. The ionospheric research of this laboratory became a very fruitful test bed for the applications of atomic physics during the 1950's and 1960's. In two respects the ionosphere has been critical to the development of atomic and molecular physics. First, it provides a laboratory for ion chemistry in which reactions continue to completion without the influence of walls. Secondly, the upper regions of the atmosphere are a rich showcase of phenomena based on non-equilibrium energy distributions for both photons and electrons. Although one could hardly call the structure of the atmosphere simple, more progress has been made in basic understanding of this environment than in the glow discharge, even though one captures the latter in one's laboratory.

The Central Radio Propagation Laboratory subsequently evolved into two institutions, the Office of Telecommunications of the Department of

Commerce and the Research Laboratories for the National Oceanic and Atmospheric Administration.

In 1951, Condon left the Bureau of Standards to became Director of Research and Development in the Corning Glass Works. There he was to apply his knowledge of solid state theory to the many glass and ceramic projects of Corning, to develop materials for use in the heat shields of spacecraft, and to help Corning move into the emerging field of electronics.

In 1956, he returned to the university world as Chairman of the Department of Physics at Washington University in St. Louis. Moving in 1963 to Boulder as Fellow of the Joint Institute for Laboratory Astrophysics and Professor of Physics, he once again returned to his original love, the theory of atomic spectra. Together with Hallis Odabasi, a graduate student who subsequently worked with Condon as post-doctoral research associate, Condon began the task of a new and up-to-date version of his treatise with George Shortley on the *Theory of Atomic Spectra*. A major portion of this work was complete before his recent death. If present hopes to bring this work into publishable form by 1976 are borne out, it would be a fitting tribute to the 50th anniversary of the Ph.D. degrees and first publication of Edward Uhler Condon, a scientist whose life has touched many fields and many people.

E.U. Condon — The Physicist and the Individual

I.I. RABI

Transcript of a talk
Columbia University
New York, New York

Having heard Dr. Branscomb's talk, I think we will turn the program around, as he gave Condon's importance for physics, and I will talk more about Condon the individual, and the times in which he lived, and something about the scientific scene in the United States at that time and some of his influence on that and through that on the world.

First let me say that I find it very astonishing that I should be here in Heidelberg at this big conference talking to you in English. When I first came to visit Germany in 1927 such a thought would have been ridiculous, incomprehensible, unbelievable. In the first place Germany and the German-speaking world were the center of physics at that time. If you wanted to learn physics, you learned German, and very few Germans knew any English — it was a very esoteric kind of language. And in particular they did not know many Americans and there was no need to know many Americans at that time.

Fifty years ago, about 1924, physics seemed to have come to an end in frustration. There had been no great experimental discovery for a few years and the quantum theory in the form of the correspondence principle of Bohr and his students had after its magnificent career come to the end of the line. The two-electron spectrum presented a clear paradox and obstructed future progress along the lines of the classical correspondence principle. It is true that towards the end of 1924 de Broglie had published his brilliant paper, but with all due apologies very few physicists in Germany or in the English-speaking world read *Annales de Physique*. Practically nobody read his paper except for provincial graduate students like myself in far-away New York. The reason I read that paper was that I read everything because we were so far away from the center.

The final blow of this period came in 1925. If you look up the *Zeitschrift für Physik* there is a long paper, not very long but 8 or 10 pages, by Bohr. This paper did not contain a single equation. It spoke very broadly of things like "Unmechanischer Zwang" and things of this sort — all very mysterious, as Bohr knew how to be when he wanted to be mysterious. From that point on one seemed to take off on wings into metaphysics. I remember listening to a paper by a distinguished American scientist who had a theory that the mass of the electron was in one place, its charge in another, and its name somewhere else. This period of frustration fortunately lasted a very short time.

The end of 1925 and the year 1926 saw the second great revolution of ideas in this century: the birth and the formulation of quantum mechanics. The first one, and perhaps as great or greater a revolution, had occured in 1905: the work of a single individual — Albert Einstein. Though an enormous development of the Bohr theory and the correspondence principle took place in the intervening years and many people had worked on it, it had come to an end. Now in 1926 a whole new world opened.

So I take this occasion of the Condon Memorial Session to make a few personal remarks which may help to communicate some of the quality of the time when Condon and I and quantum mechanics were young. Sometimes it seems that events conspire to thwart one's desires and hopes, and at other times it is happily just the other way around. In my own case, after many false steps I became a graduate student of physics at Columbia University in the period from 1923 to 1927. I call this the Golden Age of de Broglie, Schrödinger, Dirac, Heisenberg, and Pauli. In this period physics emerged from inspired groping in darkest shadows into the new universe of clarity.

Condon was also one of the fortunate students to come into the world of physics at that time. One could talk endlessly about these wonderful few years when young physicists walked around with a shining light in their eyes.

In the mid-1920's every American physics student who could get fellowship support came to Germany or to German-speaking Europe, the land of Einstein, Bohr and the young Turks like James Franck, Otto Stern, Sommerfeld, Debye, Ehrenfest, Born, and a host of others. I was one of those and Condon was too. He came a bit earlier, in 1926. I arrived in the summer of 1927.

I had just completed my Ph.D. at Columbia and set off for Europe on a minute stipend, first to Zürich with Schrödinger, Debye, Scherrer, Linus Pauling. It was Stratton who recommended a pension in München to be with Sommerfeld. It was the Pension Elvira, which unfortunately had disappeared from the face of the earth. I think of this with tremendous nostalgia because one could have room and board and all meals for six Marks a day, just exactly my size. I came to this pension, and when I came down to breakfast in the morning there was one of the most German looking men I have ever seen. He had the German hair-do of that time: "ein Millimeter". Anyway, there was this man, Condon, reading the *American Mercury*, which was a literary critical magazine of that time. We struck up a warm friendship immediately.

Now a few words about Condon before he came to Europe. As mentioned by Branscomb, he had has a varied career, although very young. He started to be an experimental physicist and he went to the stockroom and got a lot of glassware. On the way up to his room he stumbled and all the glass fell down and broke. He decided experimental physics was not for him. And thus began a career in theoretical physics which was in many ways unique. He attended the lectures by Birge there at the University of California. Birge mentioned some rules for molecular transistions which interested and impressed Condon very much. He went home and quantified them, and came back and showed the paper and it looked very good. He had the original idea of submitting it as a dissertation (it was the Franck-Condon-Principle) for his doctor's degree although he had only been a graduate student for a few months. Of course, there was tremendous opposition; "you can't do a dissertation in a weekend". But that was what he did. But there were people there at Berkeley, like Loeb, who said, "Why can't you do a dissertation in a weekend, or a day for that matter. It's the quality of the dissertation that counts, and this should be accepted." Being very aggressive he won; so Condon got his doctor's degree in one year and got a fellowship, a pretty good fellowship, a National Research Council fellowship, and he came to Göttingen and then München. There he found Oppenheimer and H.P. Robertson, and others. Actually we came there as if on a pilgrimage. Most of the people of my age that I knew in physics I first met in Germany, like Condon, Robertson, Loomis, Stratton, Oppenheimer and so on. So there he came with his Ph.D. and his Franck-Condon principle.

In the year 1926 he went to Sommerfeld in München. This is where I met him. There were Condon and Robertson and I, and we had a wonderful time there. I recall that we went to a fair in München. As we went along, the smell of beer in the air became thicker and thicker. There was a brass band guzzling the beer in at one end and blowing it out through the trumpets at the other end. The late professor Robertson, the famous relativist, and Condon got together and played the "Peas, Porridge Hot . . ." game and pretty soon we had a circle of people around us watching, and I passed the hat and got a few Pfennigs. Altogether we had a wonderful time. Although we were not German students, we lived a little bit of the life, to a point where somehow or other, we never succeeded in finishing a meal in a restaurant before we were thrown out, especially with Robertson, who was rather boisterous. We were rather sensitive about the whole thing because German is a difficult language. It was not so difficult for me because I had studied chemistry first and even undergraduates in chemistry at that time had to read German because the real literature was in German. We had to look up things in "Beilstein" and so on. At one time, Condon, being very adventurous, bought a Lederhosen outfit. So there he was very beautiful in this thing going along with the short pants and the straps. And he explained to the maid that he would not have much use for it back home in America. But she said, "Doch, doch, abends zum Spazieren!"

Jessie Dumond had written Sommerfeld about visiting him. Sommerfeld had never seen him, neither had we. Sommerfeld asked us to meet Dumond at the railroad station and Condon went in this outfit. I went along because his German was poor and he was afraid that if people talked to him and he answered in his language, with this outfit on, he would be lynched!

We came along and we recognized Dumond immediately because it was easy to recognize an American then and I suppose it is now too, because of his luggage and his shirt. Well, Dumond had married a French wife, and when she looked at Condon she was very turned off. We got together that evening but she was too "sick" to join us. It took longer for the first world war to end than the second world war, it took a very long time. In 1927 I was in Zürich, and there was the first meeting where French and German physicists came together on an informal basis.

Now just a little bit more about those days. I would like to tell about it because it may give a lot of encouragement to people in less developed countries. We felt that we came from a less developed country scientifically in 1927. For example, the *Physical Review* was taken by the libaray in Göttingen but was made accessible only at the end of the year. You can see it was not a very exciting journal even though I published my dissertation in it. And we felt this very keenly. Here was the United States, a vast and rich country but on a rather less than modest level in its contribution to physics, at least per capita. And we resolved that we would change the situation. And I think we did. By 1937 the *Physical Review* was a leading journal in the world.

Condon contributed greatly to this success because of his early journalistic career. This included writing books or having books written and, particularly, editing, which he did brilliantly through the following years. As told already by Dr. Branscomb, he had had an upbringing where his family went from place to place in the United States. So he never had any deep roots in any one place. He was able to make friends, to pick up friends and acquaintances very easily.

When he left Germany he went to Bell Telephone Laboratories. Later he recalled to me that on the way home to London he overheard an early transatlantic telephone conversation and he was struck by the extraordinary range difference between the high level of technology that developed the transatlantic telephone connection and the trivialities that go over it.

Condon left Bell Telephone Laboratories and became professor of physics at Princeton and then at the University of Minnesota for a short time, then came back to Princeton. At Princeton at that time, both the Physics Department and the Institute were at a very high level mathematically. There he gave these wonderful courses and papers on relativity. One day Condon announced an elementary course on relativity. All of these high powered people came to his lecture to find out what relativity was really about. He had this very remarkable gift of making complicated things simple and clear.

He then went to Westinghouse. He was of an age, of course, where he knew all the good people in the field who were young. He had a great capability for finding talented students and interesting them. At Westinghouse he brought a group of brilliant young people together. Of course it was not so difficult then. This was in the 1930's and a job was simply not to be had. The general idea that one gets a doctor's degree in physics and immediately moves to a job was not the case then; there were no jobs. I know when I got my degree at Columbia I did not expect a job to go with it. But I was interested in physics and I did want to do it anyhow. Under those circumstances he could pick a very good group of people, many from the then advanced subject of nuclear physics. At that time the young nuclear physicists were just as arrogant as the young people now in high energy physics. Dr. Stepian, a brilliant electrical engineer at Westinghouse, came around to converse with them about some subjects other than nuclear physics, only to be rebuffed. He said he couldn't understand why a man should be so proud of knowing so little physics. This group was very successful and some of the leaders of the Westinghouse Company came from this group.

When the war came and we all became engaged in some war activity, Condon was always called on to take a very important position. We were together at the Radiation Laboratory in Cambridge, Massachusetts in the development of radar. This very successful laboratory was organized under Lee DuBridge. It began with about 30 people in 1940 and ended with about 4000 in 1945. A very unique experience, having a laboratory of this size and spending this enormous amount of money, having the very vital connections with the army and navy and with the British and with people of other countries. Well, nobody in the government group of the laboratory had ever had any administrative experience of that sort. When it came to making the budget for the following year we always did very well but we were afraid to give away the secret. We simply plotted an exponential curve. That always did it, because if you have more and more physicists you have more and more ideas and this goes exponentially. It's as simple as that. Condon was a great help to us because of his connections with the Westinghouse Company.

Later on, Los Alamos was organized. It was natural to bring in Condon and he wrote the first primer on how to build an atomic bomb. It was a good thing to have in those days a book about how to build an atomic bomb; it explained things to the young people who came in. He didn't stay very long in Los Alamos because he and the General Groves were orthogonal. He went back to Westinghouse. This battle with Groves was to haunt him one way or another for the rest of his life.

Wherever the action was thick, there was Condon. He performed a very great job as a consultant to a U.S. Senate special committee. At the end of the war there was this big question of what to do with atomic energy. The law which was proposed was the May-Johnson bill, to put the whole thing under the control of the military. And a small group of scientists outside the government got into action to oppose the bill in the most formidable manner to the surprise of all politicians. They organized speeches, they set up an office in Washington, they distributed all kinds of documents, and in spite of the strong backing of the military and the government itself for the May-Johnson bill, this group won out. A new bill put the control over nuclear energy under a civilian administration. Simultaneously a small group of scientists was formed to advise the government about nuclear energy.

I would like to turn to Condon's later career, his unfortunate difficulties with the United States government. He was a man of independent mind, and could not be pushed around by people in the government. He suffered very greatly from this. There were times when he could not get a passport to go abroad. Anybody who could get a look at Condon or listen to one sentence of Condon's speech would have had no doubts about his basic and essential Americanism. But he was attacked and although he had a very natural and cheerful manner, this troubled his life in many respects. These battles took a great fraction of his time and were a drain on his emotions for many, many years. At the same time he did not stop in his various efforts in teaching and in writing. He was a great idealist, an idealist who could laught at himself with some bitterness.

When we came together we talked over the old times in Germany when we were young. The gaiety and spirit of fun with which we approached physics at that time remained with us mostly for the rest of our lives. Physics was great fun, a wonderful thing. But beyond that it did for us, and I think for that group, my age group, something that I sense is somewhat less important nowadays. I don't believe any of us took physics in a sense of a profession, it was not a trade, it was a way of life, it was a form of interest, it was an orientation, a philosophical orientation, not articulate but with a certain feeling that there was something exciting about this whole quest of mankind to extend its understanding, to make the universe more understandable, to clarify man's place in the universe, to develop tools to extend his power, like the radiotelescope, to help mankind in understanding itself. I think this was a basic attitude and assumption at that time. Perhaps the great success of physics and the large sums of money that were going into it have to some degree — I hope not too much — turned physicists into professionals rather than what used to be called "natural philosophers".

Condon got his Ph.D. in a year and took the rest of his life in completing his education and thus remained young and fresh in spirit and in intellect until the very end.

Foreword to *Topics in Modern Physics,* A Tribute to Edward U. Condon

FREDERICK SEITZ

President
The Rockefeller University
New York, New York

During the academic year of 1929–30, which was punctuated by a dramatic descent in the stock market, I was a sophomore at Standford University and decided to do my bit to reverse negative trends in society by becoming a professional physicist. While still enjoying the feeling of euphoria brought on by this decision, I read in the university newspaper that the visiting professor in theoretical physics for the summer quarter would be a brilliant young man, twenty-eight years old, who had discovered the Franck-Condon principle while a graduate student at Berkeley, had spent two years at the great centers of theoretical physics in Europe as a National Research Council fellow, and had held prominent posts at the Bell Telephone Laboratories, Princeton University, and the University of Minnesota. Just a year earlier, he and Ronald Gurney had given an interpretation of spontaneous alpha disintegration of nuclei in terms of quantum mechanical tunneling. To top it all, the campus paper related that he had earned his way through Berkeley as one of the more worldly reporters for the *Oakland Tribune.* In this pursuit he had, among other things, stirred up a lively public discussion of whether a birdcage would weigh more or less when the bird was flying around inside instead of resting on its perch.

The visitor, Edward Condon, was slated to give a course in modern physics which would be open to duly qualified undergraduates. I succeeded in persuading an indulgent father to provide the means to attend the summer session and, early in July, found myself perched on a chair in the front row of the lecture room waiting for the show to start. It was not a disappointment.

Precocious and crew-cut, Ed Condon exhibited even then all of the characteristics that have carried him through a lifetime near the center of the stage. He was creative, energetic, perceptive, humorous, restless, eloquent, worldly and friendly. Moreover, he knew, on a first-name basis, most of the top-billed physicists on the planet and loved to spin endless anecdotes about them. This was very rich fare for an undergraduate. Condon's lectures, which followed a notable tradition that included such physicists as Karl K. Darrow, J.H. Van Vleck, and Floyd K. Richtmyer, were then as now a wonderful combination of logic, anecdotes and humor.

He has often said that the ideal physicist-lecturer should have the ability to "make hard things look easy and easy things look hard." From the start, his greatest attribute has been the ability to achieve the first of these goals. He loves directness and clarity far too much to indulge in obfuscation for its own sake.

Perhaps I should add that Condon also gave a more advanced course that summer, which my friends and I observed with great interset and curiosity from a safe distance. Among the participants were William W. Hansen and the Varian brothers, Russell and Sigurd. Although I eventually came to know the three quite well, they seemed at the time to be a pretty fearsome group. Actually they turned out to be among the most generous, friendly and gentle individuals I have ever known, but that is another story.

In those days, long before physicists were taken very seriously by the public at large and when they still were all but unknown to congressmen and security officers, Condon was flamboyantly cheerful practically all of time, his occasional bursts of wrath being directed at the petty annoyances of everyday life which plague us all. His bouts with various prominent individuals — particularly with General Leslie Groves — lay far in the future. Not least, he made a very colorful addition to the Stanford quad that summer as he strode sturdily about in conservatively styled golf knickers.

Condon was so deeply interested in other people that he quickly came to know personally everyone in the class who managed to act reasonably alive. The small band of embryonic physicists who dominated the front row in the lecture hall became his close friends. A number of incidents that occurred during the lectures stand out vividly in my mind. Perhaps the most dramatic followed a weekend visit Condon made to Berkeley early in July. This was his first visit since leaving for Europe in 1926. He not only wanted to catch up with campus life there, but also to make peace with the counselors of his student years, Raymond Birge and Leonard Loeb, for having chosen to remain "back East" after his sojourn in Europe rather than returning to Berkeley. On the Monday morning following this visit he strode into class with a stop-the-press air and proceeded to describe an all but incredible device his old friend Ernest Lawrence was working on with the hope of accelerating charged particles to very high energies. All of us were soon busy calculating cyclotron frequencies for varying circumstances.

With Condon's ardent help, continued family indulgence, and some permissiveness on the part of the Princeton admissions committee, I followed him back to Princeton a year and a half later as a graduate student. This was a new and much more formal world than the one I had known in the West. K.T. Compton had just left Princeton to become president of the Massachusetts Institute of Technology, and the physics department was groping its way to a new equilibrium. Condon was as much at the center of stimulation as ever, and with his brilliant student, George Shortley, whom he had met in Minnesota, was just starting to write what was to be his greatest scientific treatise, *The Theory of Atomic Spectra*. However, his lectures that spring were centered on Frenkel's book about the classical electromagnetic theory of light, which he embellished in countless ways. I still cherish a carbon copy of his notes, which the departmental secretary (now Mrs. L. Bonner) typed.

Condon's western outlook emerged when he came to purchase a home. It was more or less traditional in Princeton at that time for a young faculty member on the rise to rent a house or apartment in one of the more acceptable districts until he was prepared to build a home of distinction with the aid of a well recognized architect. The Condons preferred a home providing the maximum volume at the lowest cost to take care of their growing family and many guests. As a result they purchased a large two-story frame house in what was then a less fashionable part of town north of Harrison Street. That house was a Mecca for an endless stream of students and visitors. The dinner table was always well loaded with food and drink, and the chairs with guests. Anyone with a good story to tell or in need of companionship or sympathy was a welcome guest of Emilie and Ed. Without the warm hospitality of the Condons, the winters at Princeton would have seemed far more bleak for a student from California. It would be difficult to count the number of times I enjoyed their companionship, food, fireplace, and guests.

In the fall of 1932 the Department of Physics appointed me as Condon's research assistant. However, he and Shortley were so deeply involved in their book that Condon arranged for me to work with Wigner as the latter's first American student, with the sole provision that I spend a few hours a week checking over equations in the text that was to become "Condon and Shortley."

During these years we frequently drove to scientific meetings together, often through bitter weather. I recall one all-night drive that started on Christmas afternoon and took us to a meeting of the American Physical Society in Pittsburgh. The weather was so severe that we decided as we approached Pittsburgh in our open roadster that we had finally achieved Viking status. On another occasion late at night, as we drove back through a thick storm from a colloquium at Columbia University, I noted that Condon had an unusually worried look on his face. In an attempt to relieve the situation, I said as brightly as I could, "Cheer up, we are doing great." He replied, "O.K., but I should let you know that we are riding on thin tires and do not have a spare aboard." This was at a time when the mean free path between flat tires was of the order of one thousand miles.

As the depression years ground on, Princeton, like most other universities, found itself in a fairly static position as far as posts and promotions were concerned. The greatest shifts in the part of the university with which I was concerned resulted from the creation of the Institute for

Advanced Study, to which a few individuals, such as John von Neumann and Oswald Veblen, were transferred.

By 1937 Condon was ready for a new adventure and accepted a post as associate director of research at Westinghouse Research Laboratories in Pittsburgh. Once again he became the center of a vast and lively community of stimulating individuals. He and Emilie purchased an even larger and roomier house close to Wilkinsburg. This home seemed to be undergoing continual growth and renovation and was always bursting with interesting, if occasionally unconventional, visitors. Condon not only brought into closer communication the alert and promising young scientists and engineers already employed at Westinghouse, but soon added many new faces through a system of appoinments of the postdoctoral research type. Under his leadership the laboratory quickly grew to the state where it could become a significant factor in the research that was to become necessary in World War II. In fact, the first time I saw klystrons and related microwave gear in routine use was during a visit to the laboratory in 1939.

I had joined the General Electric Research Laboratory as a summer visitor in 1935 and 1936, and became a full-time member in 1937. Condon visited Schenectady several times during this period and I arranged to have him visit the laboratory. It was interesting to note that the laboratory administration relaxed for him some of the restricting formalities which we usually observed with visitors from competing organizations. Condon always focused on issues and ideas of such importance to industry and management as a whole that proprietary themes became secondary in the resulting discussions that evolved.

In 1942 the position of chairman of the department of physics at the Carnegie Institute of Technology became open and Condon proposed that I be considered. The administration acted on his recommendation, and by the end of that year we were close neighbors again. Later in the winter of 1942–43 we made a train journey to California together on the City of San Francisco in order to follow up more or less parallel interests related to war research. Robert Oppenheimer was riding on the same train. He was then in the process of starting up the Los Alamos Laboratory and had been on a recruiting expedition. Before the trip ended he had signed up Condon to be one of his right-hand men, on leave from Westinghouse. This assignment turned out to be short-lived, however. Condon soon decided that, from the standpoint of cost effectiveness, Los Alamos was an impractical place for a great scientific laboratory. His extended feud with General Leslie Groves, which grew out of this period, needs no embellishment here. Suffice it to say that the episode is representative of the way in which matters can go when great men clash. Subsequently I came to know Groves fairly well and found that I could admire him as one of the remarkable engineering administrators of our time. It is part of Condon's own greatness that he has been willing to acknowledge Groves' capabilities.

As the program of the Manhattan District developed, Westinghouse, in cooperation with Ernest Lawrence, became involved in the electromagnetic isotope separation process. Condon took up residence in Berkeley for a period of time (August 1943 to February 1945) with the thought that he might furnish additional leadership and guidance to this work. However, Lawrence proved able to provide his own all-encompassing form of direction, and Condon rejoined the Pittsburgh community, where he remained through the rest of the war.

As the conflict drew to an end and plans began to be drawn for an organization that could take over from the Manhattan District in the time of peace we hoped would follow, Condon became deeply convinced that his organization should be entirely civilian in nature and began to direct his great energies to this issue. The goal was two-fold: first, to attempt to give the leaders in Congress some concept of the physical features of nuclear weapons and their strategic implications, and second, to support the drafting and passage of an appropriate bill. Ultimately the struggle narrowed to the so-called May-Johnson bill, introduced in the House by Representative Andrew May, and the McMahon bill introduced by the young senator from Connecticut who, as chairman of the joint House-Senate Committee on Atomic Energy, came to play a critical role in the postwar development of nuclear energy. Condon spared no energy in helping to defeat the May-Johnson bill and in supporting the McMahon bill. In the course of this struggle he became for a time a close ally of Leo Szilard who had the same objectives and pursued them in his own unique way. The two men often met with Henry Wallace, then the Secretary of Commerce. Wallace succeeded in convincing Condon that he had a more permanent role to play in Washington as director of the National Bureau of Standards — a post which he assumed in the fall of 1945.

Condon brought to Washington not only the fervent desire to make the National Bureau of Standards one of the greatest scientific laboratories in the world, which he might well have succeeded in going under proper circumstances, but also a desire to further international understanding and world peace by capitalizing on the continuance of wartime alliances. He approached these objectives with characteristic breadth and zeal and was soon giving skillfully prepared lectures on science to congressmen. On one occasion he even dared to challenge them with a final examination. He had an old mansion-like house on the grounds of the Bureau converted into an official residence and soon made it a visitors' Mecca, in keeping with traditions I remembered so well from Princeton and Pittsburgh. It was a time in which the scientists became amateur politicians on a massive scale and were looked upon as interesting allies by the professionals in Washington.

Once the Soviet Union had rejected the proposals contained in the Lilienthal-Acheson report, and once it became clear that the liberated nations of Eastern Europe were to become Soviet satellites, the climate in Washington began to change. For reasons too complicated to elaborate here, it became increasingly difficult for Condon to achieve the goals that he had established for the National Bureau of Standards in spite of the fact that he made truly remarkable progress in part by obtaining project support from various federal agencies. One is reminded somewhat of the difficulties faced by John Wesley Powell sixty-odd years earlier as he tried to advance the course of the geological survey while opposing the establishment of the types of "water companies" that eventually made it possible for cities like Los Angeles to transport water from large distances.

As one who has spent seven continuous years in Washington, I realize now that the head of any agency will eventually face such enormous difficulties in attaining the goals he establishes for his agency that he must, axiomatically, abjure the pursuit of other politically oriented goals. A few individuals have violated this principle and succeeded, but they are very rare.

In 1951 Condon left the Bureau to become director of research of the Corning Glass Works. When his detractors, with whom he had traded blow for blow as best he could, threatened the company with punitive action for using his services, he felt obliged to relinquish the directorship and seek an academic post. It is a sad comment on the temper of the time that those who opposed him did their best to prevent any university from employing him. In 1956, however, he accepted an appointment as professor of physics at Washington University in St. Louis as a result of a wise action by the great scientist-statesman Arthur Compton. Compton not only wanted to add an outstanding man to the staff, but also realized that the nation, as well as Condon, would be the loser in the long run if the irrational chain of events which had been set in progress were allowed to continue unabated. It is interesting to note that Condon gave very great stimulus to Corning Glass Works during his brief period there and has been retained as a major consulant by the company since then.

In 1963 Condon, who never really got the western blood drained from his veins, in spite of the altercations in which so much of it was spilled, joined the Joint Institute for Laboratory Astrophysics of the University of Colorado and achieved a position of longitude less than one degree east of the site of Alamogordo, New Mexico, where both he (1902) and the first atomic bomb (1945) were born. He is nominally retiring from this post as this volume goes to press. We shall see just what such retirement consists of. Unfortunately I have not yet visited Boulder to see the type of house in which he and Emilie live. His recent studies of unidentified flying objects show that his energy and sensitivity remain completely undiminished since that summer forty years ago, even though he may have a broader range of issues at which to direct his ire. To me, at least, the introductory chapter of the report* on UFO's, in which Condon describes with characteristic clarity his own view as a scientist on what constitutes worthwhile research, is a classic. It deserves to be a landmark in the journey science has taken since the days of Stevin, Galileo, and Kepler. Thank you, boss.

---

* *Scientific Study of Unidentified Flying Objects*. E.P. Dutton & Co. (hardback), Bantam Books (paperback) New York, 1969.

Preface to *Topics in Modern Physics,* A Tribute to Edward U. Condon

WESLEY E. BRITTIN and HALIS ODABASI

The papers in this volume were prepared as a tribute to Prof. E.U. Condon by his associates on the occasion of his becoming professor emeritus in the Department of Physics and Astrophysics in the University of Colorado in the summer of 1970. He has also been given the title of fellow emeritus by his colleagues in the Joint Institute of Laboratory Astrophysics of the National Bureau of Standards and the University of Colorado.

The range of topics included is, itself, a reflection of the diverse subjects to which Condon gave his attention during a long and varied career in teaching, research and public affairs. His scientific papers are listed in the bibliography. In addition, there could be listed an almost equal number of prepared papers and addresses dealing with public issues, especially on the subject of avoidance of war and the promotion of international understanding between peoples and governments. One of these (p. 349) is included in this volume.

Edward Uhler Condon was born in 1902 in Alamogordo, New Mexico. His father, William Edward Condon, was an early western railroad builder, and in consequence the son grew up "nearly everywhere west of Denver" as his father took on different railroad location and construction jobs. He attended high school in Oakland, California, was graduated in 1918 and became a San Francisco bay area newspaper reporter in the summer of 1918, continuing to work part-time in this way after he entered the University of California in Berkeley as a freshman in 1921. He entered the university in the fall of 1919, intending to become an "educated newspaperman" (newspaper reporters with college degrees were rare in the West in those days), but dropped out in a few weeks after finding that the courses in social science subjects held no interest for him. When he entered again in 1921 it was as a freshman in the College of Chemistry, but the next year he changed to the College of Letters and Science, and in 1924 received the A.B. with highest honors in physics.

Two years later he received the Ph.D. at Berkeley for a thesis (7) based on an extension of some work by James Franck, then of Göttingen, which later became widely known and recognized in physics and chemistry as the Franck-Condon principle.

The world of theoretical physics had undergone a major evolution in
1925 and 1926 with the discovery of the main principles of quantum
mechanics by Werner Heisenberg and Max Born in Göttingen, by Louis de
Broglie in Paris, and by Erwin Schrödinger in Zürich. With the help of an
International Education Board fellowship, Condon and his wife (he
married Emilie Honzik in 1922) and infant daughter, Marie, went from
Berkeley to Göttingen in the fall of 1926 so that he could study the new
subject. After a semester in Göttingen, he spent a second semester in
Munich working under Professor Arnold Sommerfeld.

The rate of discovery of new and difficult ideas in theoretical physics was
extraordinarily rapid in those years and Condon became discouraged. His
attention was attracted by an advertisement for a Ph.D. physicist with
newspaper experience to do public relations work for a major industrial
laboratory. He returned to New York in the fall of 1927 to a post in the
public relations department of Bell Telephone Laboratories. There,
however, he spent most of his time with the research physicists, where the
big new interest was the recent discovery of electron diffraction by C.J.
Davisson and L.H. Germer. He found that physicists in America were, in
the main, having great difficulty assimilating the new theoretical ideas. He
was much in demand as a colloquium speaker at Eastern universities, and
as a result he was encouraged to continue in university research. He took
his first academic post in the spring of 1928 as a lecturer in physics at
Columbia University, giving graduate courses in quantum mechanics and
in the electromagnetic theory of light. That spring he was offered assistant
professorships by six universities, and chose to go to Princeton University
to join a vigorous and energetic group under the leadership of Prof. Karl
Compton, who later became president of Massachusetts Institute of
Technology.

Condon's first year in Princeton was especially fruitful. He wrote the
paper which gave the wave-mechanical basis of the Franck-Condon
principle (20); with H.D. Smyth he gave an interpretation of the
continuous spectrum of molecular hydrogen (22); with P.M. Morse he
wrote the first English language book on quantum mechanics (25); and
with Ronald Gurney he wrote the papers (19, 23) on the barrier leakage
interpretation of radioactivity by alpha particle emission. This latter
discovery was made simultaneously with, and independently of, the work
of George Gamow, who was then in Göttingen. Many years later he and
Gamow became colleagues and close neighbors at the University of
Colorado.

In the summer of 1929 he was one of the five lecturers at the summer
institute for theoretical physics at the University of Michigan. In the fall of
1929 he became full professor of theoretical physics at the University of
Minnesota, although only twenty-seven years old. He was invited there as
successor to Prof. John H. Van Vleck who had just moved to the
University of Wisconsin. During that year he suggested (26) the work

which led to the discovery, by Walker Bleakney and others, of a mode of dissociation of molecules under electron impact in which the fragments acquire a considerable amount of kinetic energy, now regarded as an important topic in chemical physics. After a year at Minneapolis, Condon returned to Princeton, as associate professor of physics, where he remained until 1937.

In the summer of 1930, he gave two courses as a visiting professor at Stanford University. In his sophomore course on atomic physics he had as a student Frederick Seitz, who later became president of the National Academy of Sciences and is now president of Rockefeller University. Seitz presents some charming reminiscences of that summer in his contribution to this volume. In his graduate course on quantum mechanics he had W.W. Hansen and Russell Varian as students. They later achieved fame as inventors of the klystron, a type of velocity-modulated electron tube based on excitation of microwave oscillations in a cavity resonator.

Back at Princeton, Condon turned his attention principally to the theory of atomic spectra. The basic work of John Slater had opened the way for theoretical treatment of the subject. Condon's interest in it was greatly stimulated and helped by the presence, in Princeton, of Henry Norris Russell, Allen G. Shenstone, and Charlotte E. Moore, who were actively engaged in experimental work in this field. In the year at Minnesota he had arranged for a young electrical engineering student, George Shortley, to come to Princeton for graduate work. Working with Condon, Shortley soon began producing research papers on the subject (30, 32), culminating in their preparation and publication, in 1935, of the book *The Theory of Atomic Spectra* (44) which is still widely cited as the classic reference in this field.

At Princeton, Condon supervised three doctoral theses, those of Frederick Seitz, George Shortley, and Edwin McMillan, now director of Lawrence Radiation Laboratory of the University of California. He recalls how the young McMillan came to him as a graduate student and said "I would like to work with you if you will promise to leave me alone." Condon promised and kept his word and McMillan has turned out well nevertheless!

After the book on atomic spectra was completed, Condon turned his attention to two distinct directions: molecular physics and nuclear physics. In the former field he continued to work on problems involving mass spectroscopy (53); he took an interest in infra-red spectroscopy (37, 52), then being developed in Princeton by R. Bowling Barnes (now president of the Barnes Engineering Company of Stamford, Connecticut), and in the theory of optical rotatory power, working with Henry Eyring and William Altar in the chemistry department (57, 58). In nuclear physics he worked primarily with Gregory Breit, who spent the year 1936–37 at Princeton. The principal outcome of that work was the paper of Breit, Condon and Present (51) on the interpretation of proton-proton scattering experiments done by Tuve, Heydenberg and Hafstad at Carnegie Institution in Washington. This work first clearly recognized the charge independence of the strong interaction between nucleons, a result that lies at the foundation of all modern work in nuclear physics. A paper with Benedict Cassen (50) was among the first to draw attention to the extra significance gained by isospin formalism because of this property of strong interactions.

While at Princeton Condon took a strong interest in undergraduate teaching. In four of the years he gave a special introductory course for the top twenty or so of those enrolled in beginning physics. Several of those who took it have gone on to distinguished careers in science, and of those who chose other careers Condon says "most of them admit to being no worse off for having taken that course."

In the summer of 1937, Condon was offered the post of associate director of research at Westinghouse Electric Corporation. He accepted and moved to Pittsburgh with his family, which now included two sons. (Paul, born in 1932, is associate professor of physics at the University of California at Irvine, and Joseph, born in 1935, is a research physicist at Bell Telephone Laboratories.) Westinghouse wanted to strengthen its work in fundamental physics which had been curtailed during the depression years, and assured Condon of liberal support and a free hand in developing such work at the laboratories in East Pittsburgh.

Westinghouse had already started the construction of a large pressurized Van de Graaf machine for nuclear work. This project was put under Condon's direction, and other lines of work were initiated. Because of the depression there was an oversupply of young Ph.D. physicists and chemists. To build a young and eager research staff, Condon arranged to establish ten positions, called Westinghouse Research Fellowships, which gave opportunities to young men to work on problems of their own choosing. This was an unprecedented move, for most work in industrial laboratories had been closely directed toward immediate effort on applied problems.

The plan was extremely successful in bringing able young men to Westinghouse. Among the first group were William E. Shoupp, the present vice-president for research there; John A. Hipple, now vice-president for research at the North American Philips Company; William E. Stephens, now provost of the University of Pennsylvania, and Sidney Siegel, now vice-president for atomic energy development at North American Aviation. The Fellowships were awarded each year until 1940, when the need to convert facilities to defense work forced the discontinuance of the plan.

Condon's first initiation into government-sponsored defense research came in the fall of 1940 when he was invited to join the group that established the Radiation Laboratory at Massachusetts Institute of Technology under the direction of Lee A. DuBridge, then professor of physics at the University of Rochester and now scientific adviser to

President Richard Nixon. This laboratory was devoted to the development of airborne microwave radar under the auspices of the National Defense Research Committee (NDRC) which was established by President Franklin Roosevelt that summer, after the German invasion of France. Experience in the Battle of Britain had proved the vital importance of this equipment in modern aerial warfare.

It was agreed that Condon should devote as much of his group's energy as possible to this kind of work at East Pittsburgh, in cooperation with the Rad Lab at MIT, so during the winter of 1940–41 he made weekly trips between Cambridge and Pittsburgh. Westinghouse responded by making him chairman of the company committee to coordinate a rapidly expanding microwave research effort at its electronic laboratory in Bloomfield, New Jersey, and its radio systems factory in Baltimore, Maryland — activities that greatly added to his crowded travel schedule.

In this early period of NDRC work he served briefly with R.C. Tolman and C.C. Lauritsen on the NDRC committee responsible for the rocket program which led to establishment of the Jet Propulsion Laboratory of the California Institute of Technology, now a part of the National Aeronautics and Space Administration.

Uranium fission was discovered in December 1938 by Lise Meitner, Otto Hahn, and Fritz Strassman, working in Berlin. Word of the discovery was brought to a conference on nuclear physics in Washington by Niels Bohr in January 1939. Condon was among those at the conference who returned to their laboratories fired with ideas about further research in this field. At Westinghouse this enthusiasm led to two contributions: measurement of the neutron energy threshold for fission by U^{238}, which was found to be close to one million electron volts, and the discovery of photo-fission, that is, fission of uranium nuclei stimulated by absorption of gamma rays, rather than by neutrons.

During 1939 many physicists, but principally Enrico Fermi and Leo Szilard at Columbia, began to have ideas about self-sustaining chain reactions based on uranium fission. These might be arranged to proceed slowly with the controlled release of fission energy for the generation of electric power, or perhaps explosively for use in a military weapon. Now, thirty years later, when both of these results have been achieved, it is hard to realize the uncertainties which pervaded the original thinking because of the lack of quantitative fundamental knowledge about the relevant processes. And now, when the universities are being criticized so violently both internally and externally for their extensive involvement in research sponsored by the Department of Defense and the Atomic Energy Commission, it is hard to realize how slowly and ineffectively work on these two applications of uranium fission was sponsored at first by the military and other agencies of the United States government.

The story of how a secret program in this area was initiated by President Roosevelt after his receipt of a fateful letter on the subject from Albert

Einstein, written in August 1939, has often been told. This resulted in the formation of the secret S-1 committee for this subject, under the chairmanship of Lyman J. Briggs, then director of the National Bureau of Standards. But the work moved slowly and with little support, except for some basic studies under the direction of Fermi and Szilard, which proved fundamental to the development of controlled energy release for reactors; initial studies of isotope separation by the membrane diffusion method under H.C. Urey and J.R. Dunning at Columbia; the study of the physics and chemistry of the transuranium elements at Berkeley; and initial theoretical studies of bomb possibilities by J.R. Oppenheimer and R. Serber at Berkeley.

In the late summer of 1941 the S-1 committee was expanded by three members: Jesse W. Beams of the University of Virginia, Henry D. Smyth of Princeton, and Condon. Important figures in the National Academy of Sciences, in addition to Vannevar Bush and K.T. Compton of the NDRC, and including Arthur H. Compton of the University of Chicago, Ernest O. Lawrence of the University of California, Harold C. Urey and George B. Pegram of Columbia University, were impressed by the preliminary results which the small program was obtaining. They began to plan for a greatly expanded government program of uranium research. This involved a visit to England by Urey, Pegram and others, to learn what was being done there under the extremely difficult war conditions of continuing aerial bombardment, and a visit to America by leaders of the British project to tell the S-1 committee of their plans for the future.

By coincidence these plans came to a head for presentation to President Roosevelt, by Vannevar Bush, after formal action by the S-1 committee, just a few days before the Japanese attack on Pearl Harbor plunged the United States into the war. Orders were given, at once, for a large-scale secret expansion of the uranium work, which some months later was called by the code name, Manhattan District of the U.S. Army, Corps of Engineers. Although he was a member of the S-1 committee during the fall of 1941, Condon was unable to devote much attention to this work because of his heavy involvement with the Westinghouse microwave radar program.

In January 1943 Condon was asked by J.R. Oppenheimer to become associate director of the then secret Los Alamos Scientific Laboratory in New Mexico. This was opened gradually in March and April of 1943, and produced the first uranium bombs for test near Alamogordo in June 1945, and for use in Hiroshima and Nagasaki in August 1945. He went to Los Àlamos for a short time when the laboratory was being started, but returned to Pittsburgh, largely because Westinghouse felt that he was needed to direct the program of microwave radar work.

His next association with the atomic bomb project was when he went to Berkeley in August 1943 to head the Theoretical Physics Division of the Radiation Laboratory under E.O. Lawrence. He remained there until

February 1945, when he returned to Pittsburgh. The work at Berkeley was directly concerned with development of equipment which Westinghouse had the responsibility to manufacture. Lawrence's laboratory had undertaken the development of the electromagnetic mass spectrograph method of separating uranium isotopes which was the basis of one of the enormous secret plants built at Oak Ridge, Tennessee. Basically, the mass spectrographs, which were built by Westinghouse, were of the Dempster type, built on a large scale, but with a sophisticated distortion of the usually uniform magnetic field. This improved the quality of focus of the uranium ion beams and therefore improved the performance with respect to separation of U^{235} and U^{238} ions.

The entire project had been put in jeopardy because the security officers of the Manhattan District had summarily drafted for infantry training the one young physicist in the organization who knew the theory of design of the magnetic shims used to distort the magnetic field. The action was taken over the strong protests of Ernest Lawrence, who emphasized the crucial importance of this one man. Lawrence then appealed to Condon, and to the Westinghouse management, to persuade him to come and straighten out the difficulties with the shims, which Condon did.

This was Condon's first contact with what seemed to him an arbitrary abuse of power by security officers. Most anecdotes in this connection are quite grim, but he tells one that is amusing: Cooperation between Americans and British had been discontinued after organization of the Manhattan District project, but was restored after agreements between Winston Churchill and President Roosevelt were reached at their Quebec conference. A rather large group of British scientists then came to various American laboratories. Among others who came to Berkeley was Sir Mark Oliphant, now a leader of atomic energy research in Australia, and Sir Harrie Massie, now professor of physics in the University of London. The very presence of the British in Berkeley was classified as secret. So secret was it, in fact, that the British scientists had not told their wives that their own existence was a military secret. Consequently, at first the wives talked quite freely about the reason for their presence in Berkeley at social events where they were being entertained!

After the war was ended by the use of atomic bombs at Hiroshima and Nagasaki in August 1945, Condon was one of the many physicists who became deeply concerned about the possibility that American policy with regard to nuclear weapons would, by default, remain entirely in the hands of the military, with little or no factual knowledge about the awesome potentialities of nuclear fission being made available to the public. He began to write articles and to give speeches to inform the public about the policy questions posed by military use of nuclear energy.

In early September 1945 he attended a conference on the subject at the University of Chicago. There he first met Henry A. Wallace, then Secretary of Commerce, who asked him to come to Washington as director of the National Bureau of Standards. He accepted and started at once to lobby in Washington for civilian control of atomic energy, in close connection with Leo Szilard, innocently unaware of the fact that his position required Senate confirmation and he became director of NBS at the beginning of November.

During October a special committee of the Senate on atomic energy was set up under the chairmanship of Senator Brien McMahon of Connecticut. He asked Condon to serve as his scientific advisor. During 1945–46 Condon held both positions: director of NBS, and adviser to Senator McMahon. This committee, after extensive hearings, developed the legistation known as the McMahon-Douglas bill, which became law on August 1, 1946, establishing the U.S. Atomic Energy Commission.

The controversies associated with the development of that legislation aroused bitter feelings. Condon's first awareness that he had been marked for punishment came in June 1946. He was president of the American Physical Society that year and was attending the summer convention on the campus of the University of Chicago. A bus trip had been arranged by officers of the Manhattan District to show off the Argonne Laboratory, hitherto completely secret, to the physicists, without any special requirements as to security clearance. Condon was seated in a bus waiting for the trip to start when an officer announced in a loud voice that word had just come from Washington that "Condon is not cleared for this trip." So he got off and spent the afternoon reading in the Quadrangle Club, during which time he received a telephone call from the White House appointing him a member of the high-level President's Evaluation Commission for Operation Crossroads, the Naval atomic bomb tests at Bikini held in July and August, which he witnessed in this capacity.

More attacks followed. In the summer of 1947 he was mentioned unfavorably in two magazine articles signed by J. Parnell Thomas, then chairman of the House Committee on Un-American Activities. Condon wrote a letter to him offering complete cooperation in investigating any charges, with copies to all members including Richard Nixon, then a young first-term congressman from California, but received no reply.

In the fall of 1947, attacks on members of the administration, alleging vaguely that they were "soft on Communism," became a widely used tactic of partisan political warfare against President Harry Truman. Truman responded by setting up procedures for "loyalty hearings" at which a person so attacked was given the opportunity to reply before a properly appointed board. Condon asked the then Secretary of Commerce, Averell Harriman, for such a hearing, and was given the first one in the Department of Commerce. The verdict was favorable to Condon.

The Un-American Activities Committee struck on March 1, 1948, releasing to the press a sub-committee report charging that Condon was "perhaps one of the weakest links in our atomic security," which journalists quickly contracted to "the weakest link in our atomic security."

Condon replied by saying that he hoped he was the weakest link, for in that case there was nothing to worry about. Parnell Thomas promised hearings right away but kept postponing them all spring and summer, each time repeating the original charges in the newspapers. Late in the spring the Atomic Energy Commission suspended Condon's clearance and instituted its own investigation in which more than 300 persons were interviewed by FBI agents. Clearance was finally reinstated in July with a statement that no reason had been found to doubt Condon's loyalty or trustworthiness. In the fall of 1948 a second loyalty hearing was held by the Department of Commerce which resulted in a third formal clearance.

In September 1948 President Truman addressed the centennial meeting of the American Association for the Advancement of Science at Constitution Hall in Washington, with Condon on the stage. At this time Truman delivered a scathing attack on the House Committee on Un-American Activities. The hit-and-run tactics used by that committee are now associated in the public mind primarily with the name of the late Senator Joseph McCarthy of Wisconsin; actually McCarthy merely picked up and went farther with tactics initially used in the American Congress by the Thomas Committee.

Later that fall, as a result of revelations by Drew Pearson, the committee chairman, Parnell Thomas, was indicted and tried for the crime of taking salary kickbacks from the girls who worked in his congressional office. He was found guilty and served a term in the Federal penitentiary in Danbury, Connecticut.

One of Condon's most important innovations at NBS was the establishment of an applied mathematics department as a service agency to the government, and the initiation of projects for development of automatic electronic digital computers, before there was any industrial development of such computers. He also greatly strengthened the scientific research programs of the Bureau in basic physics, particularly in the fields of atomic physics and cryogenics.

With the advent of the Korean War in 1950, the military program was further expanded. Condon had visited Boulder for the first time in the summer of 1949 and had arranged for the construction here of a major NBS facility for research on radio frequency electrical measurements and on the physics of ionospheric and tropospheric propagation. In this he was greatly aided by the late Senator Edwin C. Johnson of Colorado, who was then chairman of the Senate Commerce Committee.

In the summer of 1950 Condon arranged for the Bureau to take over a surplus Naval hospital at Corona, California, to gain facilities quickly for a guided missile development project which was being conducted for the Navy by the Bureau of Standards. That laboratory is now operated by the Navy for weapons development work.

Harassment of scientists, both in and out of the government, continued in 1949–50, mostly by actions of the House Committee on Un-American

Activities. Many able persons lost positions in the government and in universities in this period. Several good physicists were driven into permanent exile from America. Condon himself was not attacked again and did all he could to help those in difficulties.

In the summer of 1951 he was offered the post of director of research and development in Corning Glass Works at Corning, New York. He accepted and entered that position in October after resigning as director of NBS. He plunged eagerly into the task of learning a whole new field of physical-chemical science and technology (78–81) and of helping that company move into electronic areas in which it was becoming involved for the first time. He soon learned that more trouble was in store for him. Navy security officers in Buffalo who had jurisdiction over Corning insisted that it was necessary for Condon to go through the clearance process again, in spite of his having had three favorable decisions.

The new process was started, but it dragged on slowly during the national political campaign of 1952 which brought Dwight Eisenhower to the White House. In that year's campaign oratory more attacks were made on President Truman in connection with his personnel security policies. The candidates promised there would be new and better procedures in this matter. In September 1952 Condon was summoned from Corning to Chicago for a hearing before a sub-committee of the House Committee on Un-American Activities. No new issues were raised. The subject matter had nothing to do with Chicago, but the sub-committee members were running for re-election from districts in and around Chicago and wanted helpful coverage for their candidacies in the Chicago newspapers by holding the hearings there.

After the Eisenhower administration was installed many months went by while his staff drafted the promised new procedures. In consequence, Condon did not receive the new formal "charges" until the week of Thanksgiving in 1953.

Preparations for Condon's response were handled by the Washington firm of Fowler, Leva, Hawes and Symington, a great deal of attention being given to it by the head of the firm, Henry H. Fowler, who later became Secretary of the Treasury under President Lyndon Johnson. Finally, in April 1954 a two-day hearing was held in New York City before the Eastern Industrial Personnel Security Board. This board finally rendered a completely favorable verdict on Condon, his fourth.

Some younger readers may not recall, or may not have known, about the atmosphere in America in the spring of 1954. This was the period when the attacks by Senator Joseph McCarthy culminated in the nationally televised army-McCarthy hearings in which the senator hurled all manner of wild charges of disloyalty and incompetence against the army. This was also the season made notorious by the hearings which resulted in final termination of the security clearance of J.R. Oppenheimer, the man who directed the Los Alamos laboratory during the war, and who served as chairman of the general advisory committee of the Atomic Energy Commission during most of the post-war period.

Thus, in the summer of 1954, Condon was again "cleared." Among the "charges" to which he had responded in one or another of these hearings was one which said that there was reason to believe he might be disloyal "in that your wife was critical of the foreign policy of the United States and you did not reprove her."

But the administration seemed to feel that there was more political mileage to be gained from the personnel security issue, despite the fact that matters were now being handled by the "improved" procedures which they had devised. In mid-October 1954 the fact of Condon's clearance was published in the newspapers. Soon after that the Secretary of the Navy, before assembled television cameras, dramatically announced that he was suspending that clearance, saying that it needed further investigation. A few days later Vice-President Richard Nixon, making campaign speeches in Butte, Montana and Cheyenne, Wyoming, implied that he had requested the suspension (*New York Times,* October 22, 1954; also Drew Pearson in the *Washington Post,* October 29, 1954). *The Providence* (R.I.) *Journal,* October 26, 1954, said editorially, "Vice-President Nixon spilled the beans when he disclosed that the Navy had acted at his request. This made it apparent that the 're-review' is not a security proceeding at all, but simply a political maneuver . . . Vice-President Nixon's avowed role in the Condon case is more subversive of the national interest than anything Dr. Condon himself was ever charged with doing."

As it was quite clear that these actions were politically motivated, nothing was done until after the election in November. Then Henry Fowler, as counsel, went to see appropriate officials in the Department of Defense. He told Condon that the Secretary of the Navy admitted to having suspended the clearance without having read the hearing record. His press statement had implied that the record was in some way incomplete. Fowler asked for an indication of what further material was needed so that Condon could supply it. He was told that the whole hearing would have to be repeated, before a new and specially appointed board. Fowler was convinced that this would be a kangaroo court proceeding and advised against reopening the case. Condon finally resigned from Corning Glass Works around Christmas time, but he has continued to serve as a consulting physicist there ever since. A decade later, after he came to the University of Colorado, the Department of Defense quietly reinstated the clearance.

Condon had been elected president of the American Association for the Advancement of Science for 1953. With Mrs. Condon he drove to California, where he gave his retiring presidential address (82) during Christmas week of 1954.

During the spring terms of 1955 · and 1956 he was invited to the University of Pennsylvania as a visiting professor of physics. In the fall of

1956 he was appointed Wayman Crow professor and chairman of the department of physics at Washington University in St. Louis. In the year 1962–63 he served as visiting professor of physics at Oberlin College in Ohio. While at St. Louis, he spent his summers in Boulder, taking an active interest in the development of the summer Institute for Theoretical Physics here.

In the years 1957 through 1968 he was editor of the *Reviews of Modern Physics*.

Finally, in the fall of 1963 he came to Boulder as professor of physics and as fellow of the Joint Institute for Laboratory Astrophysics. He plunged energetically into the project of trying to persuade the Atomic Energy Commission to locate its large 200 Bev accelerator near Denver. In this connection he was appointed chairman of the Colorado Scientific Development Commission by Governor John Love. In 1966 he ran for regent of the University, an elective office in Colorado, but did not make it.

In 1966 he allowed himself to be persuaded to head up a project requested by the Office of Scientific Research of the Air Force, to investigate unidentified flying objects, a problem with which the Air Force had been plagued for nearly twenty years. This occupied much of his time during 1967 and 1968, leading into curious experiences with people who claim to be visitors from other worlds. The report (98) on this project was published in January 1969. He gave a light-hearted account (100) of some of his experience in a talk before the American Philosophical Society. This work has been the subject of vituperative comment from persons anxious to continue to believe that flying saucers are visitors from outer space, and who wish to see the government spend vast sums on further studies of UFO's, despite the lack of any scientific evidence to support the belief.

During the years at Boulder he has resumed research on the theory of atomic spectra with Halis Odabasi, who received his doctorate here working with Condon.

He has received many honors besides those already mentioned. He was president of the American Association of Physics Teachers in 1964, and of the Society for Social Responsibility in Science in 1968 and 1969. Besides membership in the National Academy of Sciences, the American Philosophical Society of Philadelphia and the American Academy of Arts and Sciences of Boston, he holds honorary foreign membership in the Société Française de Physique, the Royal Swedish Academy of Engineering Sciences, and in the Royal Norwegian Society for the Sciences. He holds foreign honorary D.Sc. degrees awarded by the University of Delhi (India) and the New Mexico Institute of Mining and Technology in 1951, and by American University and Alfred University in 1952. In May 1970 he received notification of his election to membership in the Royal Norwegian Society for the Sciences. In September 1970 he was elected national co-chairman of SANE, a citizen's organization for a sane world.

The University of Colorado has established an annual lectureship in chemical physics in his honor, of which the first two holders have been E. Bright Wilson of Harvard (1969), and Gerhard Herzberg of the National Research Council of Canada (1970).*

A group of friends were gathered after Wilson's lecture (Wilson studied undergraduate classical mechanics under Condon at Princeton) and were discussing the many troublesome experiences Condon has had.

"Sometimes I think he looks for trouble," Wilson declared.

"It's not hard to find," Condon replied.

Ed Condon, and Emilie, your friends all wish you well, and wish for both of you many active happy years in retirement.

---

* The 1971 lecturer was Linus Pauling of Stanford University.

Measures for Progress
A History of the National Bureau of Standards

Rexmond C. Cochrane

Editorial consultant, James R. Newman

Edward Uhler Condon

On May 7, 1945, 4 months before the end of the war in the Pacific, Dr. Briggs quietly celebrated his 71st birthday. A year beyond the compulsory retirement age, he had served as Director since 1932 under five Secretaries of Commerce, Roy D. Chapin, Daniel C, Roper, Harry L. Hopkins, Jesse H. Jones, and, since the first of the year, under Roosevelt's new Secretary, Henry A. Wallace. Anxious to return to the comfort and quiet of his old laboratory in West building. Dr. Briggs submitted his resignation to Secretary Wallace.[1]

Two members of the Bureau, Dr. Eugene C. Crittenden and Dr. Hugh L. Dryden, came under consideration by the Secretary's Visiting Committee to the Bureau as Dr. Briggs' successor. Dr. Crittenden, at 65, was the senior, with 36 years of service in the Bureau. But he felt his health was not up to the task, and Dr. Briggs urged the candidacy of Dr. Dryden. Secretary Wallace, however, did not have the advice of his Visiting Committee in selecting a successor.[2] Moreover, he was strongly inclined to

---

[1] Dr. Briggs' first years of retirement were spent, at Secretary Wallace's request, compiling the report on NBS War Research (1949). Latter, Wallace to LJB, Oct. 11, 1945 (NARG 40, Box 112, file 67009, pt. 1, 7–12). See also E.U. Condon, "Lyman James Briggs (1874–1963)," *Year Book, Am. Phil. Soc.*, 1963, pp. 117–121

[2] Interview with Dr. Briggs, Nov. 1, 1961
Dr. Briggs put his request for retirement on the agenda for the meeting for the Visiting Committee on June 22, 1945, just prior to his notification to Secretary Wallace. The chairman of the Visiting Committee subsequently accepted responsibility for the failure of the Visiting Committee to submit promptly its nominations, in response to his request, for the Secretary's consideration. In turn, Secretary Wallace acknowledged that he sent in his own nomination earlier than he had originally contemplated. Reports of the Visiting Committee to the Secretary of Commerce, July 5, 1945, and Oct. 31, 1945 ("Gen. Corresp. Files of the Director, 1945–1955," Box 6).

find someone outside the Bureau for the post. He first met his new Director of the Bureau at a conference of scientists in Chicago.

The successful test of the atomic bomb at Alamogordo in July 1945 had almost at once aroused concern among scientists over the control of the weapon and the peacetime development of atomic energy.[3] Ranged against continued military control were most of those who had worked on the bomb at Los Alamos and in the universities. One of the first of the many conferences that were called to discuss the future of atomic energy was that convened by Robert M. Hutchins, Chancellor of the University of Chicago. It met in September 1945 at the opening of the university's new Institute. of Nuclear Studies. Lending his support to the conference, Secretary of Commerce Wallace attended and brought with him as special advisor, Dr. Philip M. Hauser, a sociologist on leave from the University of Chicago, then with the Bureau of Census.

Meeting Dr. Condon, Associate Director of Research of the Westinghouse Electric Corp., for the first time at the conference, Dr. Hauser found him "a most amiable and knowledgeable fellow * * * [with] broad interests in the physical sciences." Aware that the Secretary was searching for a replacement for Dr. Briggs, Hauser suggested to Wallace that "this was a man he should meet and consider for the post of Director of the National Bureau of Standards." As Wallace remembers it, he discussed the directorship with several others at the conference, but "Dr. Condon was the only one who was available and really interested."[4]

Dr. Condon's name was submitted by President Truman to the Senate and confirmed without a dissenting vote. On November 7, 1945, he was formally appointed Director.

As Dr. Condon told an Appropriations Subcommittee not long after, he was "born * * * actually in the town where the bomb was tested, but there [was] no connection between those two events."[5] Then in his 43d year, he had indeed been born in Alamogordo, N. Mex., on March 2, 1902, but had spent his early school years largely in California. Taking his doctorate in physics at the University of California at Berkeley in 1926, he went to Germany for a year's study, where the new quantum physics of Heisenberg, Born, Schrödinger, and Dirac was being taught. He returned

---

[3] One result of that concern was the publication of *One World Or None* (eds. Dexter Masters and Katharine Way, New York: McGraw-Hill, 1946), a report to the public on the meaning of the atomic bomb. Contributors to the report included Einstein, Bohr, Compton, Bethe, Langmuir, Oppenheimer, Szilard, Shapley, Seitz, Urey, Wigner, and Condon.
[4] Communications to the author from Henry A. Wallace, Jan. 7, 1964, and from Dr. Hauser, Jan. 29, 1964 (NBS Historical File). See also Wallace letter in *New Republic*, 118, 10 (1948). For Wallace's possible prior interest in Dr. Condon, see letter, LJB to H.A. Wallace, Aug. 2, 1945, sub: Standing of certain scientists (NBS Box 504, IG).
[5] Hearings * * * 1947 (Jan. 29, 1946), p. 175.

FIGURE 1. Dr. Edward U. Condon, fourth Director of the Bureau and the first theoretical physicist to head its operations. Reorganizing the Bureau in the postwar period, he cleared its attics of 50 years of accumulated lumber and began the modernization and systematizing of present Bureau operations.

to lecture in physics at Columbia University and in 1928 went to Princeton as assistant and then associate professor.

While at Princeton, he coauthored the Frank-Condon principle in molecular physics; developed the theory of radioactivity decay, with Ronald W. Gurney; a theory of optical rotary power; the theory of proton-proton scattering, with Gregory Breit; and the theory of charge-independence of nuclear forces, with B. Cassen. His definitive treatise on the theory of atomic spectra, with George H. Shortley, established his reputation as an outstanding theoretical physicist.[6]

In 1937, Dr. Condon went to the Westinghouse Electric Corp. at Pittsburgh as associate director of research and there developed a program of nuclear research.[7] Appointed a consultant to the National Defense Research Committee in 1940, he helped organize the Radiation Laboratory at MIT, where America's microwave radar program was started, and wrote a basic textbook on the subject of microwaves for the laboratory. During the war he introduced and directed the microwave radar research program at Westinghouse.

While setting up the radar program, he served on Dr. Briggs's S-1 Committee, meeting monthly at the Bureau. In April 1943 he went ot Los Alamos at the request of General Groves as associate director under Dr. Oppenheimer. Later that year he was called to the Radiation Laboratory at the University of California to head the theoretical physics group working on the electromagnetic (mass spectrograph) separation of uranium isotopes. Toward the end of the war he started the nuclear reactor program at Westinghouse which later produced the power plant for the Navy's atomic submarine.

Dr. Condon was no stranger to the Bureau laboratories when he became their Director. Actually, his acquaintanceship dated back to the late 1920's, when as a Princeton professor he attended the annual meetings of the American Physical Society, regularly held for many years at the Bureau. But Dr. Condon had no sooner seated himself in the Director's chair in South building, to learn something of the dimensions of his office, when he was called to Capitol Hill as scientific adviser to the Special Senate Committee on Atomic Energy. The hearings of Senator Brien McMahon's

---

[6] Biographical note, "About Edward U. Condon," What is Science? ed. James R. Newman (New York: Washington Square Press, 1961), pp. 105–108; interview with Dr. Condon, Oct. 27, 1963. With P.M. Morse, Condon writes Quantum Mechanics (1929) and with G.H. Shortley, The Theory of Atomic Spectra (1935), both standard works in their fields.

[7] Time, 35, 44 (Feb. 12, 1940), called him "king of the atomic world at Westinghouse," where its new Van de Graaff generator, the only one in industry, was being used to make artificially radioactive substances for studies of nuclear structure.

committee on the question of civilian control of atomic energy began on November 27, 1945, and lasted until April 8, 1946.[8]

In the interim, Dr. Crittenden served as Acting Director and Dr. Condon contended himself with brief visits to the Bureau to acquaint himself with its operations and activities. With only his Sundays free, he came with his master key and toured the unpeopled laboratories looking at work in progress, read the reports of current research left on his desk, and studied reports on operational procedures at the Bureau.[9]

Late in January 1946, Dr. Condon appeared for the first time before the House Appropriations Subcommittee for the annual hearing on the budget. Unaware of the deep affection of the committee members for Dr. Briggs and their long-standing interest in the Bureau under his direction, Dr. Condon brought up the subject of Bureau administration. The immediate order of business, Dr. Condon told the committee, was "to modernize and systematize the entire administrative activity of the Bureau, which has just grown up over the years without any special organization unit to coordinate and supervise the work.[10] Dr. Briggs and two division chiefs acting as Assistant Directors had borne the responsibility not only for all research at the Bureau but for the work of the 141 members of the administrative staff.[11] It seemed to Dr. Condon an impossible task.

Dr. Condon asked for funds for three full-time Assistant Directors to administer the professional and scientific functions of the Bureau, and an Executive Director to supervise business management functions. These four, he said, would "do what Dr. Briggs was doing before." As for the Director of the Bureau, he should not have 13 division chiefs and 4 or 5 administrative heads reporting directly to him for decisions and policy determinations. The greater part of his time should be devoted to

---

[8] As a result of the hearings, Congress established the Atomic Energy Commission on Aug. 1, 1946, with complete civilian control over all atomic affairs of the United States, peaceful and military. All Manhattan District facilities, including the Los Alamos weapons laboratory, the isotope separation plants at Oak Ridge, and the plutonium piles at Hanford, were turned over to the AEC. It became responsible for procuring ores of the fissionable heavy metals, uranium and thorium, for converting them into concentrated pure metal, for manufacturing weapons as well as radioactive isotopes, electric power reactors for ship propulsion, and generators for electricity. The AEC was also charged with conducting all research necessary to keep the United States ahead of the world in atomic development. Finally, the act authorized free international exchange of basic scientific information when an international arrangement and techniques of inspection made that possible. See James R. Newman and Byron S. Miller, *The Control of Atomic Energy* (New York: McGraw-Hill, 1948).

[9] Interview with Dr. Condon, Oct. 27, 1963.

[10] Hearings * * * 1947 (Jan. 29, 1946), p. 183.

[11] The assistants were Dr. Crittenden, chief of the electricity division, and Dr. McAllister, chief of codes and specifications. The latter retired in the spring of 1945 and had not been replaced when Dr. Condon took over.

"maintaining appropriate relations with the Secretary's Office, other activities of the Department and other Federal agencies, and commercial concerns and educational and scientific societies and institutions with which the Bureau is associated in cooperative or allied work."[12]

Asked by Congressman Louis C. Rabaut, chairman of the subcommittee, if the increased staff would promote greater efficiency at the Bureau, Dr. Condon replied: "That is my hope, and if it does not we will have to do something about that. It is my own feeling * * * that we have a great many overlapping operations and practices there that have just grown up over the years * * *." It was not a diplomatic note and Mr. Rabaut, and many at the Bureau hearing it later, reacted to it.[13] Steeped in an academic rather

---

[12] Hearings * * * 1947, pp. 183–184

[13] Ibid., p. 184. For Chairman Rabaut's great affection for and delight in Dr. Briggs, see Hearings * * * 1945 (Jan. 11, 1944) and Hearings * * * 1946 (Feb. 2, 1945), passim. For his reactions to Dr. Condon's criticism, see Hearings * * * 1947, passim.

The House subcommittee seems to have resented Dr. Condon's remarks on the state of Bureau facilities and equipment, his observations that there was serious duplication and overlapping in laboratory equipment and in shops, but that "with a complete reorganization of the administrative functions * * * we can introduce many simplified practices"; that the laboratories had become storehouses of obsolete records and equipment, "housing * * * useless items which should be disposed of"; and that despite its famed safety code experts, "the Bureau itself is probably one of the worst violators of its own safety codes" (Hearing * * * 1947, pp. 190–191).

The Congressmen queried Dr. Condon on his choice of speech and efforts at explanation. Despite his acknowledged unfamiliarity with Bureau statistics, they sought from him breakdowns in appropriations, work loads, expenditures and other data that neither Crittenden, Parsons, Thompson, Dellinger, nor other administrative officers at the hearing with Condon could answer offhand.

Three years later the House subcommittee sent up a group of investigators, including inspectors from the Public Buildings Administration, who over a 6-months' period surveyed Bureau grounds and buildings maintenance, the shops and laboratories, efficiency of operations and activities, use of personnel, and administration of research and testing.

The questioning of Dr. Condon on the line-by-line details of the resulting House survey report, which everywhere found "the administration of the Bureau * * * weak and timid," occupied almost 75 pages of the hearings for 1951 (Feb. 23, 1950, pp. 2179–2230, 2242–2246, 2249–2260, 2288–2293). Midway in the quizzing, Congressman Daniel J. Flood of Pennsylvania interrupted to ask: "What is the most exciting thing that has happened in the Bureau of Standards in the year outside of this investigation by the Appropriations Committee?" Dr. Condon could only deny it had been exciting; it had been rather depressing (p. 2237).

Prior to that questioning, Dr. Condon had talked steadily for over 2 hours (pp. 2158–2181) on the scope of activities of the Bureau, in answer to the repeated queries of the subcommittee: "What does a 'Bureau of Standards' mean?" "Does the Bureau's work embrace all of science and technology?"

At the next year's hearing, in March 1947, Congressman Karl Stefan of Nebraska replaced Rabaut as chairman. Stefan requested that Dr. Condon use layman's language before the committee, and raised again the joke about the scientist and

than industrial or even bureaucratic tradition, the Bureau, with almost a hundred on the staff who had been there since Stratton's time, braced itself for the shock.

Dr. Condon was not to project a father image as had Stratton, softening the severity of his strictures. He was not to capture cooperation by his appeal for help, as had Burgess, or to inspire devotion by his presence, as had Briggs. Genial, gracious, and the world's best company away from his desk, Dr. Condon brought to an organization largely staffed with experimental physicists the new-broom outlook of the theoretical physicist. Perhaps more than most at the Bureau, he was aware that the war years had revolutionized science and scientific thought and, always a prolific writer, he had for sometime expounded the new physics in a steady stream of articles in the periodicals.[14]

---

the plumber, alleging that in reply to a New York plumber who had asked the Bureau about the use of hydrochloric acid for clearing drainage stoppages, a Bureau physicist had answered: "The efficacy of hydrochloric acid is indisputable, but the corrosive residue is incompatible with metallic permanence." Assuming that meant it was all right, the plumber wrote thanking the Bureau. The Bureau supposedly replied "We cannot assume responsibility for the production of toxic and noxious residue with hydrochloric acid and suggest you use an alternative procedure." The plumber wrote that he agreed with the Bureau: hydrochloric acid worked fine. Frightened at what might happen to the drainage of New York skyscrapers, the Bureau was alleged to have resorted finally to simple speech: "Don't use hydrochloric acid. It eats hell out of the pipes." (Hearings * * * 1948, p. 289). The joke was brought to Dr. Condon's attention in each of the next 2 years. (Hearings * * * 1949, p. 538; Hearings * * * 1950, p. 493).

Representative Walt Horan of the State of Washington quizzed Dr. Condon about the purpose of the Bureau: "The title 'Bureau of Standards' should have some meaning. Otherwise we are going to get lost in a maelstrom of scientific research. What does 'Bureau of Standards' mean?" Continuing the questioning at the next hearing, Congressman Stefan advised Dr. Condon: 'Give it to us as Dr. Briggs used to do * * * so that we can understand." At that and subsequent hearings, Dr. Condon was told, "Remember, we are laymen" (Hearings * * * 1948, p. 299; Hearings * * * 1949, p. 526; Hearings * * * 1950, p. 485).

Few men have written more clearly and simply about the complexities of modern physics or are more lucid in general exposition on any subject than Dr. Condon. His sole public rejoinder to his "problem of relations with Congress" occurred in a speech on Sept. 25, 1951, wherein he urged at some length the establishment of a committee of Congress concerned exclusively with science and scientific research in the Government (*Physics Today*, 5, 6, 1952).

[14] See "Making new atoms in the laboratory," *Sci. Am.* 158, 302 (1938); "Sharpshooting at the atom," *Pop. Mech.* 74, 1 (1940); "Physics in industry," *Science*, 96, 172 (1942); "Tracer bullets of science," *Pop. Mech.* 77, 170 (1942); "Physics gives us nuclear engineering," Westinghouse Eng. 5, 167 (1945); "Science and our future," *Science*, 103, 415 (1946); "Is war research science?" Sat. Rev. Lit. 29, 6 (1946); "Science and the national welfare," *Science*, 107, 2 (1948); "60 years of quantum physics," *Physics Today*, 15, 37 (1962). See also files of his speeches and addresses on electronics, nuclear physics and other fields of Bureau research in NBS Historical File.

The Bureau as presently established, Dr. Condon told the Appropriations Subcommittee, is "one of the finest scientific laboratories in the country, and it would be wise to maintain and extend its functions at this time, when there seems to be a disposition to recognize the importance of pure science in the Government's activities more than ever before."[15] As one who had made important contributions to pure science and at Westinghouse brought it to bear on industrial work, he was determined to advance pure science at the Bureau and to move the Bureau rapidly into the postwar world. "Think big!" he repeatedly told the Bureau staff. There was no alternative, and he challenged the staff with his cry, "Are you going to think in terms of peanuts or watermelons?"[16]

Dr. Condon himself thought big. His outstanding characteristic, it proved unnerving to some of the older members of the Bureau, and frightening to congressional appropriation committees. At his second appearance on the Hill, in March 1947, he was to stagger the committee members with a proposed $25 million budget, up from $5 million the previous year.[17] He talked of acquiring not one but three mass spectrometers for the Bureau, not one but two giant betatrons. He requested a fourfold increase in publication funds, to expand the regular series of Bureau reports and prepare and publish multivolume tables of atomic energy levels, tables of the thermodynamic properties of chemical compounds, and a new and comprehensive handbook of physics. The Bureau had lately become the central agency in the Federal establishment for radio propagation research and service. Dr. Condon proposed that it also assume direction of all Federal research in synthetic rubber and in mathematical analysis and machine computers.

Was all this, the committee asked, contemplated in the act that created the Bureau? What about the present program? "Are all of your tremendous, gigantic activities out there carried on under a two-page

---

[15] Hearings * * * 1947, p. 178.
In the Steelman report to the President in 1947 on the role of the scientific agencies of the Government in the Nation's total scientific effort, the National Bureau of Standards was described as "* * * the principal Federal agency for research in physics, chemistry, and engineering; it acts as custodian of the Nation's standards of measurement, carries on research leading to improved mesurement methods, determines physical constants and properties of materials, develops and prescribes specifications for Federal supplies and generally serves the Government and industry as adviser in scientific and technical matters and in testing, research, and development in the physical sciences." (The President's Scientific Research Board, Science and Public Policy, II, The Federal Research Program, Washington, D.C., 1947, p. 151.)
The statement reflected the view of Dr. Condon, who served as an alternate on the President's Scientific Research Board that prepared the report.
[16] Interview with Dr. John Hoffman, Apr. 28, 1964.
[17] Only 15 years later the Bureau's operating budget, exclusive of construction appropriations and transferred funds, would rise to $28.5 million.

law?'' Congressman Stefan asked. Did the Bureau actually intend to "spend about nine or ten million dollars during the next fiscal year on the basis of a two page law?''[18] The committee began vigorously debating with Dr. Condon on what he thought the phrase "bureau of standards" meant and what such a bureau was really supposed to do. He explained point by point how the new science, enormously stimulated by the war, had changed the Bureau and the Nation.

In many ways Dr. Condon was the very man for the Bureau in the years after the war, sparking new ideas and impulses among his associates and energetically recruiting a new scientific staff.[19] He acknowledged that recent technological developments demanded continuance of the Bureau work on rubber, plastics, textiles, liquid fuels and lubricants, on structural metarials, ceramic and electroplated coatings, metallic alloys, electronic devices, and new ranges of radio wave frequencies. But "it would be a serious mistake * * * to let these projects in the fields of applied science interfere with the Bureau's work on fundamental problems of physics and chemistry and on methods of measurement and the standards and instruments which provide the basis for measurements of every kind," as primary responsibilities of the Bureau.[20]

New industries and wholly new technologies were to make unprecedented demands upon the laboratories. Perhaps no one at the Bureau comprehended better than the new Director the implications of nuclear technology, just emerging from its pioneer state, or the need for new instruments, materials, and processes spawned by that technology. More than administration and organization, the thought at the Bureau needed redirection, and as the cold war and then the Korean war came and research for defense intensified, Condon's redirection paid off in the years that followed.

New direction required new men, and Dr. Condon's arrival happened to coincide with an almost complete turnover of the top echelon. Age had begun to make its claims and many, like Dr. Briggs, past the retirement age, had waited only for the war to end. The five division chiefs who retired

---

[18] Hearings * * * 1949 (Jan. 20, 1948), p. 526.
[19] As he told the committee, in addition to the prewar cuts in staff, budget, and services, during the war much of the Bureau's basic research had been reduced and its best men put into war work, from which they had not yet been released. The Bureau was therefore very shorthanded in the field of fundamental research, and it was that area he sought to rebuild and expand. He hoped "to be allowed to do for peacetime fundamental research [in the Bureau] something of the sort that [had] recently been announced as part of the Navy's research plans, involving a high degree of collaboration, and intimate cooperation at the working scientists' level with universities throughout the country" (Hearings * * * 1947, pp. 178–179). Dr. Condon referred to the Office of Naval Research, organized later that year.
[20] Hearings * * * 1947, p. 176.

in 1945 had been with the Bureau since World War I or earlier.[21] Submitting requests for retirement with them were two section chiefs and a number of nonadministrative scientists and technicians with long years of service.[22] Still other division chiefs and 14 additional section chiefs reached retirement age over the next 4 years.[23] By 1950 the top echelons of the working force was essentially new, and the average age level at the Bureau had plummeted by some 20 years.[24]

In most instances division chief replacements were found among senior heads of sections. Continuity was further maintained by appointing Bureau-bred members to top administrative positions. The redirection of the Bureau was carried out principally through changes in organization, through new men that came in to head new fields of research, and the special assistants that Dr. Condon brought in from outside.[25]

Appointed Associate Directors early in 1946 were Dr. Crittenden and Dr. Dryden, the latter, upon going to NACA as director of research in 1947, replaced by Dr. Wallace R. Brode, organic chemist and spectroscopist from Ohio State. From the Navy Bureau of Ships that spring came Dr. John H. Curtiss as assistant to the Director, to take charge of mathematical and statistical research and analysis. From Westinghouse came two other assistants, Dmitri I. Vinogradoff, as liaison between the Bureau and foreign scientific and engineering laboratories, and Hugh Odishaw, to oversee scientific and technical information and Bureau publications.[26] And in a reorganization of housekeeping elements, budget and management, personnel, plant, and shops became formal divisions.

Changes in Secretary Wallace's Visiting Committee to the Bureau included the appointment in 1945 of Harold C. Urey, research chemist at the University of Chicago and Nobel laureate, and in 1946 of Eugene P. Wigner, physicist at Princeton and director of research at the Oak Ridge laboratories, who was to receive the Nobel Prize in 1963. The appointment of two theoretical physicists resulted in a significant change in the

---

[21] They were Bearce of weights and measures, Dickinson of heat and power, Rawdon of metallurgy, P.H. Bates of silicate products, and Fairchild of trade standards.

[22] The section chiefs were Acree in chemistry and Stutz in mechanics.

[23] Retiring section chiefs were Curtis and Dellinger in electricity, Miss Bussey, Wensel, Van Dusen, and Ingberg in heat and power, Bridgeman, Brooks, and Peters in optics, Smither in chemistry, Tuckerman and Whittemore in mechanics, Wormeley in organic materials, and McAdam in metallurgy.

[24] Dr. McPherson of organic materials was to say that in 1943 he was the youngest division chief in point of service; by 1950 he was the oldest. Interview, Dec. 5, 1961.

[25] In a few instances, senior section chiefs were made assistant division chiefs as areas of the Bureau research were phased out or several sections were combined.

[26] A third special assistant, Nicholas E. Golovin, trained in physics but then a management specialist from Naval Ordnance, arrived in the spring of 1949 to take over the analysis and planning of Bureau technical programs.

composition of the Visiting Committee, long dominated by representatives of industry. Urey and Wigner joined long-time members Gano Dunn of the J.G. White Engineering Corp., Karl T. Compton, president of MIT, and William D. Coolidge, director of research at General Electric.

In place of the informal notices and occasional memoranda on administrative matters that previous directors had issued were the numbered Bureau Orders, Administration Procedural Memoranda, and Bureau Memoranda introduced in December 1945. They were timely, for the next decade was to see more changes in organization, policies, and staff than in all the previous years put together. For one thing, the wartime influx of workers that raised the staff above the 2,000 level for the first time in Bureau history did not recede with the end of hostilities but increased steadily. Administration grew proportionately more complex.

Between serving on the McMahon committee and familiarizing himself with the Bureau establishment, it was May 1947, a year and a half after assuming the directorship, before Dr. Condon completed his initial reorganization of the Bureau structure.[27] In the new order, divisions were merged to bring related interests or functions together,[28] new divisions and new sections were created,[29] and still other sections wre relocated as a matter of logic. Several sections, some of them one- or two-man units, were absorbed in larger units elsewhere.[30] Two divisions saw little more than a name change as weights and measures became the metrology division, and clay and silicate products became the mineral products division.[31] And as Dr. Stratton had once headed his own optics division,

---

[27] Announced in NBS BuOrder 47–14, May 19, 1947.

[28] The new electricity and optics division included three sections from optics (photometry and color, optical instruments, and photographic technology) that depended upon electrical standards. Simplified practices and trade standards were combined as the commodity standards division.

[29] The atomic physics division grouped all Bureau facilities and activities relating to atomic and molecular physics and also certain phases of optics and of electronic physics. The Central Radio Propagation Laboratory stemmed from the radio section in electricity. Building technology division took over the fire resistance and heat transfer sections of heat and power, the masonry section (renamed structural engineering) from silicate products division, and the whole of the codes and specifications division. The applied mathematics division had its origin in the New York mathematical tables project.

[30] Underground corrosion went to metallurgy. The huge special projects section (i.e., guided missiles) in mechanics became part of the ordnance development division, and a ballistics group in electricity was transferred to the new division. Transferred to chemistry and no longer separate units were the polarimetry, radiometry, and interferometry sections of optics. Combined with the temperature measurements section of heat and power were the division's thermometry and pyrometry sections. One section in heat and power, aircraft engine research, was discontinued in 1948 when the work was taken over by the NACA laboratory at Cleveland.

[31] Weights and measures administration, for a time a section in metrology, became

so Dr. Condon for a time doubled in brass, as chief of his new atomic physics division.

Laboratory space became critical even before the President's decision in 1950 to construct the hydrogen bomb and the onset of the Korean war. Under the shadow of atomic war, talk of dispersal or military installations and defense facilities was translated into policy. The pressure of space and Truman's refusal to permit expansion of facilities in Washington led to the establishment of two Bureau stations far from the Nation's Capital, the Corona Laboratories in California and the Boulder Laboratories in Colorado.

Two major Bureau projects stepped up when the Korean war began were those in nonrotating proximity fuzes for Army Ordnance and guided missiles for the Navy. Additional temporary structures across Van Ness Street were sufficient to accommodate the augmented fuze group, but the missile staff was approaching a hundred members and its development mission had been accelerated by the requirement for an expanded series of production models for possible use in the Pacific. The project needed space quickly and there was no time to build.[32] On June 1, 1951, the project left Washington and moved into surplus Navy hospital structures, idle since the war, at Corona.[33]

Still another cooperative project, for the Atomic Energy Commission, called for large-scale assistance from the Bureau and required facilities for which space was lacking in Washington. The year before, in 1949, a 220-acre tract had been donated by the citizens of Boulder, Colo., at the foothills of the Rockies, on what was then the outskirts of the city, for new radio facilities for the Bureau. On the slope back of the site marked out for the radio laboratories ground was leveled for the erection of new Bureau cryogenic laboratories.[34]

---

a separate Office of Weights and Measures in October 1947, and was later joined by an Office of Basic Instrumentation. All of these organization changes are shown in app. J.

[32] Letter, EUC to Secretary of Commerce, Dec. 13, 1949, and letter, EUC to Director, Bureau of the Budget, Sept. 13, 1950 ("General Correspondence Files of the Director, 1945–1955," Boxes 4 and 6).

[33] BuOrd 51–18, June 1, 1951; Hearings * * * 1952 (Apr. 10, 1951), pp. 497–502; interview with Dr. Condon, Oct. 27, 1963.

[34] The site was acquired in mid-December 1949 and construction began in the summer of 1951 (Department of Commerce records, NARG 40, file 83583; NBS BuOrd 52–7, Aug. 15, 1951).

Edward Uhler Condon, 1902–1974

PHILIP M. MORSE

Department of Physics
Massachusetts Institute of Technology
Cambridge, Massachusetts

The middle third of the twentieth century was the era of hegemony of physics in American science. During that whole period Edward Uhler Condon was a leader in physics, in research of his own, in stimulating research in others, in applying physics, and in calling attention to the effects on all of us of its indiscriminate and irrational application. When he made his first contribution to theoretical physics in 1926, the word physics was not in the vocabularies of most Americans and the revolutionary concepts of quantum mechanics and relativity were just being worked out in Europe; by 1960 the applications of electronics and solid state physics had begun to change our lives irreversibly, and the implications of nuclear physics were manifest to everyone. Ed Condon contributed to each part of this explosive evolution.

Condon's father, William Edward Condon, was a builder of railroads in the west. He and his wife, Carolyn Uhler Condon, moved from place to place as the construction jobs require. When Ed was born, on March 2, 1902, they happened to be in Alamogordo, New Mexico, an ironic concidence not apparent until July 16, 1945. By the time he was ready for high school the family had settled down in Oakland, California. Ed's rival interests, science and journalism, pulled in different directions. In the turbulent year of 1918, when he graduated, rather than going on to college he became a reporter for the *Oakland Enquirer*. His experience in the ensuing three years had a lasting effect on his attitude toward government and society.

In his own words, "On the *Enquirer* I specialized in news of organized labor. The dock and timber workers and the migratory farm laborers were drawn to communism. The California State Legislature had passed a strong bill defining criminal syndicalism and making it a felony. The politicians were looking for a place to use it. On November 9, 1919, I was the only reporter from a conservative paper to cover the organization meeting of the Communist Labor Party of California, as it was then called. I wrote sensational stories about this small group of persons, which resulted in indictments against them, and which required that I had to testify against them, in trial after trial, over the next several years; there I watched police

FIGURE 1. Edward Uhler Condon. Courtesy of Photographic Department, University of Colorado.

framing some of the defendants in matters where I knew the facts to be otherwise. The effect of this involvement on me was to wipe out any desire to be a newspaperman; so I entered the university and went in to physical science largely as a means of escape from the corruption of the world, in addition to the fact that I was genuinely interested in physical science" (Condon, 1973).

He entered as a freshman in the College of Chemistry of the University of California at Berkeley in 1921, but when he learned that his high school physics teacher, W.H. Williams, had joined the physics faculty at Berkeley he switched from chemistry to take Williams' courses in theoretical physics; thus his choice of career was determined. In that same year, 1922, he married Emilie Honzig, a tiny bundle of energy who encouraged Ed in his scientific work and actively supported his extracurricular activities.

At that time, as Condon has commented, "the physics department was rather weak in a research way, except for the recent addition to the faculty of Raymond T. Birge, who concentrated on the early development of the quantum theory of interpretation of diatomic molecular band spectra, and of Leonard B. Loeb, who spent his life making important contributions to the conduction of electricity through gases" (Condon, 1973).

Ed did well, for he received his AB degree in three years with highest honors, went on directly to graduate work in physics, and received his

Ph.D. in 1926. Birge was then making great progress in measuring and analyzing band spectral intensities. Condon put these observations together with a suggestion of James Franck concerning the photodisintegration of diatomic molecules, to come up with an explanation of the regularities in the intensities. He wrote it up over a couple of weekends and presented it to Birge as his Ph.D. thesis. The combined suggestion-explanation later became known as the Franck-Condon principle, when Condon reworked it later in the language of the newer quantum mechanics.

In those years an education in physics was not complete without a year or two spent in Germany. Condon got an NRC fellowship and he and Emilie, with infant Marie (now Mrs. Wayne Thornton, Jr.), spent the fall of 1926 in Göttingen, and the spring of 1927 in Munich. He imbibed the probabilistic interpretation of quantum mechanics from Max Born and, under Arnold Sommerfeld, began the wave mechanical formulation of the Franck-Condon principle.

Again the rivalry of interest between the gaining and the imparting of new knowledge intervened. Quoting Condon again, "By spring and summer of 1927, papers on quantum mechanics were appearing at a great rate. In those days a young theoretical physicist was supposed to keep abreast in every area of physics. I became discouraged and decided that if this were the normal pace of work in my chosen field (which it was not!) then I was not equal to the task. About this time there appeared a help-wanted advertisement in the *Physical Review* for a man to write popular write popular science for an industrial laboratory, the requirement being that the candidate must have newspaper writing experience as well as a Ph.D. in physics. I may well have been the only person in America with that combination at that time. At any rate I applied and was accepted. It turned out that the position was in the public relations department of the Bell Telephone Laboratories, then at 463 West Street in lower Manhattan."

"We returned to America and found an apartment near Columbia in October 1927. At Bell Laboratories, C.J. Davisson and L.H. Germer had just done the experimental work on scattering of low-energy electrons by single crystals of nickel, which led to the discovery of electron diffraction. The importance of this work was not at first appreciated in the business management side of the Laboratories, and I devoted a good deal of attention in the fall of 1927 to explaining to such people that the work was destined to win for the Bell Labs, the first Nobel prize to be awarded to an industrial organization."

"In that fall I found that the American physicists on the Atlantic coast were having as much trouble understanding and assimilating quantum mechanics as I had had in Germany. The profession of theoretical physics was much smaller then than now. As I remember it, Gregory Breit, John Slater, John Van Vleck, and Edwin Kemble were about the only ones in America who were really active in research in quantum mechanics then. I soon found myself in demand as a colloquim speaker at various universities and my boss, R.W. King, encouraged me to accept such invitations, even though they bore little relation to work I was supposed to be doing for the telephone company."

"I was asked by George Pegram to be a lecturer in physics at Columbia University in the spring of 1928. I accepted and started on my first regular university appointment by giving two graduate courses, one in quantum mechanics and the other on electromagnetic theory of light. Besides giving these courses I travelled around giving colloquium talks on quantum mechanics and on the Franck-Condon principle. So great was the demand for young faculty who could deal with these subjects that in the spring I was offered six assistant professorships for the fall of 1928. I ended up by taking the offer from Karl Compton to go to Princeton."

The chronicler of this biography first met Condon at Princeton in the fall of 1928. He was a new kind of professor. A close-cropped brush to black hair accentuated the roundness of his head, his broad face was usually adorned with a grin, his brown eyes looked steadily but somewhat skeptically at one through rimless glasses. The western vocabulary, the proletarian outlook, the rough-edged kindliness, all contrasted with the eastern establishment manners then the Princeton norm. He was only a year older than the chronicler, but while his greater experience and maturity made a great deal of difference to the student it made no difference to the professor.

Condon has remarked that this first year at Princeton, 1928–29, was the most productive in his life. He has said (Condon 1973), "For teaching I gave a course in quantum mechanics again, improving the notes of the previous Columbia course, and a junior course in classical mechanics, of which the most outstanding student was E. Bright Wilson, now Mallincrodt professor of chemistry at Harvard. Philip Morse, who received a doctorate under K.T. Compton that year, took my course and we worked up the lecture notes into the book *Quantum Mechanics* (Condon and Morse) which was published in the fall of 1929, the first text in the new subject in America."

"I personally wrote the paper that give a fuller statement of the quantum mechanics of the Franck-Condon principle. But by far the most important piece of work done that year was the development of the barrier leakage picture of alpha-particle radioactivity, done with R.W. Gurney. The same idea was developed almost simultaneously by George Gamow, then a postdoctoral fellow in Göttingen. This was the first application of quantum mechanics to details of inner structure of atomic nuclei, and at the same time its success gave a big boost to the probability interpretation of the intensity of the Schrödinger wave, that was only being reluctantly accepted in some quarters."

He still was footloose. He accepted an offer of full professorship at the University of Minnesota for the fall of 1929. But within a year he decided he preferred the stimulation of congenial colleagues to the kudos of the full professorship, so after giving summer courses at Stanford, he returned to Princeton in 1930, where he remained until 1937. During that decade he began to show his ability to spot, energize and guide emerging leaders in the next generation of theoretical physicists. Two of them have reported how he did it.

George Shortley, who became professor of physics at Ohio State University and went into the field of operations research during and after World War II, writes: "I was a senior at the University of Minnesota, taking a physics minor in my electrical engineering program. I signed up for both of his courses. His appearance was quite different from that of any professor I had ever seen. He was jovial, chubby, black-haired, crew-cut and boyish in appearance, wearing cream-colored plus-fours, after the fashion of the day for students, but decidedly not for faculty. One of his courses was the theory of atomic spectra, taught in the quantum-mechanical technique of Dirac before any useable text was available. The other was a course in classical methods of mathematical physics. The two courses meshed perfectly because the same mathematical functions were used in both. Condon was a beautiful lecturer; he had the facility of 'making a difficult subject sound easy' whereas other professors often had the opposite tendency to 'make a simple subject sound difficult'."

"These courses aroused my interest in the theory of atomic spectra and led eventually to my collaboration with Condon on the well-known book on this subject. In fact, later in this same senior year, Condon and I wrote and published our first joint research paper in this field."

"Early in 1930 Condon decided to leave Minnesota and return to Princeton. With considerable difficulty he arranged for me to go with him to Princeton as a graduate student. He also arranged for me to be his research assistant, at a salary that would enable me to support myself. After teaching at Stanford in the summer of 1930, he picked me up in Iowa for the drive back to Princeton with his wife Emilie and their little child, Marie, called Mädi; in fact Mädi sat on my lap for most of the trip."

"When I reported for my duties as research assistant he proposed the collaboration on the monograph on atomic spectra, and we proceeded to outline the chapters then and there. As indicative of the energy he expected of himself and of his students, he asked me the next morning how much I had written. Fortunately I had applied myself the previous afternoon and evening and had the draft of half the introductory chapter to show him."

And Frederick Seitz, president of the National Academy from 1962 to 1969 and since then, president of Rockefeller University, writes (Seitz, 1969): "I was a sophomore at Stanford University and decided to do my bit to reverse 1929 trends by becoming a professional physicist. While still enjoying the euphoria brought on by this decision, I read in the university newspaper that the visiting professor in theoretical physics for the summer quarter would be in brilliant young man, 28 years old, who had discovered the Franck-Condon principle while a graduate student at Berkeley, had attended the great centers of physics in Europe as a National Research Fellow, and had held prominent posts at the Bell Telephone Laboratories, Princeton University and the University of Minnesota. Just a year earlier, he and Ronald Gurney had given an interpretation of alpha disintegration of nuclei in terms of quantum mechanical tunneling. To top it all, the campus paper related that he had earned his way through Berkeley as one of the more worldly reporters of the Oakland Tribune [sic]. In this pursuit he had, among other things, stirred up a lively public discussion of whether a birdcage would weigh more or less when the bird was flying around inside instead of resting on its perch."

"The visitor, Edward Condon, was slated to give a course in modern physics which would be open to duly qualified undergraduates. I succeeded in persuading my father to provide means to attend the summer session and, early in July, found myself perched on a chair in the front row of the lecture room waiting for the show to start. It was not a disappointment."

"Precocious and crew-cut, Ed Condon exhibited even then all the characteristics that have carried him through a lifetime near the center of the stage. He was creative, energetic, perceptive, humorous, restless, eloquent, worldly and friendly. Moreover he knew, on a first-name basis, most of the top-billed physicists on the planet and loved to spin endless anecdotes about them. This was very rich fare for an undergraduate. Condon's lectures were then as now a wonderful combination of logic, anecdotes and humor. In those days, long before physicists were taken seriously by the public, and when they were still all but unknown to congressmen and security officers, Condon was flamboyantly cheerful practically all the time, his occasional bursts of wrath being directed at the petty annoyances of everyday life which plague us all. His bouts with various prominent individuals — particularly with General Leslie Groves — lay far in the future."

"Condon was so deeply interested in people that he quickly came to know personally everyone in the class who managed to act reasonably alive. The small band of embryonic physicists who dominated the first row became his close friends. With Condon's ardent help, continued family indulgence and some permissiveness on the part of the Princeton Admissions Committee, I followed him back to Princeton as a graduate student a year and a half later. His lectures that spring were centered on Frenkel's book on the classical electromagnetic theory of light, which he embellished in countless ways. I still cherish a carbon copy of his notes."

With the completion of *The Theory of Atomic Spectra*, Condon's interest returned to atomic nuclei. He collaborated with Gregory Breit on a paper on the photodisintegration of the deuteron. But, as he has written, "much more important was the work done jointly with Breit and R.D. Present on the theoretical interpretation of the experimental results obtained by Tuve, Hafstad and Heydenberg at the Carnegie Institution of Washington on the scattering of protons by protons at energies up to about one million volts. These results showed clearly the charge independence of the strong nuclear force between nucleons on which all modern nuclear theory is based."

Between 1928 and 1938 Condon published two books; "Quantum Mechanics" and "The Theory of Atomic Spectra", both with co-authors; nine papers on general quantum mechanics; six papers on atomic spectra, all but one with co-authors; eight papers on the quantum mechanics of molecules, all but two with co-authors; two papers on solid state theory, one with a co-author; and two papers on the biological effects of radiation. In addition there were three articles in the *American Physics Teacher* on simple ways to understand physical concepts, two on semi-philosophical topics and one, published in the *Proceedings of the U.S. Naval Institute*, that can be considered either as an early example of operations research or as an example of Ed's sense of humor. He had come across, in his omnivorous reading, a set of heuristic rules for the amount of food a shipboard cook should prepare, as a function of the number of men to be served. Assuming that the rule represents a balance between satisfying the men's shipboard appetites and reducing the amount of food left over, he determined the parameters of the normal distribution of the men's appetites that the rule inferred and then embellished it with comments on the implications of the distribution and on the validity of the conclusion that there was a nonzero fraction of the men with negative appetites. The conclusions seemed to puzzle some commentators in later issues of the *Proceedings*. This is by way of illustrating that, in spite of his earlier noted complaint at keeping abreast of progress in physics, Condon did, in fact, read and understand an unusually large sample of scientific literature.

Princeton could not hold him long. In 1937 he accepted the post of associate director of research of the Westinghouse Electric Corporation. He moved his family, now increased by two sons, Paul Edward (now on the physics faculty at the University of California at Irvine) and Joseph Henry (now with the Bell Telephone Laboratories) to Pittsburg. Westinghouse wanted to strengthen its work in fundamental physics and assured Condon of liberal support and a free hand in developing such work at the laboratories in East Pittsburg. Construction had already been started on a large pressurized van der Graaff machine for nuclear work. The project was put under Condon's direction and other lines of work were initiated.

Once again Condon became the center of a lively community of stimulating individuals. He purchased a roomy house close to Wilkinsburg that seemed to be undergoing continual growth and was usually bursting with interesting, if occasionally unconventional, visitors. Condon not only brought into closer communication the promising young scientists and engineers already employed at Westinghouse, but soon added new faces, through a system of postdoctoral research fellowships. Under his leadership the laboratory quickly grew to the state where it could become a significant factor in the research and development that was to be necessary in World War II.

The approaching war broke in on those developments. In the fall of 1940 the National Defense Research Committee was authorized by President Roosevelt. It soon established the MIT Radiation Laboratory, to develop microwave radar. It was agreed that Condon should devote as much of his group's energy as possible to radar work at East Pittsburg, in cooperation with the MIT Radiation Laboratory. So, during the winter of 1940–41 he communted weekly between Cambridge and Pittsburg. Westinghouse made him chairman of the company committee to coordinate the expanding microwave research effort at its electronic laboratory in Bloomfield, New Jersey, and its radio systems factory in Baltimore, adding to his crowded travel schedule. He also served briefly with R.C. Tolman and C.C. Lauritsen on the NDRC committee responsible for the rocket program that led to the establishment of the Jet Propulsion Laboratory of Caltech.

Parallel to these developments was the work on nuclear fission that grew slowly at first, but by 1942 expanded into the huge complex of the Manhattan District. Condon worked for a while with the S-1 Committee, coordinating the start of this work. Later he spent a little time with Robert Oppenheimer, planning the establishment that was to become Los Alamos. But his duties to the microwave work at Westinghouse prevented his participating further, beyond preparing a text on nuclear physics that became known as the *Los Alamos Primer*. In 1943 he spent some time at E.O. Lawrence's laboratory at Berkeley, arranging for Westinghouse to build the huge magnets to be used for the electromagnetic separation of uranium isotopes. These multiple contacts with the Manhattan District stengthened Condon's aversion to the military control of scientific research and development. He had many anecdotes, some grim and some humorous, about the military attitude and the consequences of the paranoia for secrecy. For example, he would recall the time he, Oppenheimer, and General Groves were discussing the site of what was to be the Los Alamos Laboratory. Ed inserted the question, "As a western boy, I am wondering how we are going to supply this place with water?" General Groves brusquely said that that was his own problem and that Condon should concern himself with physics. "Yes, General," came back Condon, "but just how are we to get the water?" The fact that Condon's worry was justified and that the problem later had to be solved at enormous expense by trucking water up the mesa probably did not endear either of these strong characters to the other.

The end of the war pushed Condon onto the national stage. He had been elected to the National Academy of Sciences in 1944, and in 1945 he became Vice President of the American Physical Society and became its President the following year. The many physicists who were concerned at the military control of nuclear weapons looked to him for leadership, and his penchant for action set him to writing articles and giving talks about the dangers as well as the potentialities of nuclear power. These came to the attention of Secretary of Commerce Henry Wallace, who persuaded President Truman to appoint Condon Director of the National Bureau of Standards, Condon accepted and, before he was confirmed by the Senate, he came to Washington to work with Leo Szilard and others to lobby for the civilian control of atomic energy. The fight was violent and bitter, and affected the rest of Ed's life.

His appointment to the Bureau was confirmed by the Senate in November 1945, but before that Senator Brien McMahon asked Condon to serve as his scientific advisor for the special committee on atomic energy that McMahon chaired. For several months Condon gave a course for legislators on the atomic nucleus, its implications in war and peace. Until the summer of 1946, when the McMahon-Douglas bill established the Atomic Energy Commission, under civilian control, Condon held two jobs. With the establishment of the AEC he felt able to turn his undivided attention to the Bureau of Standards.

Condon was the first Director of the NBS to be appointed from outside the Bureau ransk, the first Director to be recruited from industry, the first theoretical physicist to head the Bureau and the first and only Director to live in a house on the Bureau grounds. As his collegue at the Bureau, Hugh Odishaw, has written: "The NBS had had a long and honorable history of scientific and technical contributions, but the depression years had seen its budgets slashed. Instruments and facilities were wearing out; there was little if any new gear; no significant opportunities to enter into new areas of research and negligible funds to attract young scientists. Condon was determined to change this."

"The struggle for greatly increased appropriations was limited in success, but Condon drew much larger funds from other agencies. With these he strengthened sound on-going activities and initiated new ones, in mass spectroscopy and betatron studies, for example, and through the creation of new divisions, as in applied mathematics and electronics. These latter two collaborated in a pioneering computer program — SEAC in the East and SWAC in the West. These were the first automatically-sequenced, high-speed digital computers, and much of subsequent computer technology stems from this endeavor." In addition, he and Odishaw assembled the highly useful "Handbook of Physics," finally published in 1958.

Condon was also interested in administrative problems. He simplified the Bureau's organization, initiated the first complete restatement (Public Law 81–619) of the Bureau's functions since its founding, and he presided over the establishment of major new facilities at Boulder, Colorado. The results of his initiation of new programs and recruiting of young blood are still quite apparent at the Bureau, 23 years later.

Condon believed in removing obstacles, not going round them. This chronicler remembers being castigated for commending the formation of not-for-profit corporations as a means of providing technical assistance to government agencies without becoming enmeshed in civil service redtape. Condon felt this was a cowardly evasion of the Augean task of revising civil service.

Such direct action, of course, makes enemies. He had already roused the ire of the House Unamerican Activities Committee (HUAC) by his opposition to the military control of atomic energy. In 1948 the Committee's chairman, J. Parnell Thomas, proclaimed that "Dr. Condon is one of the weakest links in our atomic security." Privately Condon described the impossibility of refuting such a charge as follows: "If you say I've got a wart on my nose, I can deny it. But if you just say I'm one of the ugliest men in town, all I can do is to argue that I'm really quite pretty." The verbal duels at the hearing reached heights of invective and illogic. Condon once alleged that someone actually asked how it had come about that Dr. Condon had been born so near the site of the first atomic bomb test. Time and again his security clearance status was reviewed and re-established, only to be challenged again, long after Congressman Thomas had been put in jail for taking kickbacks from his staff.

In 1951, the year that a star-chamber hearing had removed the clearance of J.R. Oppenheimer, Condon regretfully decided that the Bureau would fare better if he left. He had expected to stay at the Bureau for much of the rest of his life, devoting his energy and skill to making it one of the greatest scientific laboratories in the world, but it was clear that the continuing attacks on him were hindering further support of the Bureau by Congress. So he accepted an offer to become director of research and development for the Corning Glass Works.

Here again Condon recruited new scientists and initiated new research on the structure of glass and new applications of its properties. He published a highly useful sequence of four papers on the physics of the glassy state in 1954. At the Corning laboratories he initiated a number of new projects. Unfortunately one of them, on missile nose cones, was supported by the Navy, and thus clearance was required.

In September 1952 Condon was again called before the HUAC to answer further charges, one of which (Britten, 1971) was that there was reason to believe he might be disloyal "in that your wife was critical of the foreign policy of the United States and you did not reprove her." After a long delay, during which the strong support of his scientific colleagues was shown by his election to the presidency of the American Association for the Advancement of Science, Condon had his fourth hearing before the clearance review board and again was given a completely favourable verdict. Within four months, however, the Secretary of the Navy demonstrated the irrelevance of the semijudicial clearance hearings by arbitrarily suspending Condon's clearance. A few days later Vice President Nixon implied, in a campaign speech, that he had requested the suspension.

As Condon has written: "The Republicans were still bent on smearing the Truman record by pretending to a concern over the loyalty. The campaign promise in 1952 was that they would 'clean the reds out of Washington'. They kept their campaign promise to the extent that the procedures were revised and a number of persons were subjected to long and tiresome hearings. One of these was J. Robert Oppenheimer, who was finally deprived of his clearance; another was myself, where the outcome was favourable, as it had been in three previous hearings. However this verdict was arbitrarily suspended by the Secretary of the Navy. The Corning organization had proved to be the most satisfactorily, scientifically and humanly, of any with which I have had the good fortune to be associated. But I had been under intermittent harrassment in this way since 1947 and I decided I would subject myself to it no longer. So I arranged to become a consultant of Corning Glass Works (a position he held till his death). In the spring of 1955 I was offered professorships by the faculties of two major universities, but in both cases the trustees refused to confirm the appointments, under pressure from Washington. Finally I was allowed to become chairman of the physics department at Washington University in St. Louis, and later to come to the University of Colorado as professor and fellow of the Joint Institute for Laboratory Astrophysics (joint with the Bureau of Standards, thus formally reestablishing Ed's relationship with the Bureau that had never really been broken). As the cold war slowly died down the Department of Defense finally reinstated my clearance, but this, I am proud to say, I have never used."

The appointment to Washington University was the result of the efforts of Chancellor Arthur Compton, who not only wanted to add an outstanding physicist to the staff, but also realized that the nation as well as Condon would be the loser if the irrational chain of events were allowed to continue. At Colorado he could finally settle down again to research in atomic theory with Halis Odobasi (now at the University at Istabul) with the intend of rewriting *The Theory of Atomic Spectra*, and in further work on the properties of glass. He continued to write and lecture on the need for peaceful, worldwide cooperation; he took on the job of Editor of the *Reviews of Modern Physics*; he actively participated in the research of the Joint Institute of Laboratory Astrophysics and he found time to be president of the American Association of Physics Teachers in 1964.

And, in an incautious moment, he agreed to head a project, supported by the Office of Scientific Research of the Air Force, to investigate the many reports of unidentified flying objects (UFO's), with which the Air Force had been plagued for nearly twenty years. This occupied much of his time during 1967 and 1968. The report of this project was published in 1969. Condon gave a light-hearted account of some of his experiences as a talk before the American Philosophical Society. The report has been the subject of vituperative comment from persons anxious to continue to believe that flying saucers are visitors from outer space and who wish to see the government spend vast sums on further studies. Despite the views of many of his colleagues that the investigation was a waste of Condon's time, Seitz has said, "The introductory chapter of the report on UFO's, in which Condon describes with characteristic clarity his own view as a scientist on what constitutes worthwhile research, is a classic; it deserves to be a landmark in the journey science has taken since the days of Stevin, Galileo and Kepler."

Edward Uhler Condon died on March 26, 1974. Two comments may serve to close this survey of his life. One is by one of his colleagues at the NBS, Churchill Eisenhart: "Edward Condon was a brilliant scientist, with highly original ideas, a wide range of interest, a restless probing mind with voluminous information indexed for instant retrieval. He could meet with scientists of diverse specialities and stimulate each with fresh enthusiasm and new insights. He could elucidate scientific intricacies to non-scientists with clarity in layman's language. Whatever he knew he saw with crystal clarity; he could summarize it in a nutshell on a moment's notice or discuss it in detail with experts, with equal ease. He had an exuberant sense of humor, a gift of repartee and could be wittily caustic when provoked. He was a cordial, genial, straightforward individual; fond of people, mathematics, science, chamber music and conversation; allergic to formality, fuzzymindedness, pomposity and all forms of physical exercise. He was an active Quaker, a firm believer in human dignity, an outspoken liberal and anti-isolationist. He gave freely of his counsel and his time, generously of his finances and his home."

The other comes from Lewis M. Branscomb, a colleague at Boulder, now with IBM: "Watergate came as no suprise to Edward Condon, nor did its aftermath. I imagine he would have liked to see the outcome of the impeachment inquiry. But Condon understood and paid his share of the price of liberty. Somehow his idealism, his sense of humor and his inexhaustible energy made his relentless quest for a better world look like optimism. He was elected president of the AAAS during the height of his troubles with HUAC. He was president of the Society of Social Responsibility in Science (1968–69) and co-chairman of the National Committee for a Sane Nuclear Policy (1970). He was appropriately honored, on his retirement from JILA and the University of Colorado in the summer of 1970, by the volume edited by Brittin and Odabasi. Brittin relates a comment about Condon by E. Bright Wilson: 'Sometimes I think he looks for trouble'; and Condon's answer, 'It's not hard to find'."

Unfortunatly it is not easy to find a brilliant scientist who is willing to speak out on questions of public policy, often with humor but always with determination, even in the face of official persecution.

References

Branscomb, Lewis M., 1974, Phys. Today, **27**, 68.

Brittin, W.E., 1971, Preface in Topics in Modern Physics, a Tribute to Edward U. Condon, edited by W.E. Brittin and H. Odobasi, (University of Colorado Press, Boulder, 1971)

Condon, E.U. 1969, "UFO's I have loved and lost" Proc. Am. Philos. Soc., **113**, 425–427.

Condon, E.U., 1973, "Reminiscences of a Life in and out of Quantum Mechanics," Proceedings of the International Symposium on Atomic, Molecular, Solid State Theory and Quantum Biology, 7th Sanibel Island, Florida, 1973.

Eisenhart, C., 1974, "Edward Uhler Condon," in Dimensions, the Technical News Bulletin of the National Bureau of Standards, **58**, 151.

Seitz, F., 1971, Foreword in Topics in Modern Physics, a Tribute to Edward U. Condon, edited by W.E. Brittin and H. Odobasi (University of Colorado Press, Boulder, 1971).

Other quotations in this memoir are taken from personal letters to Philip M. Morse.

Obituary: Edward U. Condon

LEWIS M. BRANSCOMB

Vice-President and Chief Scientist
International Business Machines Corporation

The extraordinary career of Edward Uhler Condon, president of the American Physical Society (1946) and of the American Association of Physics Teachers (1964), ended with his death in Boulder, Colorado, on 26 March 1974.

Born in Alamogordo, New Mexico, on 2 March 1902, Edward Condon was one of the young Americans who made the pilgrimage in 1926 to Göttingen and Munich and grasped immediately the significance and power of the new quantum theory. As an undergraduate, Condon had worked as a reporte for the Oakland *Inquirer*, thinking he might pursue a career in journalism. But the intellectual challenge of physics, after a brief flirtation with chemistry, caught his fancy. When he returned from Göttingen, he worked briefly as a public-relations man for Bell Labs, lectured at Columbia and then embarked on an academic career that took him to Princeton, Minnesota, and back to Princeton, where he taught until 1937.

Like most great scientists, Condon made important contributions while still a student. The basis for his papers on the separability of electronic and vibrational motions in molecules (the Franck-Condon Principle) was in his Berkeley thesis. With R.W. Gurney, he was an early explorer of quantum-mechanical tunneling, applied to the phenomenon of alpha-particle radioactivity. In 1937, with Gregory Breit and Richard Present, he interpreted proton-proton scattering data and established the importance of charge independence in the strong nuclear interaction. His early solid-state theory work was the explanation of optical rotatory power, and later he studied semiconductor-contact potentials.

With Philip M. Morse he wrote the first English-language text on quantum mechanics (1929). With G.H. Shortley he wrote the *Theory of Atomic Spectra* (1936), still the primary treatise in the field. In later years the *Handbook of Physics*, which he edited with Hugh Odishaw, and his editorship of *Reviews of Modern Physics* demonstrated once again his facility for dealing with the full range of topics in physics.

These brief notes are an inadequate tribute to the side of Ed Condon's career that will have the most lasting value — his great contributions to knowledge throught discoveries in physics. This side of his career is much more fully documented in *Topics in Modern Physics: A Tribute to Edward U. Condon* by Wesley Brittin and Halis Odabasi (Colorado Associated UP, 1971). Younger physicist who may wish to emulate Condon's courageous public record as an outspoken defender of truth, civil liberties and peace may lose sight of the monumental research contributions that won him the admiration of his fellow scientists and the respect of the public, which permitted him to make a major impact on public affairs.

The second phase of Condon's career began with his move to Westinghouse as associate director of research, just two years before the beginning of World War II in Europe. He brought Westinghouse into the nuclear age and earned an accolade from *Time* as "king of the atomic world." He served on the National Defense Research Committee during World War II, but was not present at his birthplace in Alamogordo when the Trinity explosion gave that small New Mexico town its second claim to fame.

With the war over, Condon became director of the National Bureau of Standards and, concurrently, science advisor to Senator Brian McMahon, chairman of the special Senate committee on atomic energy. McMahom was leading the forces for civilian control of the nuclear-weapons program and with Condon's active help saw success in the McMahon-Douglas bill, passed in August 1946. Condon believed deeply that civilian control over nuclear-weapons development and production was essential to avoidance of nuclear war. In the year of Condon's death, this issue may be reopened as Congress considers the Energy Research and Development Administration proposal which restructures the AEC.

At NBS, as he had at Westinghouse, Condon concentrated his attention on good science, stripping away administrative encrustations of the past, hiring the next generation of scientific leadership, pulling together programs (like building technology) of great potential benefit to the public. He built the NBS Boulder Laboratories. But soon these accomplishments were dwarfed in the public eye by the relentless attacks of Congressman J. Parnell Thomas and the House Un-American Activities Committee, which Thomas headed. The press picked up the phrase in a HUAC report (published on Condon's birthday in 1948) that stated, "It appears that Dr Condon is one of the weakest links in our atomic security." Privately, Condon described the impossibility of refuting such a charge with characteristically colorful language: "If you say I've got a wart on my nose, I can deny it. But if you just say I'm one of the ugliest men in town, all I can do is argue that I'm really quite pretty." Time and again, his security clearance status was reviewed and re-established, only to be challenged again. Finally, in 1951, with his record cleared and with Parnell Thomas in Danbury Prison, convicted of taking kickbacks from his office staff, Condon left government to become head of research and development for the Corning Glass Works.

E.U. Condon. Reproduced with permission of AIP Niehls Bohr Library. Photo by Heka Davis.

In October 1954, Condon's Navy clearance was again re-established in connection with government contract research at Corning. When the clearance was dramatically suspended by intervention of the Secretary of the Navy, the press reported that Vice President Nixon, a former member of HUAC, implied in campaign speeches that he had requested the suspension.

Ten years later, after Condon had taught at Oberlin two years and at Washington University for seven, he moved to Boulder, Colorado, as professor of physics and fellow of the Joint Institute for Laboratory Astrophysics. His security clearance was quietly restored, clearing his record once again.

What kind of man was he? Grace Marmor Spruch's profile in *Saturday Review* (1 February 1969) says it well: "The composite Condon is a moral, impassioned man, with a depth of concern for mankind not common in scientists; a man fiercely principled and anti-diplomatic; a man who believes and feels in sharp contrasts, who will let the world know his position without ambiguity. Fuzzimindedness is an anathema to him and he insists on saying so at every opportunity. But this rasping trait is wedded to an extreme generosity and kindness. Throughout his life he has given freely of his time, his counsel, his finances, and his home."

Watergate came as no surprise to Edward Condon, nor did its aftermath. I imagine he would like to have lived to see the outcome of the impeachment inquiry. But Condon understood and paid his share of the price of liberty. Somehow his idealism, his sense of humor and his inexhaustible energy made his relentless quest for a better world look like optimism. He was elected president of the American Association for the Advancement of Science during the height of his troubles with HUAC. He was president of the Society for Social Responsibility in Science (1968–69) and co-chairman of the National Committee for a Sane Nuclear Policy (1970). He was appropriately honored on his retirement from JILA and the University of Colorado in the summer of 1970 by the volume edited by Brittin and Odabasi mentioned earlier. Brittin relates a comment about Condon by E. Bright Wilson: "Sometimes I think he looks for trouble," Wilson said. Condon's comment: "It's not hard to find."

Sadly, brilliant scientists — who serve their country and principles, their love of truth and their fellow citizens with relentless determination and delightful good homour — are hard to find indeed.

The Age of the Stars

EDWARD U. CONDON

Department of Physics
University of California
Read before the Academy November 11, 1924

It is well-known that the supply of energy made available by the gravitational contraction of the material of a star is insufficient, in view of the Stefan law of total radiation, to allow the time-life of a star to be more than a small fraction of the age demanded by geological considerations. Discussions of the problem of maintenance of stellar energy have been given by Shapley[1] and Russell.[2] The former discusses an hypothesis of asymmetrical radiation flow in which the rate of radiation to empty space is much less than the radiation toward other matter. Shapley also considers the destruction of mass as a possible source of stellar energy. He concludes that " . . . it now appears that the disagreement between the long and short time scales must be decided the favor of an exceedingly prolonged history for sidereal systems, permitting a relatively slow evolutionary development for stars and planets." Russell's paper is a very general sketch of the main outlines of the problem and is concerned mainly with the conditions under which a nuclear "unknown" source of energy comes into action.

In this paper, it is shown that the relativistic relation between energy and mass leads directly to a means of estimating the age of the stars, the methods being independent of the atomic processes whereby mass as "matter" is changed into mass as "radiant energy." This relation,

$$\Delta E = c^2 \Delta m \tag{1}$$

in which E is the energy of the system in ergs, c the fundamental constant of space-time in centimeters per second, and m is mass in grams, has been given an interpretation that is too narrow, at least in discussions of its possible applications to stellar problems. It is necessary to bear in mind that, from the manner of its derivation, the equation necessitates that loss of mass by a system accompany loss of energy as a result of radiation. We do not need to inquire into the source of the energy lost by radiation.

---

[1] Shapley, Harlow, *On Radiation and the Age of the Stars. Pub. Astron. Soc. Pac.*, **31**, 178 (1919).
[2] Russell, H.N., *On the Sources of Stellar Energy. Ibid.*, **31**, 205 (1919).

The point is brought out clearly in the original work of Einstein.[3] He says, "Gibt ein Körper die Energie L in Form der Strahlung ab, so verkleinert sich seine Masse um L/V^2." And again, "Die Masse eines Körpers ist ein Mass für dessen Engergieinhalt; ändert sich die Energie um L, so ändert sich die Masse in demselben Sinne um $L/(9 \times 10^{20})$ wenn die Energie in Erg und die Masse in Grammen gemessen wird."

Let us assume that the amount of mass which is gained by a star during its life by means of the actual accumulation of meteoric matter is negligible and further, that there is no explusion to infinity of matter by the star during its life. If a star is born by the accumulation of dispersed matter, the first of these assumptions is probably inadmissible in the early stages of its life, but will be true with increasing accuracy as the star gets well along in years. The second assumption is probably true at all times as the explosive disruption of a star is considered to be an exceptional occurrence.

If we grant these assumptions, at least after the period of infancy, it is evident that since there is a continual loss of energy by radiation, the mass of the star will decrease throughout its life.

The rate of radiation of energy may be taken to be in accordance with Stefan's law. This allows the equation,

$$dE/dt = 4\pi r^2 \sigma T^4 \tag{2}$$

in which dE/dt is the rate of loss of energy, r the radius of the star, σ the Stefan constant, T the effective temperature, all in G.C.S. units. Equation (1) in the derivative form is

$$dE/dt = c^2 \, (dm/dt). \tag{3}$$

Combing (2) and (3) we may write,

$$dt = \frac{c^2 \cdot dm}{4\pi\sigma r^2 t^4} \tag{4}$$

which is the fundamental equation of this paper. It expresses the lapse of time corresponding to a differential decrease of mass in terms of known quantities.

For the computation it was convenient to use other units. Let τ be time in years, M be mass with the Sun's mass as unit, and R be the star's radius with the Sun's radius as unit. Then we have

$$\Delta t = 3.15 \times 10^7 \, \Delta\tau, \qquad m = 1.99 \times 10^{33} \, M, \qquad r = 6.95 \times 10^{10} \, R.$$

Also we have

$$\sigma = 5.72 \times 10^{-5}, \qquad c = 3.00 \times 10^{10}.$$

---

[3] Einstein, A., *1st die Trägheit eines Körpers von seinem Energienhalt abhängig? Ann. Physik,* **18**, 639 (1905).

Using these figures, (4) gives the relation,

$$d\tau = 1.63 \times 10^{28} \cdot dM/R^2 T^4 \qquad (5)$$

The division of stars of given spectral type into two classes according to absolute magnitude (giants and dwarfs) presented by Russell[4] is the basis of modern theories of stellar evolution. It is usually inferred that the course of evolution of individuals is along the lines of maximum frequency on the diagram of spectral type against absolute magnitude; the individual makes his debut as a giant of M-type and roughly zero absolute magnitude proceeds to B-type at approximately constant absolute magnitude, then declines from B to M while losing in luminosity so that at M-type its absolute magnitude is about 9 or 10. But there is considerable question whether each individual actually follows this course.

Eddington[5] has shown on theoretical grounds that the maximum temperature attained by a star will depend on the mass of the star and that only the more massive stars will attain B-type. Those of lesser mass will start to decline after having reached only A or F type.

If for some reason the stars all start with approximately the same mass and this mass is sufficient to carry them to B-type, the path of individuals on the Russell diagram will be simply that of the maximum frequency on the diagram. But if there is dispersion in the initial values of the mass then some individuals will short-circuit the main path, appearing on the dwarf branch without having attained B-type. Such a short-circuiting accounts for the presence of a considerable number of individuals between the giantline and the dwarf-line of the diagram.

In the very comprehensive discussion of the present state of knowledge of masses of the stars by Seares,[6] it is brought out that the observational data on the masses of the giants are meager. The mean masses of stars on the dwarf-line, however, seem well-defined. We may ignore, in the absence of an exact method of allowing for it, the effect of short-circuiting by small stars. This short-circuiting will affect the data in such a way as to make the decrease of mass of an individual in going from B to M type appear greater than it really is and so to make the time scale too long. Ignoring this effect, Seares' data allow the evaluation of a rough time scale along the dwarf branch.

From table XII of Seares' paper were taken the values for the effective temperature of the dwarfs of each spectral type, assuming this to be the same as for the giants in the range B_0 to F_0 inclusive. From Seares' table

---

[4] Russell, H.N., *Relations between Spectra and Other Characteristics of the Stars.* Pub. Amer. Astron. Soc,. **3**, 22.

[5] Eddington, A.S. *On the Radiation Equilibrium of the Stars. Zeit. Physik,* **7**, 351 (1921).

[6] Seares, F.H. *The Masses and Densities of the Stars. Astrophys. J.,* **55**, 165 (1922).

XIV were obtained the values of R $(=D)$ for each type and from his table XXIII were obtained the mean values for the mass of each type.

From these data was computed the value of the integrand in (6) for each spectral type. From the values so obtained was formed a column giving the arithmetic mean of the values of the integrand for each adjacent pair of types. These means were multiplied by the corresponding ΔM to effect an approximate integration by the trapezoidal rule. Taking the origin of time as the instant when the star is of type G_0, to correspond with the choice of the Sun as unit, the integrated time lapses were computed by summing the separate time intervals.

The results of the computation follow:

TYPE	Y	M	Δm	$\dot{y}\Delta m$	$\Sigma\bar{y}\Delta m$	time (yrs.)
B_0	1.78	10.	1.7	7.4	−564.	-9.2×10^{12}
B_5	6.92	8.3	2.3	28.1	557.	9.06
A_0	17.5	6.0	2.0	57.8	529.	8.62
A_5	40.1	4.0	1.5	108.	471.	7.67
F_0	104.	2.5	1.0	165.	363.	5.92
F_5	226.	1.5	0.5	198.	−198.	−3.23
G_0	568.	1.0	0.24	210.	0.	0.
G_5	1180.	0.76	0.08	136.	+210.	+3.42
K_0	2220.	0.68	0.06	217.	346.	5.64
K_5	5030.	0.62	0.03	492.	563.	9.18
Ma	27900.	0.59			+1055.	+25.2

In the table, y is equal to $10^{18}/R^2t^4$. The time in years in the last column is obtained by multiplying the quantities $\Sigma\bar{y}\Delta M$ by 10^{-18} and also by 1.63×10^{28}, the factor of equation (5).

In figure 1, M is plotted as a function of the time on the basis of the third and seventh columns of the table.

There is a correspondence between the results here and some given by Shapley and Miss Cannon[7] that is worth mentioning even though it leads to considerations which cannot be developed further at present. One might suppose, at first sight, that the number of stars in a given volume of space of each spectral type would be proportional to the amount of time that each individual spends in the several spectral classes. This is the case, at least roughly, for Shapley and Miss Cannon,[7] from a study of the recently-completed Draper catalog, find the following figures for the number of dwarfs of the several types in a million cubic parsecs of the space around the solar system:

$$B, 4.4; \quad A, 250; \quad F, 680; \quad \text{and} \quad G, 7,600.$$

Although further figures are not given, they remarks, "Dwarfs stars of classes K and M are probably much more numerous per unit volume than

---

[7] Shapley, H., and Cannon, A.J. *Summary of a Study of Stellar Distribution. Proc. Amer. Acad. Arts Sci.*, **59**, no. 9 (1924).

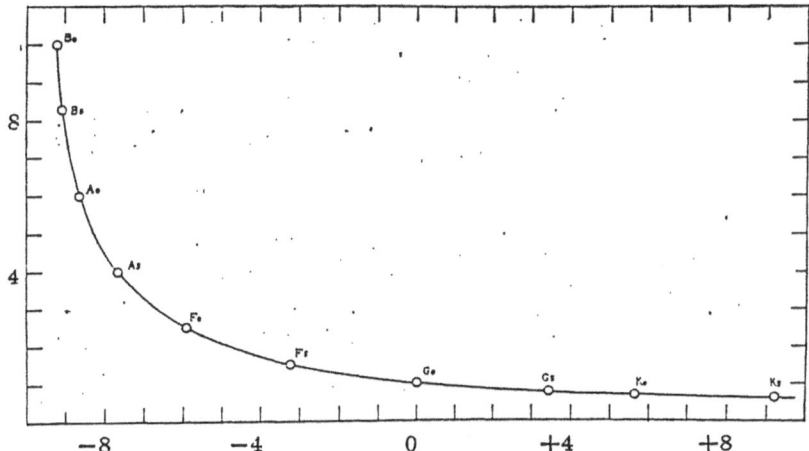

FIGURE 1. Mass of dwarf stars as a function of the time, the unit of the time (abscissæ) is 10^{12} years; that of the mass (ordinates) is the Sun's mass.

dwarf stars of class G ... The most important deduction from the table above is that the great majority of the stars are extreme dwarfs apparently indicating that a star spends most of its life in the later stages of the spectral series."

It is really doubtful whether this important deduction can be made from these data alone. The relative frequencies in the sky today depend on three things: 1, the relative duration within each spectral class; 2, the distribution in space of the birthplaces of the stars and in time of the birthdays of the stars; and 3, the motions of the stars with respect to the solar system in the course of their evolution. The exact analysis of the effect of these three causes does not seem to have been given. If the chaos of the motions makes the third factor ineffective and if we can assume a constant "birth-rate" among the stars, then (and only then, I think) is the deduction of Shapley and Miss Cannon allowable.

The relative duration of the stars within the various spectral types must be related in some way to the relative frequency of stars of each spectral type as found in the sky today. If we suppose that new stars are being born at a steady rate and old stars are dying at a steady rate so that the population of the stellar universe is approximately constant, then we should expect to find the actual number of stars of each spectral class in a given volume at any time to be exactly proportional to the relative duration of an individual within each class.

Comparison with observation, however, will neither support nor discredit any particular calculation of relative duration in each spectral class. From exact knowledge of relative duration and from exact observation of the relative frequency in the sky today, we can only infer the distribution in time of the stellar birthdays.

While, therefore, no great weight is to be attached to the comparison, it is of interest to note that the frequency of the dwarfs of various types in a million cubic parsecs around the solar system, corresponds roughly to the relative durations as computed here.

Statistics of Vocabulary

E.U. CONDON

Bell Telephone Laboratories
New York, New York

While studying some data on the relative frequency of use of different words in the English language, I noticed a rather interesting functional relationship which is here communicated. The note which follows, admittedly incomplete, is published in this form because the subject is one which I can not pursue and which may be of interest to those who are actively enagaged in the study of language.

Suppose one takes a large representative sample of written English, counts the number of times each word appears and arranges the words in order of decreasing frequency of occurrence. The nth word in such a list will then occur with an observed frequency which is a function of n, call it f(n). This function is clearly a monotonicly decreasing function of n, from the way the data have been arranged. But what is its form?

Two large published word counts are available. One is that of L.P. Ayres, "A Measuring Scale for Ability in Spelling," Russell Sage Foundation, 1915, and the other is that of G. Dewey, "Relative Frequency of English Speech Sounds," Harvard University Press, 1923. Each of these writers analyzed samples of 100,000 words of written English.

In the accompanying figure is plotted the logarithm of the observed frequency of the nth word against the logarithm of n. The circles are based on Dewey's count while the crosses are based on that of Ayres. The close approximation of the points to a straight line with unit negative slope is at once remarked. This suggests that there is something about the way in which man uses his language (Is the relation true for other languages?) which makes the frequency of occurrence of the nth word be given by a formula of the form,

$$f(n) = \frac{k}{n}$$

On this form some comments will be made at the close of this letter. An interesting question concerns the value of the constant, k. Supposing the law to be valid over the entire range of the language, k must have such a value that the result of summation over all the different words in the sample will equal unity. That is, k is determined by the equation,

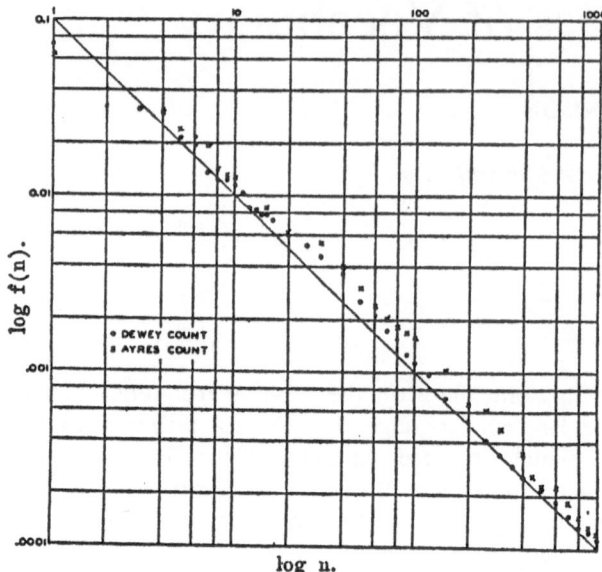

FIGURE 1.

$$k \sum_{n=1}^{m} \frac{1}{n} = 1.$$

Since for large values of m (the number of different words in the sample) the summation can be replaced by $\lambda + \log_e m$, where $\lambda = 0.5772$, is Euler's constant and the logarithm is to the natural base, one has a ready means of computing the value of k from the total number of different words in the sample. Dewey found 10,161 different words in his sample, accordingly the value of k is 0.102.

A sort of check on the accuracy of this representation is given by assuming that the most infrequent word occurred just once and inferring from that fact and the value of k the total size of the sample. The value would clearly be 10,161 divided by k, or 99,500, which checks well with the actual size of sample counted, i.e., 100,000.

On the figure the solid line has been drawn to represent the function with k = 0.100. It is seen to fit the data quite well, although there are systematic deviations.

Supposing the law to be substantially correct, the writer ventures to point out that it is perhaps a quantitative appearance in language of the Weber-Fechner law of psychology. In the language of the economist, it is a quantitative law of diminishing utility in vocabulary. The frequency of use of a word measures in some way its usefulness in transmitting ideas between individuals. Considering a vocabulary of n words it appears that the marginal increase in ideatransmitting power which can be accomplished by the addition of another word to the vocabulary is smaller the greater the value of n, according to the same law which governs the relation between the psychological increase in sensation accompanying an increase in the total intensity of the physical stimulus.

Recent Developments in Quantum Mechanics[1]

EDWARD U. CONDON

Palmer Physical Laboratory
Princeton University

The fundamental fact for quantum physics which has emerged from experimental investigations in the field is a double duality of wave and particle concepts. The wave theory of light, firmly grounded by the wave explanations of interference and diffraction and by Maxwell's electromagnetic theory of light waves, is now recognized as giving but one aspect of the nature of radiation. The corpusclar aspect of radiation is revealed in the photoelectric effect and in the Compton effect. In these effects Planck's constant appears as a universal constant which connects the wave and particle modes of description of the same radiation.

The radiation which functions as a wave disturbance of frequency ν waves per unit time, and as having a wave number of σ waves per unit length, when interference and diffraction phenomena are being discussed, also functions as a mechanical corpuscle of definite energy, E, and momentum, P, when effects like the photoelectric and Compton effects are considered. The connection between the quantities, ν, σ, E, and P, given by experiment is contained in the equations:

$$E = h\nu, \quad P = h\sigma.$$

(Note that the equations are made more symmetrical by using σ, the wave number, instead of its reciprocal, the wave-length, which is more commonly used.)

The duality is made double by recent discoveries concerning the electron. The most direct are those of Davisson and Germer, reported in the *Physical Review* for December, 1927. In their experiments a beam of electrons is scattered from the direction of normal incidence on a single crystal of nickel. It is found that there are certain directions of strong scattering and that the location of these directions can be quite closely represented by assuming the scattering to be governed by laws of wave interference, exactly parallel to the wave interpretation of the diffraction

---

[1] Based on a lecture delivered on July 19, 1928, before the summer school for engineering teachers of the Society for the Promotion of Engineering Education at the Massachusetts Institute of Technology, Cambridge.

patterns formed when X-rays are scattered by a crystal. The association of wave and particle concepts which holds for radiation is also seen to hold for electrons in these experiments, for when the mechanical momentum is varied the wave-number or wave-length effective for diffraction of the electron beam varies in just the manner given by the equation, $P = h\sigma$.

Instead of trying to go back of this duality, quantum mechanics is attempting to give a self-consistent mathematical development to the phenomena of quantum physics, which recognizes the duality as fundamental. The recognition of a wave-length associated with the motion of electrons really antedates the Davisson and Germer experiments, going back to de Broglie's wave interpretation of Bohr's rule for picking out the stationary states in atoms. In Bohr's theory of the hydrogen atom, it is postulated that circle orbits of just certain discrete sizes are possible, the rule being that the angular momentum of the orbit must be equal to an intergral multiple of h/2π. De Broglie recognized in this rule an analogy with the discrete modes of vibration of a stretched string. Such a string when fastened at the two ends may vibrate only in ways which make the wave-number of the standing waves be an integral multiple of the length of the string or in a superposition of several of these modes of vibration. If there is some kind of wave motion associated with a moving electron, it is then natural to suppose by analogy that the wave-number of the waves must be an integral multiple of the circumference of the orbit. One may readily see that this leads to the same choice of orbits as did Bohr's rule, when the fundamental connection, $P = h\sigma$, is assumed. It has the advantage that it gives some inkling of the reason for the existence of stationary states.

Schrödinger's contribution lay in the mathematical development of this idea. Working by analogy with the theory of propagation of elastic waves he set up a partial differential equation for the propagation of the waves associated with an electron which has had extraordinary success in dealing with quantum phenomena. His work was in no way a derivation by strict logical steps from previous knowledge of what this equation must be. In fact, it is now recognized that his equation is a simplified form of other equations for the waves discovered by Dirac this year. The correction given by Dirac is a small one so far as numerical magnitudes go but an important one in that it puts the magnetic effect of the electron spin into the theory in a rational way.

Whenever there are waves having a wave amplitude propagated in space and time one must have a physical understanding of the wave amplitude or quantity which does the waving. In the electromagnetic theory of light this wave amplitude consists of two vectors, the electric and magnetic forces, which are recognized physically by the forces which they exert on electrified or magnetized bodies placed in the field. When physical reality is attributed to the corpuscular light quanta, one must still find a way in which the wave field is important. Einstein suggested that the electromagnetic field determines the relative probability that the quanta go to different places. In a set of interference fringes, the wave amplitude is strong at some places and weak at others. The quanta go to the different places with a relative probability that is given by the wave measure of the intensity, namely, the square of the wave ampitude.

In attempting to provide an interpretation for the wave amplitude Schrödinger supposed that when in an orbit an electron does not remain in a little particle of negative electricity whose radius is of the order of 10^{-13} cm, but spreads out and fills a larger region of space with a volume distribution of charge. The total amount was just the charge on an electron, but the electricity was thought of as being actually smeared out over a region of space which is of the same order of magnitude as the orbits on Bohr's theory, therefore more like 10^{-8} cm. Against this view can be urged that in dealing with the free particle, the wave associated with a freely moving electron on Schrödinger's theory fills all space uniformly: there is not much left of an electron which is so smeared out that it fills all space uniformly.

Recognizing this difficulty in connection with problems involving the motion of electrons free from atoms, Born carried over Einstein's idea concerning the electromagnetic field as a probability field for the positions of the light quanta. He supposed the square of Schrödinger's wave function to give not the actual charge density arising from a smeared-out electron, but the probability of finding the corpuscular electron in the different parts of space. This view has been the basis of the developments due to Dirac and Jordan which have given a much more general form to the mathematics of quantum mechanics.

This interplay of wave and particle concepts which makes the wave field serve as giving the probability of different positions of the particle or particles governed by the wave field implies a fundamental limitation on the precision of certain physical quantities, a point which has been emphasized by Heisenberg and by Bohr. (See especially Bohr's article in *Nature* for April 14, 1928.) A plane wave of infinite extent is to be associated with a particle of which the momentum is exactly known. The infinite extent of the wave in turn means that the particle is equally likely to be anywhere in space. That is, exact knowledge of the momentum implies absolute ignorance concerning the position of the particle. By superposing waves of different wave-lengths, it is possible to have the different waves interfere everywhere except in a certain small region of space. Such a group of waves is taken as the wave representative of a particle *of which it is known* that the particle is in this region of space. It follows from the laws of wave interference that the smaller the region in which the waves do not destructively interfere, the greater the range of wave-lengths which must be represented in the different plane wave constituents which are superposed to make up the group. Recalling the connection between wave-length or wave number and momentum, it is seen that such a group of waves, which represents a particle known to be in a certain region, implies a range of values of the momentum or a lack of precision in the knowledge of the momentum. The size of the region in which the particle is known to be located may be thought of as the uncertainty of our knowledge of the position of the particle. The range of wave-numbers in the constituent waves of the group measures the associated uncertainty in the momentum of the particle. Calling Δx, Δy, Δz the uncertainties in positional coordinates and Δp_x, Δp_y, Δp_z the uncertainties in the momentum components, the laws of wave interference, together with the quantum law of association between the concepts of wave number and momentum, give the equations:

$$\Delta x \Delta p_x \gtreqqless \frac{h}{2\pi}, \ \Delta y \Delta p_y \gtreqqless \frac{h}{2\pi}, \ \Delta z \Delta p_2 \gtreqqless \frac{h}{2\pi}$$

If the laws of wave interference really do govern the motion of particles this implies that our simultaneous knowledge of the position and momentum of a particle may never be so precise that the product of the uncertainty in a coordinate multiplied by that in the associated momentum component that the product of the two uncertainties is less in order of magnitude than Planck's constant, h. A consideration of various methods of measuring simultaneously the position and momentum of a particle has indicated that all physical measurements are really subject to this fundamental limitation. This point is likely to prove of considerable interest to philosophers. It appears that in the concepts of position and momentum we are confronted with two quantities, either of which may be given a precise definition when considered alone, but when considered together there is a correlated vagueness about their magnitudes which appears as a fundamental law of nature.

In conclusion, it may be well to point out that the reason that the classical laws of mechanics prove to be so satisfactory for macroscopic things is that the wave-lengths of the wave phenomena for them are so small that diffraction effects are negligible, just as in many problems concerning light it is admissible to ignore the wave nature of light, as is done in geometrical optics. The laws of classical mechanics bear the same relation to those of quantum mechanics as the laws of geometrical optics bear to the wave theory of light.

Wave Mechanics and Radioactive Disintegration

RONALD W. GURNEY and EDWARD U. CONDON

Palmer Physical Laboratory
Princeton University

After the exponential law in radioactive decay had been discovered in 1902, it soon became clear that the time of disintegration of an atom was independent of the previous history of the atom and depended solely on chance. Since a nuclear particle must be held in the nucleus by an attractive field, we must, in order to explain its ejection, arrange for a spontaneous change from an attractive to a repulsive field. It has hitherto been necessary to postulate some special arbitrary 'instability' of the nucleus; but in the following note it is pointed out that disintegration is a natural consequence of the laws of quantum mechanics without any special hypothesis.

It is well known that the failure of classical mechanics in molecular events is due to the fact that the wave-length associated with the particles is not small compared with molecular dimensions. The wave-length associated with α-particles is some 10^5 smaller, but since the nuclear dimensions are smaller than atomic in about the same ratio, the applicability of the wave mechanics would seem to be ensured.

In the classical mechanics, the orbit of a moving particle is entirely confined to those parts of space for which its potential energy is less than its total energy. If a ball be moving in a valley of potential energy and have not enough energy to get over a mountain on one side of the valley, it must certainly stay in the valley for all time, unless it acquire the deficiency in energy somehow. But this is not so on the quantum mechanics. It will always have a small but finite chance of slipping through the mountain and escaping from the valley.

In the diagram (Fig. 1) let O represent the centre of a nucleus, and let $ABCDEFG$ represent a simplified one-dimensional plot of the potential energy. The parts ABC and GHK represent the Coulomb field of repulsion outside the nucleus, and the internal part $CDEFG$ represents the attractive field which holds α-particles in their orbits. Let DF be an allowed orbit the energy of which, say 4 million volts, is given by the height of DF above OX. Approximately, we can say that with this orbit will be associated a wave-function which will die away exponentially from D to B. Again, corresponding to motion outside the nucleus along BM, there will be a

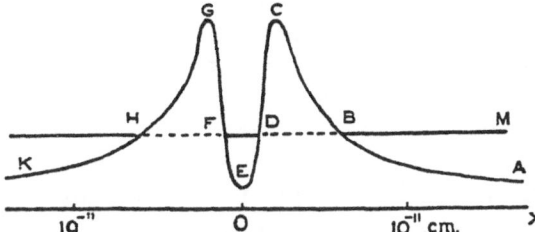

FIGURE 1.

wave-function which will die away exponentially from B to D. The fact that these two functions overlap in the region BD means that there is a small but finite probability that the particle in the orbit DF will escape from the nucleus along BM, acquiring kinetic energy equal to the height of $DFBM$ above OX, say 4 million volts. This occurrence will be spontaneous and governed solely by chance.

The rate of disintegration, that is, the probability of escape, depends on the amount of overlapping of the wave-functions in the regions DB and FH, and this is extremely sensitive to the height to which the potential curve at C rises above BDF. By varying this height through a small range we can obtain all periods of radioactive decay from a fraction of a second, through the 10^9 years of uranium, to practical stability. (In considering the transmutation of a molecule into its isomer, Hund found a similar vast range of transformation periods, *Zeit. f. P.*, **43**, 810; 1927.) If the potential curves for the interaction of an α-particle with the various radioactive nuclei are similar, we can obtain a qualitative understanding of the Geiger-Nuttall relation between the rate of disintegration and the range of the emitted α-particles. For the α-particles of high energy the wave function for outside motion will overlap that for the inside motion more, and the rate of disintergration will be greater.

Besides obtaining a general idea of the mysterious instability of the nucleus, we can visualise in this way one of the most puzzling results of recent experimental work. An α-particle having the same range (**2.7 cm.**) as those emitted by uranium should, if fired directly at the uranium nucleus, penetrate its structure; while faster α-particles should do so, even when not fired directly at the nucleus. It was therefore disconcerting when, on examining the scattering of fast α-particles fired at uranium, Rutherford and Chadwick (*Phil. Mag.*, **50**, 904; 1925) could find no indication of any departure from the inverse square laws. But from the model outlined above, this is what would be expected. For if the height of BM above OX represents the energy of the uranium α-particles, then a faster particle fired at the nucleus will simply run part way up the hill ABC and return without having encountered any change in the repulsive field or any nuclear particles (which are describing orbits within the region GEC).

The peculiar property of the wave mechanical equations which finds application here has also been applied to the theory of the emission of electrons from cold metals under the action of intense fields (Oppenheimer, *Proc. Nat. Acad. Sci.*, **14**, 363; 1928; and Fowler and Nordheim, *Proc. Roy. Soc.*, **A**, **119**, 173; 1928). Ordinarily, an atom does not lose its electrons because the attractive field of the atom remains attractive to all distances. But when an intense field is applied, then the attractive field is reversed in sign a short distance from the atom. This makes the resultant potential energy curve similar to that in the diagram, and so the atoms begin to shed their electrons.

Much has been written of the explosive violence with which the α-particle is hurled from its place in the nucleus. But from the process pictured above, one would rather say that the α-particle slips away almost unnoticed.

Note on the Velocity of Sound

E.U. Condon

Palmer Physical Laboratory
Princeton University

In any of standard textbooks of college physics we may read that the velocity of sound is given by $(E/\rho)^{1/2}$ where E is the modulus of elasticity for compression and ρ is the density of the medium. It is usually explained that the compressibility modulus is the one associated with adiabatic compression which is equal to $\varkappa p$, where $\varkappa$ is the ratio of the specific heat at constant pressure to that at constant volume, and p is the pressure.

That one uses the adiabatic modulus rather than the isothermal is then usually attributed to the fact that the sound vibrations are so rapid that the temperature inequalities do not have time enough for equalization. But with regard to the effect of heat conduction the contrary is the case — the vibrations are much too slow for approach to the isothermal condition. This fact is not altogether new but is certainly not well known, at least in the field of elementary physics teaching.

To get a clear view of the matter let us think that the wave is propagated adiabatically and consider how heat conduction works toward making the process isothermal. The maxima and minima of temperature occur in different wave-fronts which are spaced at intervals of $\lambda/2$ where λ is the wave-length. The change from maximum to minimum temperature at any particular place occurs in the time $\lambda/2u$ where u is the speed of wave propagation. From the theory of heat conduction we know that the distance d in which a considerable alteration of temperature penetrates during the time t is

$$d \sim (kt/sp)^{1/2}$$

(here $\sim$ means "of the order of magnitude of"), where k is the thermal conductivity, s the specific heat and ρ the density. Therefore if heat conduction is to cause appreciable departures from adiabatic behavior, conditions must be such that the distance a temperature inequality penetrates in time $\lambda/2u$ is comparable with $\lambda/2$, the actual distance apart of the places of maximum and minimum temperature. This leads to the result

$$\lambda \sim 2k/s\rho v$$

This result is more illuminating if we substitute for k its value as given in the kinetic theory of gases. There it is shown that the heat conductivity is of the order of the product of the specific heat, the density, the mean speed of the molecules and the mean free path, l. It is well known that the velocity of sound is of the order of the men speed of the molecules. Therefore we have the result that

$$\lambda \sim l.$$

In other words, the effect of heat conduction comes in for waves of such very high frequency that the wave-length of the sound waves is comparable with the mean free path of the molecules; that is, about 10^{-5} cm. Of course the ordinary wave theory which treats the gas as a continuous medium breaks down at such frequencies but even this rough argument makes the point that ordinary sound is of too low rather than too high frequency for conduction to be effective.

Looking now to the literature of acoustics on this topic I find that Rayleigh[1] has treated the effect of conduction on the propagation of sound waves and finds the main effect to be an absorption. His formula (18) shows that the amount of absorption introduced through the operation of heat conduction becomes appreciable only at the frequencies of the order of those given by the rough considerations of the preceding paragraph. But the result does not seem to have found its way into any of the textbooks. Herzfeld and Rice[2] have noted the point in connection with studies on supersonics and commented on the general prevalence of the erroneous view of the matter.

The point is not generally known, however, and the interesting connection of the wave-length with the mean free path of the gas seems not to have been pointed out before; these are the excuses for writing this note.

[1] Rayleigh, *Theory of Sound*, II, 28.
[2] Herzfeld and Rice, Phys. Rev. **31**, 691 (1928).

Food and the Theory of Probability

Edward U. Condon

Associate Professor of Physics
Princeton University

The United States Navy has found by experience that the proper amount of food to prepare per man for a mess of N men is not a linear function of N. As this is not in accord with what one might expect at first sight, it was thought worthwhile to consider the problem from the standpoint of the theory of probability. The *Navy Cook Book*[1] says,

All of the following recipes are intended for 100 men. When the number in any one mess is in excess of this amount, the following reductions are recommended:

300 to 500 men, reduce recipe by 8%
500 to 1,000 men, reduce recipe by 10%
More than 1,000 men, reduce recipe by 12%

In a large mess, considerable saving will result by carefully following the reductions recommended.

The object of this paper is to subject this point to analysis in terms of the theory of probability. It will be shown on three different assumptions that the curve which shows the proper amount of food per man for a mess of N men should have a fractional decrease from the amount per man for a mess of 100 men which is a function of N of the form

$$k \left(1 - \frac{1}{\sqrt{N\ 100}} \right) \tag{1}$$

where k is a constant. In Fig. 1 is shown a plot of the per cent reduction as a function of the size of the mess in hundreds of men. The jagged curve (a) corresponds to the recommendations given in the *Navy Cook Book*, while curve (b) corresponds to that given by probability theory assuming that the adjustable parameter k has the value (10.16).

We interpret the effect as due to the statistical fluctuation in the amount of food desired by any particular man from day to day. The theory is

---

[1] *The Cook Book of the United States Navy* (Government Printing Office), 1932, p. 9.

carried through for three different assumptions concerning the variation in individual demand of the individual man:

(1) Assume the service perfectly continuous and that the probability that one man will want an amount of food between x and x plus dx is given by

$$P(x) = \frac{1}{\sqrt{2\pi\sigma^2}} e^{-(x-m)^2/2\sigma^2} dx \tag{2}$$

where m is the mean amount consumed and σ is the dispersion.

(2) Assume the service to be such that a man must always take an integral number of units called "portions" or "servings" and that the probability that he take n servings is given by the Poisson law,

$$P(n) = \frac{m^n}{n!} e^{-m}, \tag{3}$$

where m is the average number of portions per man.

(3) Assume that the service is in portions but that a man has only the right to pass up that course altogether or to take one serving, no more. Let p be the probability that he take the serving.

Of the three assumptions we shall see that probably the last is best in accord with the facts.

We turn now to the analysis of the first hypothesis. Form (2), assuming the appetites of the individual men uncorrelated, the probability that the first man wants between x_1 and $x_1 + dx_1$, the second between x_2 and $x_2 + dx_2$, and so on, is evidently,

$$\left(\frac{1}{2\pi\sigma^2}\right)^{N/2} e^{-1/2\sigma^2 \sum_{i=1}^{N}(x_i - m)^2} dx_1 dx_2 \ldots dx_N \tag{4}$$

We want to know what is the probability that the mean amount per man, $\bar{x}$, lies between $\bar{x}$ and $\bar{x} + d\bar{x}$, where

$$\bar{x} = \sum_{i=1}^{N} x_i. \tag{5}$$

The solution of this problem is best carried out by introducing an N-dimensional mess space, in which there is one Cartesian axis for the appetite of each man.

A particular mode of consumption of the food in which the first man gets x_1; the second, x_2; etc.; is then represented by a point in this space. We omit the details. The result that is easily obtained from an appropriate algebraic mess is

$$P_n(\bar{x})d\bar{x} = \sqrt{\frac{N}{2\pi\sigma^2}} e^{-N(\bar{x}-m)^2/2\sigma^2} dx. \tag{6}$$

Since this function gives a finite, though small, probability for $\bar{x}$ as large as we please, it is plain that no finite amount of food will be enough to

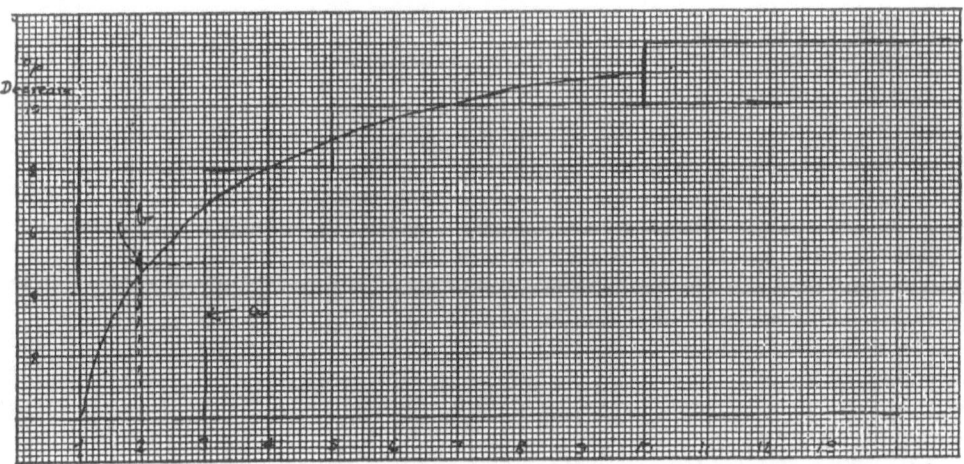

FIGURE 1. Per cent decrease in food per man as a function the size of mess in hundreds of men. The broken line *a* shows the recommendations of the *Navy Cook Book*, while the curve *b* is derived from theory.

satisfy the men completely on all occasions.[2] At this point we have to decide arbitrarily on a working limit. A good working limit is to adopt a policy of providing an amount of food per man for a mess of N men equal to

$$m + \frac{3\sigma}{\sqrt{N}}. \qquad (7)$$

The dispersion of the distribution of $\bar{x}$ is $\sigma/\sqrt{N}$ and by preparing enough up to three times the dispersion in excess of the mean, one is assured that there will be enough on all but 0.27 per cent of the meals, which is as high standard of service as anyone could ask. This figure is readily obtained from a table of values of the integral of the Gauss error curve.

According to this result the amount per man in a mess of 100 is $m + 3\sigma/10$, so that the amount per man in a mess of N is less than the amount per man in a mess of 100 by the amount

$$\frac{3\sigma}{10} \left(1 - \frac{1}{\sqrt{N/100}}\right), \qquad (8)$$

or the fractional decrease in the amount per man in a mess of N as compared with 100 relative to the amount per man in a mess of 100 is

$$\frac{3\sigma/10}{m + 3\sigma/10} \left(1 - \frac{1}{\sqrt{N/100}}\right) \qquad (9)$$

---

[2] This seems to be in accord with experience of those who have charge of the mess.

which is of the form of equation (1) and shows us a definite relation between the empirically determined parameter k and the quantities σ and m. From this relation we find

$$\frac{\sigma}{m} = 0.63.$$

Suppose we introduce a new variable $y = x/m$, and ask for the probability that a man wants an amount between y and $y + dy$, which is now the fraction of the amount which he wants on the average. Using the value we have found for σ/m, it is readily found that

$$P(y)dy = \frac{1}{0.63\sqrt{2\pi}}e^{-1.25(y-1)^2}dy. \tag{10}$$

In Fig. 2 we have plotted $P(y)$ as a function of y.

A somewhat disquieting feature of Fig. 2 is the area which has been shaded, corresponding to negative values of y. This area corresponds to a man's contributing to the mess instead of taking from it, a situation not pleasant to contemplate. This phenomenon probably does not occur, at least where there is good discipline, so we must regard it as a defect in the original assumption for $P(x)$.

Next we turn to the second assumption. From it the probability that the first man take n_1; the second, n_2; and so on; portions is

$$e^{-Nm}\frac{m^{n_1+n_2+\ldots n_n}}{n_1!n_2!n_3!\ldots n_N!} \tag{11}$$

If the total number of portions taken by the N men be called M, then

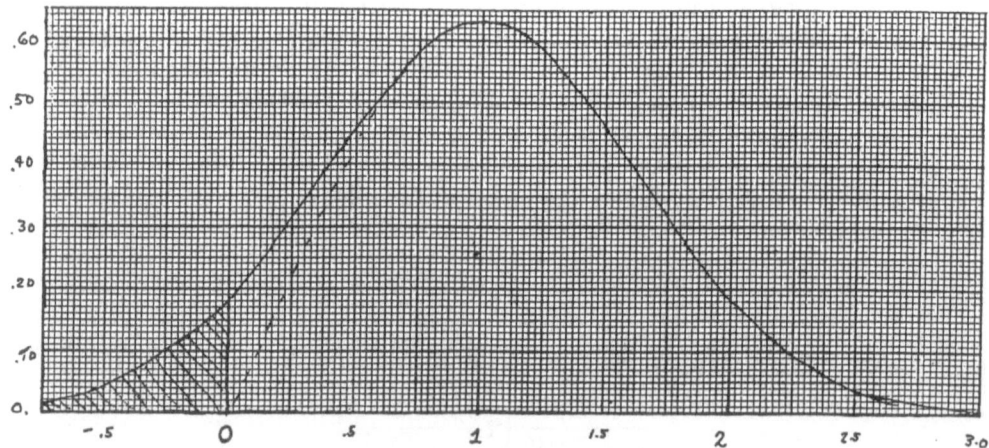

FIGURE 2. Probability distribution of appetite of one man in terms of average consumption of food as unit, calculated on assumption (1)

$$M = n_1 + n_2 + \ldots n_N$$

and the probability of a given value of M, which we call $P(M)$ is

$$P(M) = e^{-Nm} \sum_{n_1 n_2 \ldots n_N} \frac{m^M}{n_1! n_2! \ldots n_N!}$$

where the sum extends over all values of $n_1, n_2 \ldots$ whose sum is M. This sum is readily evaluated as follows: Consider the function $e^{\Sigma x_i}$. Expanded in a Taylor series in the N variables x_i, the terms of the Mth degree in the N variables are

$$\sum_{n_1 \ldots n_N} \frac{x_1^{n_1} x_2^{n_2} \ldots x_N^{n_N}}{n_1! n_2! \ldots n_N!}$$

with the same condition on the n_i in the summation as before. This evaluated for $x_1 = x_2 = \ldots = x_N = m$ becomes

$$\sum \frac{m^M}{n_1! n_2! \ldots n_N!}$$

which is just e^{Nm} times $P(M)$. But if we write $X = \Sigma x_i$, the terms of the Mth degree in e^x are given by the Taylor expansion in one variable. The value is $X^M/M!$ which for $X = Nm$ becomes $(Nm)^M/M!$. Hence,

$$P(M) = \frac{(Nm)^M}{M!} e^{-Nm}. \tag{12}$$

In other words, $P(M)$ is a Poisson distribution for an expected number of portions Nm. For large values of Nm we can approximate to the Poisson distribution with the normal Gauss error curve,

$$P(M) = \frac{1}{\sqrt{2\pi}} e^{-(M-Nm)^2/2Nm}. \tag{13}$$

In other words, the distribution of M/N, the portion per man, is normal around m with a dispersion given by $\sqrt{m}/N$.

Comparing this with the data as before, we find that to fit the Navy recommendations we must take m equal to 2.5. This assumption does not have the disadvantage of the first, namely, of leading to finite probabilities for negative consumptions of food, but it does lead to a rather spread-out set of probabilities of $P(n)$. The values given by the Poisson law for m equal to 2.5 are shown in the table below.

In other words, this interpretation leads us to the conclusion that something like a fifth of the men eat four or more portions. This is unlikely, and so we do not regard this second assumption favorably.

The analysis for the third assumption is very simple. The probability that exactly M portions will be taken in this case is

No. of portions	Probability a man wants this No.
0	8.2%
1	20.5%
2	25.6%
3	21.4%
4	13.4%
5	6.7%
6	2.8%

$$P(M) = \frac{N!}{M!(N-M)!} p^M (1-p)^{N-M}, \tag{14}$$

by an elementary proposition in the theory of probability. For large values of N it is well known that this can be approximated accurately by a Gauss error curve,

$$P(M) = \frac{1}{\sqrt{2\pi}} e^{-(M-pN)^2/2Np(1-p)}; \tag{15}$$

so that the mean amount of food per man is found to be

$$P + \frac{3\sqrt{p(1-p)}}{\sqrt{N}}, \tag{16}$$

whereas before we set our standard of service so that there will be enough to satisfy all demands in all but 0.27 per cent of the cases. This formula for the amount per man in a mess of N men again leads to an equation of type (2) and from the empirical value of k we may obtain a value of p. The result is 0.71. In other words, it appears that about 30 per cent of the men do not eat any particular course, on this hypothesis. This seems the most reasonable result so far obtained.

The foregoing analysis is intended merely to illustrate how statistical theory may be applied to this problem.

In conclusion, it should be noted in Fig. 1 that curve (a) according to the theory represents an empirical approximation to curve (b). The curve (b) is rising very rapidly between 100 and 300 men. This suggests that if mess sizes between 200 and 300 men are of very common occurrence in the service, appreciable savings could be effected by instructing the cook to discount the recipe by 5 per cent for messes between 200 and 300 men. This would replace (a) by the curve including the dotted portion, which is a much better fit to the theoretical curve.

Where Do We Live?
Reflections on Physical Units and
the Universal Constants

Palmer Physical Laboratory
Princeton University

Physicists are always busy measuring things. Measurement implies methods of measurement and the adoption of arbitrary units in terms of which things are measured. Our arbitrary units have been chosen from a definitely practical standpoint and thus have no direct relation to the fundamental quantities of nature; that is, to the so-called universal constants, the quantities which are not properties of special substances but are fundamental characteristics of the physical world. Let us therefore have a look at the unit systems of physics.

First let me make some general remarks on the c.g.s. system. The discovery of the theory of dimensions of physical quantities was a truly great forward step in our science. I will not dwell on it long. By its means the units of all physical quantities are fixed in a natural way in a manner designed to give an especially simple form to certain physical equations. In this way to each physical quantity is assigned a dimensionality, and an absolute unit for it is fixed in terms of the arbitrarily chosen fundamental units of mass, length and time.

For example, let us consider the unit of temperature. (I purposely choose this example because for some strange reason the rational methods of the absolute unit system have not been applied to this quantity.) We choose some simple law of universal validity in which the new quantity, temperature, enters together with other quantities whose dimensions have already been fixed by the same procedure. We may take the Wien displacement law which says:

$$\lambda_{max}T = \text{constant};$$

that is, that the wave-length of black-body radiation for which the spectral energy density is a maximum varies inversely as the temperature. This fixes the unit and dimensionality of temperature. Analogous to the texbook definition of the dyne or of the electrostatic unit of charge we have: *Unit temperature is that temperature for which the maximum spectral energy density of black-body radiation comes at a wave-length of one centimeter.* This fixes temperature as a quantity whose dimensions are L^{-1} and whose c.g.s. unit is cm^{-1}. Its numerical magnitude is also of convenient size,

namely 0.288 degrees so that the ordinary (slightly chilly) room temperature of 15°C is just equal to 1,000 cm^{-1} on this scale.

Besides illustrating the most striking case in which physicists have failed
to apply their absolute method of constructing units, the example also
illustrates a point which should be observed in applying the method. In
Planck's radiation formula the combination $hc\sigma/kT$ occurs, where σ is the
wave number of the radiation, T the absolute temperature and h, c and k
are respectively, Planck's constant, the speed of light and the Boltzmann
constant. We could define the unit of T in such a way that $hc\sigma/kT = 1$ when
$\sigma = 1$ cm^{-1}; that is, unit temperature would be hc/k cm^{-1} or 1.4317
degrees which makes 15°C have a value close to 200 cm^{-1}. This definition
of unit temperature is as good as the other from the dimensional
standpoint. Their ratio is the transcendental number equal to about 4.965
and defined as the root of the transcendental equation,

$$(1 - x/5)e^x = 1,$$

which appears when one calculates the value of $hc\sigma/kT$ for which the
Planck formula has a maximum spectral energy density on the wave-length
scale. As there are a greater variety of calculations in which $hc\sigma/kT$ occurs
than in which $\lambda_{max}T$ occurs, it is advisable to choose the second unit of
temperature rather than the first in building a convenient absolute system.[1]

---

[1] It is perhaps worth while to emphasize that there are two questions involved in
the ordinary method of constructing derived units. One is that of the arbitrary
choice of the fundamental law which shall serve as the basis for defining the derived
unit. Thus unit force may be defined as that force which gives unit mass unit
acceleration, giving force the dimensions MLT^{-2}. Or we may define unit force as
the force of gravitational attraction between unit masses at unit separation, which
makes the dimensions of force be M^2L^{-2}. In the first case we have $F = ma$ without
a universal constant, and have such a constant in $F = kmm'/r^2$. In the second case
we have $F = mm'/r^2$ without a universal constant and then must write $F = kma$. I
am trying to make the point that there is a fundamental entity here which can be
made to appear in different aspects according to the particular arbitrary scheme
selected. Likewise by using the Stefan law to define unit temperature we have unit
temperature such that at unit temperature there is unit total energy of black-body
radiation in unit volume. In this case the dimensions of temperature are not L^{-1} as
in the text but $M^{1/4}L^{-1/4}T^{-1/2}$ owing to the fact that energy density varies as the
fourth power of the absolute temperature.

The other arbitrary element is that which changes the choice of size of units
without altering their physical dimensions. Examples: the two temperature units
each of dimensions L^{-1} presented in the text; or the well-known case of the
Heaviside-Lorentz electromagnetic units which are dimensionally the same as
ordinary units but differ in regard to the way 4π appears in them.

For general discussions of dimensional theory see: Bridgman, *Dimensional
Analysis*, Yale University Press, 1922; Wallot, *Handbuch der Physik*, Vol. 2, Chap.
1; Campbell, *Physics the Elements*, Cambridge, 1920, Chap. 14; Campbell,
Measurement and Calculation, Longmans, 1928, Chap. 13; Lenzen, *The Nature of
Physical Theory*, Wiley, 1931, Chap. 1; Bond, Phil. Mag. 9, 842 (1930); Porter, *The
Method of Dimensions*, Methuen, 1933; Stansfield, Nature 131, 59 (1933).

The next point which I wish to make concerns the manner in which the arbitrary choice of basic units of mass, length and time is to be made.[2] The physicist wants to have a system such that quantities describing ordinary physical apparatus have convenient magnitudes. What dictates the choice of the size of "ordinary physical apparatus"? Clearly not so much fundamental physical considerations but considerations of a psycho-physiological nature. Ordinary apparatus is clearly apparatus which a human being manipulates readily. Its spatial dimensions are of the order of magnitude of the size of the human body; hence we have the meter and the yard and, in other metrological systems, units of this general size. The details of its construction are considerably smaller, about the size of the human hand, say; at least a lower limit is set by the scale of detail which the unaided human eye can readily perceive. This gives us the practical small units like the centimeter and the inch.

Similar remarks hold for mass and time. The order of magnitude of the practical unit of mass is determined by the amount of weight which one human conveniently manipulates, giving the kilogram and the pound. Smaller units like the gram get down toward the limit of sense discrimination in our unaided sense estimation of weight differences (estimation by "hefting" the weight held in the hand). This is more determined by the weight concept than the mass concept — but, as every physics teacher knows, these two concepts are not separated in the untrained mind and weight is generally regarded as the more direct one by beginning students. For time likewise our psycho-physiological structure suggests two orders of magnitude. One is the average waking interval during which the normal individual experiences a continuous stream of sense perceptions. This gives us the day. The other is the reaction time — the interval between receipt of a sensory stimulation and response to it. This gives us the second. The reaction time is clearly fundamental to physics, as limiting the experimentalist's ability to manipulate the apparatus with which he studies nature.

This, then, is where we live, in the sense in which the question is meant in the title of this paper. Our most direct experience concerns distances between the centimeter and the meter, masses between the gram and the kilogram and times between the second and the day. The order of magnitude of all these quantities is not so much a property of inorganic nature as it is of our psychological and physiological selves. As physicists we must seek more fundamental units in nature for dealing with our science and leave to the physiologist and psychologist the task of explaining why our life exists on the particular scale that it does.

This anthropomorphic element in our measuring systems is not to be deplored. It belongs there, for quantitative science is a unification by rational means of the quantitative aspects of our sense perceptions.

---

[2] For interesting accounts of ancient and modern units see: Glazebrook, Proc. Phys. Soc. **43**, 412 (1931); Gliozzi, Atti di Torino **67**, 29 (1931).

Therefore a practical system of units should be such that our directly observed data come to us in convenient magnitudes. But, since we know that the trend of theoretical physics has been toward greater abstraction from direct observation, it is worth while to attempt such an examination of our metrology. Recognizing the human element involved in the choice of our practical units, we do not expect them to be especially adapted to the simplest description of the fundamental facts of nature.

Our physical theories attempt to relate all the facts of experience to a very few fundamentals. These fundamentals are universal in character and their measures are called universal constants. The natural thing for metrology is to define a system of units in terms of which the universal constants have convenient small values. I believe Planck was the first to devise such a system of units the G.N. Lewis has given them some attention. Lewis calls them "ultimate rational units."[3]

In setting about to build such a system the first thing which strikes us is that there are more universal constants than we need. Therefore we cannot set them all equal to preassigned numerical magnitudes. This is just another way of saying that there are in physics several dimensionless, pure numbers of a physical character. Let us set down one fundamental system of units and see how these pure numbers come in. In my experience, which is mostly in the field of atomic physics, the most convenient system is that of Hartree[4]:

Hartree's atomic units

Mass: μ, mass of the electron $\sim 9 \times 10^{-28}$ gram
Length: a, radius of the first Bohr orbit in hydrogen $\sim 0.52 \times 10^{-8}$ cm
Time: τ, time required for an electron to describe $(2\pi)^{-1}$ of a revolution in
    the first Bohr orbit in hydrogen $\sim 2.42 \times 10^{-17}$ sec.
With these units many of the physical quantities of atom-theoretic interest have simple values. Thus, unit charge is $\mu^{1/2}a^{3/2}\tau^{-1}$ and is equal to the charge on the electron; unit angular momentum, $\mu a^2\tau^{-1}$, is equal to $h/2\pi$ in the usual notation, which is the angular momentum of the electron in the first Bohr orbit. All the derived units may be readily built up in this way. Some may object to the universality in that the length and time definitions refer to the particular substance hydrogen: to that I would reply that hydrogen is universal in being the prototype of all atoms, and so has a more

---

[3] Planck, *Theory of Heat Radiation*, Blakiston, 1914, p. 173; Lewis and Adams, Phys. Rev. **3**, 92 (1914); Lewis, Phys. Rev. **18**, 121 (1921); Lewis, *Contributions from the Jefferson Physical Laboratory*, Cambridge, 1922, Vol. 15; L.L. Whyte, *Critique of Physics*, Norton, 1931.
[4] Hartree, Proc. Camb. Phil. Soc. **24**, 89 (1928). See also Ruark, Phys. Rev. **38**, 2240 (1931); Clark, Phil. Mag. **14**, 291 (1932). As to the experimental values of the constants the best critical survey is that of Birge, Rev. Mod. Phys. **1**, 1 (1929). See also Bond, Phil. Mag. **10**, 994 (1930); Birge, Phys. Rev. **40**, 228 (1932).

general significance than it would hae if it were just a particular element. In other words, when we are in possession of a complete atomic theory then it will be possible to phrase definitions of these units accurately in terms of any atom; hence they are universal.

The first of the pure numbers[5] presents itself as the magnitude of the speed of light in these units, $c = 137a\tau^{-1}$. Another important one is the mass of the proton, $M = 1840\mu$. Still another is the ratio of the force of electrostatic repulsion of two electrons to the force of gravitational attraction at any distance of separation. This is $e^2/k\mu^2$, where k is the gravitational constant; its value is 4.18×10^{42}. In the relativistic theory of cosmology there is another, the "radius of the universe," whose empirical value we do not know so well. This length is of the order of $10^{35}a$. Still another example, so simple as to almost elude our attention, is the ratio of the proton charge to that of the electron, -1.

There are others which are obviously combinations of these, such as the ratio of the gravitational force between two protons and their electrical repulsion which can evidently be expressed in terms of 1840 and 4.18×10^{42}. The existence of relations of this sort raises the obvious question whether the ones listed may not be related in some fundamental way as yet unrecognized. Thus we see that $1840 \sim (40/3) \cdot 137$ but the discrepancy is outside the experimental error. Somehow $1840 \sim (4\pi^2/3) \cdot 137$ is more attractive but it does not fit quite as well! Similarly, $4.18 \times 10^{42} \sim (137)^{20}$ has a stimulating effect on the imagination but unfortunately here also the discrepancy is somewhat larger than the experimental error. It is a pleasant arithmetical game to hunt for simple relations of this sort but it is hardly likely that this method alone will find the true relations if any such exist.[6] The problem of unifying these numbers in terms of more fundamental theory than any we possess at present is thus an open one at present.

Another attack on the problem raised by these numbers is that of the attempting to give a theory which fixes the value of any one of them. Of course if we fix several of them by a unified theory we have also determined relations between them. Eddington[7] has made the most definite attempts

---

[5] Of course, the *magnitude* of any physical quantity in any system of units is a pure number, the ratio of the magnitude of that quantity to that of the arbitrarily chosen unit. But it is convenient to restrict the term somewhat to the case of the ratio of the magnitudes of two fundamental physical quantities. That is done here, although it must be admitted that the exact meaning of fundamental is not very precise.

[6] A number of relations of this type have been published recently: Witmer, Nature **124**, 180 (1929); Phys. Rev. **42**, 316 (1932); Perles, Naturwiss. **16**, 1094 (1928); Clark, Naturwiss. **21**, 182 (1932); Rojansky, Nature **123**, 911 (1929); Mills, Science **75**, 243 (1932) and criticism by Birge, p. 383.

[7] Eddington, Nature **124**, 840 (1929); Proc. Camb. Phil. Soc. **27**, 15 (1931); Proc. Roy. Soc. **A126**, 696 (1930); Fürth, Zeits. f. Physik **57**, 429 (1929); Bond, Proc. Phys. Soc. **44**, 374 (1932); Flint, Proc. Phys. Soc. **42**, 239 (1930); Beck, Helv. Phys. Acta **6**, 309 (1933); Schames, Zeits. f. Physik **81**, 270 (1933); Narlikar, Nature **131**, 134 (1933).

of this sort that I know. It is obviously out of place to discuss the details of his interesting hypotheses here.

There is a practical sense in which these numbers are of importance as they give us the means of passing from the particular Hartree system of units to all others which may with equal right be called fundamental. The great range of the numerical magnitudes occurring in physical theory makes it necessary to extend our unit system to include multiples and sub-multiples of the basic units of each kind. This is clearly recognized in the metric system where auxiliary units are provided by multiplying the basic units by powers of ten. Even this ten has an anthropomorphic origin, for it is the base of our system of numeration which is related to our possession of ten fingers. The powers of 137 give us a set of very convenient multiples for extending the Hartree system over the field of atomic problems. Thus with regard to length we have

$137a$: $(4\pi)^{-1}$ times the wave-length of the limit of the Lyman series in hydrogen

a: Radius of the first Bohr orbit

$a/137$: $(2\pi)^{-1}$ times the amount of the shift in wave-length of radiation scattered through $\pi/2$ in the Compton effect

$a/137^2$: Electromagnetic radius of the electron.

For energy the quantities of interest in atomic physics proceed according to the even powers of 137:

$137^2\mu a^2\tau^{-2}$: Relativistic energy equivalent of the rest-mass of the electron

$\mu a^2\tau^{-2}$: Twice the ionization energy of the hydrogen atom

$137^{-2}\mu a^2\tau^{-2}$: A quantity of the order of the relativistic fine structure of the hydrogen spectrum and of the energy due to magnetic interaction of spin and orbital angular momentum in hydrogen

$137^{-4}\mu a^2\tau^{-2}$: A quantity of the order of magnitude of the natural breadth of ordinary spectral lines on the energy scale.

These four units of energy are adapted for convenient discussion respectively of (1) energies involved in radioactive transformations of an atom, packing energies, and problems in nuclear physics generally, (2) the main energy terms in atomic spectra, (3) the more detailed energy terms in the structure of atomic spectra and (4) the limit on the applicability of Bohr's rule connecting the frequency of light emitted during a quantum jump with the difference in energy of the two states involved. Similar tables may be made for other quantities although for them the range of powers of 137 is not usually so great.

It is of some interest to note that the quantity 137 is the ratio of transformation from electrostatic Hartree units to electromagnetic Hartree units. This brings clearly to the fore the real reason why all magnetic susceptibilities (except for ferromagnetics) are very much smaller thin the corresponding electric susceptibility. If we apply an electric field to an atom, the induced dipole moment P is given by $P = \alpha E$ where E is the applied field and α is the atomic polarizability. It has the dimensions of a volume and for all atoms is of the order of a^3. Correspondingly, if we apply a magnetic field to the atom, we get an induced magnetic moment that is proportional to the magnetic field. The coefficient of proportionality may be called the magnetic polarizability and in electromagnetic units it also has the dimensions of a volume. But it is always much smaller than a^3. This is because of the difference in the mechanism involved. In the electric case, electric forces act on electric charges to displace them and cause an electric moment. In the magnetic case, however, the magnetic field has to induce an electric current, which is small owing to the factor $(1/c)$ in the Maxwell equation for electromagnetic induction, and this in turn has to produce a magnetic moment which is smaller still by the factor $(1/c)$ occurring in the Maxwell equation for Ampere's law. Thus we expect diamagnetic susceptibilities to be of the order of $(137)^{-2}$ times the corresponding electric susceptibilities. This is in fact the case. The same is true for the permanent moments of molecules. The permanent electric moments of molecules, as measured from temperature variation of the dielectric constant, are all conveniently measured in terms of the Hartree unit, ea. Now ea is of the correct dimensions to measure magnetic moment on the electromagnetic system, but the actual convenient unit is the Bohr magneton which is $ea/2.137$, or twice 137 times smaller.[8] This again comes from the fact that electric moments are produced directly by the charges while magnetic moments arise from motions of charges whose speeds relative to the fundamental speed of electromagnetic theory are of the order 1/137. While we live on a scale of space and time for which the speed of light is enormous and relativistic effects remote, the magnitude domain of atoms sees the speed of light as a smaller magnitude relative to its own speeds of common occurrence.

So much for the extensions of the Hartree units by powers of 137. These are obviously the extensions needed when dealing with electrons and their relation to the electromagnetic field as modified by the existence of Planck's quantum of action. If we are interested in molecular problems, extensions by powers of the quantity 1840 obviously are of interest. For example, the frequency ν of vibration of a harmonic oscillator is given by

$$\nu = (2\pi)^{-1}(k/m)^{1/2},$$

and so the radiation wave number corresponding to this is

$$\sigma = (2\pi c)^{-1}(k/m)^{1/2}.$$

For a molecule the binding forces are principally electrostatic in character so we may expert k, which has the dimensions of force per unit length, to be of the order $\mu\tau^{-2}$. The moving masses in a molecule are

---

[8] Since it is the square of the permanent moment which occurs in the temperature-dependent part of the susceptibility in the Langevin-Debye formula the relative magnitude of the polarizability and the permanent moment terms in the susceptibility is the same in the magnetic as in the electric case. See Van Vleck, *The Theory of Electric and Magnetic Susceptibilities*, Oxford, 1932.

nuclear in magnitude and so are of the order 1840μ. Also $c = 137a\tau^{-1}$. Thus it appears that our natural unit for molecular vibration frequencies is

$$\sigma = (2\pi \cdot 137)^{-1}(1840)^{-1/2}a^{-1} \sim 5100\,\text{cm}^{-1}.$$

In point of fact this is a very convenient unit for the purpose. The vibrational energy levels of H_2 in the normal electronic state have a spacing of $4260\,\text{cm}^{-1}$. For heavier molecules the frequency is lower, partly because their masses are large compared to the proton mass (a factor of the order $100^{-1/2}$ is easily introduced in this way) and partly because the electrostatic forces are weaker when the electron orbits are considerably larger than a as in the heavier molecules. Similarly for the rotational energy of molecules. The rotational energy states are characterized by the molecule's possessing an integral number of units of angular momentum. The energy is $p^2/2I$ where p is the angular momentum and I is the moment of inertia. Since the equilibrium position of the nuclei is determined by a quantum-dynamical treatment of the electrostatic forces, the length factor in I is of the order a but the mass factor is of the order 1840μ. Hence for $p = 1\hbar$ the natural unit of rotational energy becomes[9]

$$\hbar^2/(1840\mu a^2) = (1/1840)\mu a^2\tau^{-2},$$

or 2/1840 of the ionization energy of hydrogen. The wave-number equivalent of this is $119\,\text{cm}^{-1}$ which is of convenient size for the purpose in hydrogen although for other molecules the rotational energy is a small fraction of this unit rather than a multiple because generally the nuclear mass is greater than 1840μ and the inter-nuclear distance is greater than a.

A simpler example is that of density. Since the volumes of atoms are determined by their electronic structures they are of the order a^3. The mass is principally nuclear so the natural unit of density is not μa^{-3} but $1840\mu a^{-3}$. This is about 11.3 gram $\cdot$ cm^{-3}, rather larger than most densities of solids or liquids because the molecular volumes are more than a^3 by somewhat more than the molecular weights exceed 1840μ. But the main point is that this is a natural unit in order of magnitude and is obviously close to all solid densities. It is of interest to note that here we have a case where our natural unit agrees in order of magnitude with the practical c.g.s. unit. That is because density is an intensive property of matter and so is the same for matter in bulk as for single atoms.

Let us next consider temperature from the standpoint of natural units. The natural unit following the method introduced at the beginning of this paper is a^{-1} or $1.89 \times 10^8\,\text{cm}^{-1}$ which is 286 million centigrade degrees. This is clearly not a convenient unit for discussing our direct perceptual experience. But it is the natural unit for discussing temperatures in stellar interiors because at unit temperature on this scale we have the temperature so high that the quanta of most common occurrence are those which have

---

[9] $\hbar = h/2\pi$.

energies comparable with the ionization energies of atoms; there is almost complete ionization there.

To find a derived unit more appropriate for dealing with the temperatures of our common experience, we must single out the most important fundamental characteristic of such temperatures. It is that they are such that ordinary molecules in thermal equilibrium are distributed over a small number of rotational states. This suggests the use of $119\,cm^{-1}$, the natural unit of rotational energy on the wave-number scale as the unit of temperature. This is 170 degrees which makes room temperature (300 degrees) have the value 1.76 and shows that we have adopted a unit of convenient size by this method.

The argument which led us to this unit is not completely fundamental because it did not present the fundamental reason why it is that temperatures of common experience have the property that molecules are distributed over a small number of rotational states. The fundamental reasons for this are of two characters. One is biochemical and is connected with the fact that the biochemical equilibria fundamental to our existence require temperatures of this order. The other is cosmological and is connected partly with the fundamental reason why stars have the intrinsic luminosity which they do have (an atomic phenomenon because it is the scattering of radiation by matter which limits the flow of radiation out from the stellar interior), and partly with the fundamental reason why the radius of the earth's orbit is what it is (this is the problem of the dynamical theory of the genesis of the solar system which when completely solved will presumably relate the question to several pure numbers, such as the total number of particles in the universe, and perhaps the cosmological radius of the universe or the mean density of matter in our galaxy expressed in atomic units, and the ratio of gravitational forces to electrical forces). Such fundamental considerations determine the temperature of the earth's surface. Their coincidence with the biochemical criteria will afford part of the fundamental physical theory of the fitness of the earth as an abode for life.

This completes our survey of the Hartree units and their extension to other natural units of physical interest by means of the quantities 137 and 1840. Many other applications of the same kind suggest themselves, such as the natural unit of heat capacity or specific heat, but enough has been said to show how the method provides convenient natural units in every case in which the fundamental physics of the situation involved is clearly understood.

These systems obviously do not provide us with convenient units for astronomical purposes. These are provided by moving off from the atomic units with appropriate use of the pure number 4.18×10^{42} which takes us from the part of physics where electrodynamic effects predominate to that part in which gravitational effects are important. I will not attempt to do this in detail as it would take us rather more deeply into cosmological theories than is appropriate here. I merely mention as an example that, since we know the size of stars to be due to a balance of gravitational pressure and radiation pressure when matter exists at temperatures of the order of one unit on the atomic energy scale, this balance fixes a natural unit for stellar diameters which is of convenient size.[10] To fix a natural unit in this way for the size and mass of the earth would require a definite view of the mechanics of the formation of the solar system. Since we know how to fix the natural unit of stellar sizes and if we accept a tidal fission theory of the genesis of the solar system,[11] I think it is clear that we may in this way construct natural units for the sizes involved in the solar system. Our knowledge is rather too incomplete at present to do this in explicit form. Hence, provisionally, we need to make an *ad hoc* introduction of the earth's radius in terms of atomic units as a stepping stone from these units to some of the familiar phenomena of terrestrial experience. This radius is $R = 1.21 \times 10^{17}a$.

With this unit and our natural unit of density we find that the obvious natural unit of mass for the whole earth is

$$1840\mu a^{-3} \cdot (4/3)\pi R^3 = 12.6 \times 10^{27} \text{ grams}.$$

This is, of course, of the right magnitude, the actual value being 0.485 of this unit owing to the earth's mean density being about half our natural unit of density. To get back to the first thing in the freshman physics course, let us next consider the natural unit of acceleration of gravity. It is evidently the acceleration produced at a distance R by a body of mass equal to our unit terrestrial mass. This is $2,010 \text{ cm} \cdot \text{sec.}^{-2}$ and of course bears the same ratio to the value 980 as does our mass unit to the mass of the earth. In the same way the obvious terrestrial unit of time is, analogous to the atomic unit, the time in which a body describing a circular orbit of unit radius about an earth of unit mass describes one radian of its path. This is about 5,700 sec., a little over an hour and a half.

I will not go on with the further development of explicit units appropriate for terrestrial and astronomic uses. My object was simply to illustrate the method. Those who are familiar with the work of G.N. Lewis on "ultimate rational units" will recognize that there is a considerable similarity between what he did and the subject of this paper. The program here is much more modest, however. Nothing more is attempted than to show how the universal constants determine not one system of fundamental units but a family of systems fully adapted for convenient numerical discussion of the different parts of physics. "Convenient" is meant in the sense that the units are such that the numerical magnitudes actually occurring in nature are in each case of the order of unity so that there are no large powers of ten to be handled in the calculations. The ordinary practical c.g.s. system was exhibited as occupying a place among these

[10] Eddington, *The Internal Constitution of the Stars*, Cambridge, 1926, Chap 1.
[11] Jeans, *Astronomy and Cosmogony*, Cambridge, 1928, Chap. 16.

families that is conditioned by the physical nature of human being themselves.

Lewis' program was more ambitious in that he hoped to discover the actual numerical magnitude of new physical quantities through use of the postulate that such magnitudes would be "simple" numbers. The weakness of the postulate lies in its indefiniteness as to the meaning of a simple number. I use the same notion here in a less restricted form by supposing that in the appropriate fundamental units all physical quantities have numerical magnitudes of the order of unity. If one restricts simple numbers to mean numbers of the form $2^a 3^b 5^c 7^d \pi^e$, where a, b, c, d, e are themselves positive or negative numbers of magnitude less than 10 or rational fractions whose numerator and denominator are each less than 10, the field of simple numbers while finite still includes quite a variety of numbers and this method leaves us with no rational way of choosing which simple number applies in a particular case. Moreover there are cases in which this field is too restricted as, for example, the appearance of the transcendental number x, satisfying $(1 - x/5)e^x = 1$ in the Wien displacement law.

That the numbers occurring are of the order of unity in the appropriate fundamental units seems however to be quite generally true. This much weaker principle is generally attributed to Einstein.[12] He remarked that purely mathematical processes used in physics never produce large numerical coefficients. This is a purely empirical observation for which no reason is given. It is nevertheless an important supplement to the ordinary argument from dimensions. For example, we all know the classic example of dimensional analysis which gives the result that the period of swing of a pendulum is proportional to $(l/g)^{1/2}$ where l is its length and g the acceleration of gravity. The pure number factor of proportionality is not given by dimensional analysis. Yet the principle used here makes us confident that that factor will not be some great thing like 10^{10} or anything small like 10^{-10}. It is actually 2π which is, in this case, not only of the order of unity, but simple according to the above definition of simple.

I conclude these reflections with the hope that they may serve to clarify the relation of the unit systems of physics to the fundamental entities of physics. I am sure that adoption of fundamental units along the lines here indicated can do much to simplify the calculations involved in all branches of physics and to bring out the nature of the fundamental interrelationships of the subject.

---

[12] Einstein's only publication on this subject occurs in Ann. d. Physik **35**, 687 (1911). He gives the pendulum example and obtains $T = C(l/g)^{1/2}$ as usual and says: "One can, as is known, get a little more out of dimensional considerations, but not with complete rigor. The dimensionless numerical factors (like C here), whose magnitude is only given by a more or less detailed mathematical theory, are usually of the order of unity. We cannot require this rigorously for why shouldn't a numerical factor like $(12\pi)^3$ appear in a mathematical-physical deduction? But without doubt such cases are rarities . . . ."

Three Catch Questions

E.U. CONDON

Department of Physics
Princeton University

By "catch questions" I mean questions in physics to which all of us are somewhat tempted to give the wrong answers until we have thought them over carefully. They are of the greatest pedagogic value and so it seems to me that this department of the journal might serve well as a forum for their presentation. To start the ball rolling, I will contribute three that have been "sprung" on me by various friends.

(1) First we have a perpetual motion machine that should interest those who have followed the discussions on the force between charges immersed in a dielectric. In Fig. 1, line LL' represents the horizontal boundary between air and some good insulating oil with a dielectric constant of 6, say; by dielectric constant, I mean specific inductive capacity, and *vice versa!* A spherical body A having a charge e is half submerged in the oil. Near it is a wheel with its axis in the boundary surface and carrying four arms on each of which are equal negative charges e'. The force between A and the upper arm is ee'/r^2 while the force between A and the lower arm is $ee'/6r^2$; hence the wheel will be made to spin in the counterclockwise direction and may be used to do work.

(2) Now that we have disposed of the first law of thermo-dynamics, we may proceed to violate the second. Starting with two small specks of matter A and B we may surround them with mirrors in such a way that A radiates to B at a faster rate than B radiates to A when they are at the same temperature. Hence the state of equal temperatures for A and B does not correspond to equilibrium, as the second law demands; instead, B will build up to a temperature enough in excess of that of A to make the rates of radiation equal.

In Fig. 2 use the locations of A and B as foci for an ellipse indicated by the curve C. Draw AD anywhere in the general location shown. With B as center draw a circular arc from the intersection of the ellipse C with AD to the other point E where this circle intersects AD. Through E pass another ellipse with A and B as foci. The broken curve made of the elliptical arc C, the circular arc and the second elliptical arc may now be rotated about the axis AB to make a figure of revolution. Imagine the inner surface so formed to be a perfect reflecting wall.

94

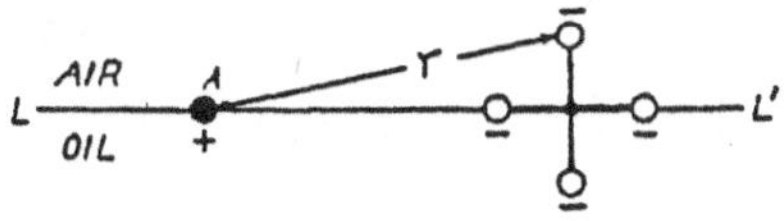

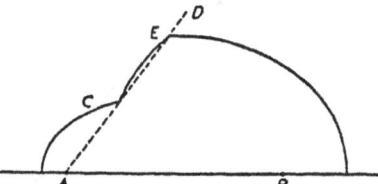

FIGURE 1.

FIGURE 2.

If now A and B are at the same temperature they will radiate at the same rate. But clearly *all* the radiation leaving A will be reflected at one or the other of the ellipsoidal surfaces and so focussed on B. On the other hand, of all the radiation emitted by B a finite fraction will strike the portion of the spherical surface produced by rotating the circular arc and thus will be returned directly to B. The rest is focussed on A. Therefore A sends all its radiation to B while B sends only a fraction of its radiation to A. Hence, for equilibrium, we must have the temperature of B enough greater than that of A to produce equality in the rate of transfer of radiant energy in the two directions.

(3) Consider a mercury barometer made in the usual way with a glass tube sealed at one end but with the tube held up by a spring balance. Will the downward pull on the balance be just the weight of the tube, or will it be the combined weight of the tube and the column of mercury? There will be, of course, a slight correction for the buoyancy of that part of the glass tube which is immersed in the mercury reservoir at the bottom but neglect this. It would seem that one must support just the weight of the glass for have we not been taught that the mercury is held up by the pressure of the atmosphere? Besides if we lift the glass tube up a bit we do not lift the mercury higher.

The argument advanced in connection with each of these three questions is incorrect; but where is the catch?

Mathematical Models in Modern Physics*

E.U. Condon

Associate Director
Westinghouse Research Laboratories
Pittsburgh, Pennsylvania

"All is fair in love and war" and, I might add, in theoretical physics. As in those activities painful adjustments are often necessary. Until a few years ago physics has been so definitely tied to Newtonian mechanics that, for many persons, progress in physics has come to mean understanding of phenomena within that frame. So much has been accomplished within that frame that it is naturally tempting to believe that mathematical physics should be thus restricted and that a phenomenon is not "explained" unless it is fully described in terms of classical mechanics.

Recent trends have emphasized the movement away from this restricted point of view by development of a non-Newtonian scheme of quantum dynamics within the field of dynamics itself. To be sure it had previously been recognized that classical dynamics alone will not give us everything. I refer especially to the inclusion of statistical ideas to explain the second law of thermodynamics, and to the difficulties of combining mechanical models of the electromagnetic field with the requirements of the Lorentz transformation. But in quantum mechanics we find the field of dynamics itself being invaded and revised, instead of simply being supplemented.

These trends raise the general questions: what *is* fair in theoretical physics? Shall we be satisfied with non-Newtonian explanations? What constitutes a satisfactory solution of a problem in theoretical physics?

It is not my desire to relate these questions to any profound presentation of technical philosophy. I shall simply try to expound a position which seems to me satisfactory enough to live in from day to day while doing one's work.

I accept the fact of man's intellectual curiosity, motivated directly or indirectly by the biological necessity of adaptation to the environment — of

---

* From "The Philosophical Concepts in Modern Physics," four addresses delivered at a symposium held Thursday afternoon, December 9, 1937.
(Note — The Franklin Institute is not responsible for the statements and opinions advanced by contributors to the JOURNAL.)

getting along in the world. John Dewey has said:[1] "Reflection is an indirect response to the environment, and the element of indirection can itself become great and very complicated. But it has its origin in biological adaptive behavior and the ultimate function of its cognitive aspect is a prospective control of the conditions of the environment. The function of intelligence is therefore not that of copying the objects of the environment, but rather of taking account of the way in which more effective and more profitable relations with these objects may be established in the future."

This purposive curiosity leads man to gather information about the environment in many ways, both natural and highly artificial and to reflect on it. He endeavors to build up some logical scheme into which his observations will fit for convenience in remembering them and transmitting them to others. Characteristic of modern times is the large scale on which activity of this kind is being done and its socialization. Formerly the spare time activity of a few gentlemen of culture, scientific study is now principally in the hands of professionals supported socially for full-time occupation at such work. Also characteristic of the times is the high degree of specialization which has resulted. Among the specialists we find physicists and among them more specialized yet a special breed of beings called mathematical physicists.

What are they doing and what is their function? Everybody knows what they do — they study carefully the results obtained by experimentalists and rewrite that work in papers which are so mathematical that they find it hard to read them themselves. What for? What values do they contribute? Or, as deKruif would ask, why keep them alive?

The answer is plain enough. *I take it to be the object of physics so to organize past experience and so to direct the acquisition of new experience that ultimately it will be possible to predict the outcome of any proposed experiment which is capable of being carried out — and to make the prediction in less time than it would take actually to carry out the proposed experiment.* When this shall have been done I will say that man has a complete understanding of his physical environment. Others may ask more, with this I am satisfied.

I lay great stress on the requirement that we be able to make the prediction in less time than it takes to try the experiment. The requirement is not strictly formulated: in some cases a loss of time could be tolerated to effect a great saving in the expense of performing an experiment.

The accomplishment of this goal obviously calls for a vast amount of experimental work but it also calls for a vast amount of organization of the results of experiment into concise forms which permit of ready application of the knowledge contained in them to new situations. Every experimental observation is a particular instance and is without value unless we may safely generalize from it. A certain amount of the generalizing is done by

---

[1] "Philosophy and Civilization," Minton, Balch, 1931, p. 30.

the experimentalist who makes sure of the definiteness of his conditions with control experiments and criteria of reproducibility. But usually that level of generalization covers only the critical evaluation of the data obtained in a particular experimental set-up or class of set-ups. After the experimentalist has provided good data in this sense someone must study its relation to the general body of knowledge previously known in order to fit it in the most consistent and concise way possible. There must be someone who is constantly watching the growth of that body of knowledge to see that no dangerous holes or weak places are left in it and above all to see that we are really getting a body that is functional: it must meet the test of "prospective control of the conditions of the environment." The theoretical physicist is the architect of that body of knowledge which the experimentalist is building for us.

Who will say what style of architecture he should employ?

The given data are usually in the form of particular quantitative statements of relation between numbers obtained in experimental observations. From these particulars one must try to infer a general relation structure embracing the expected result of all similar observations which might have been made but were not, and suggesting an expected result of other experiments which might be made in the future. Pure mathematics is the science of abstract relation structures, of the relationships which are capable of being stated in exact and quantitative form. It follows that pure mathematics is the chief tool of the theoretical physicist — so much so that he is often referred to as a mathematical physicist, a name that characterizes him by his tools rather than his function.

From this point of view it is clear that we have no a priori right to prescribe that only certain parts of the general theory of relations are to be used in coordinating experimental facts. Historically the methods of infinitesimal calculus as used more particularly in Newtonian dynamics have played the greatest role in the science of physics up to this century. So great has been the progress by these methods alone that it is no wonder that a school of thought grew up which identified progress in theoretical physics with the obtaining of results in this way. Naturally when certain leaders of theoretical physics found it convenient to employ more general mathematical methods followers of this school were greatly disturbed.

They were disturbed partly because they were more or less consciously committed to the restricted point of view which requires a physical theory to be Newtonian in form. Also, of course, they were disturbed because we all have our limitations and it really is hard to have to learn whole new branches of mathematics in order to extend one's "prospective control." I think this latter reason has, naturally enough, carried the most real weight, with the former reason being offered up as the rationalization because it sounds better.

If we look back over the history of modern theoretical physics we will see that the first kind of mathematical structure beyond the calculus of

Newtonian mechanics to be used was the theory of probability. With the founding of the kinetic theory of gases by Maxwell and Clausius in the middle of the nineteenth century, the statistical method entered physics. Regularly predictable phenomena of nature were correlated with statistical averages of undetermined and basically chaotic features of the model. The method was non-Newtonian in that it introduced supplementary statistical principles — to be sure the model, which was considered statistically, itself was governed by Newtonian mechanics.

This development never seems to have caused much of a stir philosophically. Yet it represents a radical extension of the method of mathematical physics beyond the Newtonian scheme. Why was there no stir? Probably because the model of the gas did obey Newtonian mechanics so the statistical features could be regarded as temporary and provisional stages of development to be removed by the time the theoretical edifice is completed.

But as time went on the statistical features became more and more essential to our understanding of thermal phenomena. The work of Boltzmann and Gibbs gave us a clear account of the fundamental meaning of the second law of thermodynamics in a form in which that great principle was essentially connected with the theory of probability. With the acceptance of their work, which was general a quarter century ago, it should have been clear that a wider base than afforded by Newtonian mechanics was being used for the construction of theoretical physics.

Paralleling these developments we see a great deal of effort being expended to provide a Newtonian mechanical foundation for the laws of the electromagnetic field as formulated by Maxwell. Since 1905 we have realized, however, that progress in this direction could not be made. This means that the laws of the electromagnetic field have to be accepted as a set of empirical equations, self-consistent, not inconsistent with mechanical principles, but also not explained in terms of a detailed mechanical model of the medium.

The middle period of development of quantum theory, in the years from 1912 to 1925 was characterized by the use of dynamics in the traditional mathematical form to which some additional ideas were attached. This was clearly recognized as a provisional stage by Bohr and others who played an important part in those developments — a partially successful scheme serving only to point the way to a more complete form of theory.

This more complete form came, as everyone knows, in the years 1925–27 with the work of deBroglie, Heisenberg, Schrodinger, Born and Dirac. In the decade which has followed there has been a flourishing growth of the theory of atomic structure. By means of the principles of quantum mechanics all of the known phenomena of atomic and molecular structure have found qualitative, and in many cases quantitative, interpretation. In the last three or four years this work has slowed down somewhat partly because all the relatively easy problems are done, partly because this work is outshone by the vigorous growth of experimental nuclear physics. It is interesting to note that definite theories of nuclear structure are so new that they have never existed in any other form than as quantum mechanical theories.

The new and unusual mathematical forms employed by the developments have caused a great stir in our physical thinking. The new methods have called for a complete revision of our basic ideas in dynamics in such a way as to leave no doubt about the fact that Newtonian methods are completely outgrown. Whether this is regarded as a satisfactory state of affairs dpends, of course, on whom we are trying to satisfy. Naturally if one believes that theoretical physics must ultimately be constructed wholly by Newtonian methods he will feel that the progress of the past decade is progress only in the sense that we have a scheme which gives many empirical correlations but whose deeper meaning has not yet been sensed. However, if he does not hold to such a more restricted view of the form of the ultimate development but merely requires internal consistency and agreement with observation then most of the developments of the past decade are indeed highly satisfactory. I have tried to indicate in the first part of these remarks a point of view according to which this second attitude seems to be tenable.

By that I mean that we have a reasonable view of science according to which we can regard the developments to recent years as a possible ultimate theory without feeling dissatisfied. Of course I do not mean to insist that the present form of quantum mechanics *is* ultimate nor that possible future developments which *go* beyond present ideas would be unsatisfactory or superfluous from the point of view taken here. Personally I believe that the theory is not ultimate, that the theory calls out for development in a direction which makes the Pauli Exclusion Principle be an inevitable rather than merely a possible consequence of its mathematical structure. But I also believe that these developments will come not by seeking for classical mechanical models of the universe but by more extensive exploration of the possible forms of mathematical model which are capable of coordinating all our past and future experience of the physical world.

In conclusion I want to make it clear that I am not attempting to act as spokesman for the scientific movement. The views expressed seem reasonable to me and I hope they will seem so to others but no one is asked to accept them on any other basis than their appeal to his own reason.

A Simple Derivation of the Maxwell-Boltzmann Law

E.U. Condon

Westinghouse Research Laboratories
Pittsburgh, Pennsylvania

The Maxwell-Boltzmann law is derived in a direct and simple way from the usual postulates that the mechanical system has a discrete system of allowed states and that each of these states has equal a *priori* weight in calculation of statistical averages.

The Maxwell-Boltzmann distribution law plays such a fundamental role in the entire theory of thermal phenomena that it is important to have as simple an approach to it as possible. In this paper, a method of derivation is presented which is believed to be essentially new[1] and which shows in an extremely simple manner the use of the assumptions involved.

There are only two basic assumptions needed: (1) The dynamical system in question is governed by quantum mechanics and, being a closed system, has, therefore, a discrete spectrum of allowed energy levels, (2) in calculating statistical averages, each state corresponding to one linearly independent wave function orthogonal to all the others, that is, consistent with known features of the problem, is to be given the same weight.

Let us first consider the case of a large number N of noninteracting molecules in a cubical box of edge L and volume $V = L^3$. The wave function for an allowed state of a single molecule is

$$u_{lmn}(x, y, z) = \left(\frac{2}{V}\right)^{1/2} \sin\frac{l\pi x}{L} \sin\frac{m\pi y}{L} \sin\frac{n\pi z}{L} \qquad (1)$$

and the corresponding energy is

$$W(l, m, n) = (h^2/8\mu L^2)(l^2 + m^2 + n^2). \qquad (2)$$

---

[1] *Note added in proof*: — Since this was written, I have noticed that a similar derivation is given by Kennard in his new book *Kinetic Theory of Gases*, p. 390. I have also had an interesting discussion with Professor G.E. Uhlenbeck, who remarks that the approach followed here is essentially a modern version of part of Maxwell's work as presented, for example, in Jeans' *Dynamical Theory of Gases*, Chapter V, p. 119 *et seq.* He also remarked on its connection with the discussion of statistical distribution of energy among a small number of particles as worked out for a problem in nuclear physics in a paper by Uhlenbeck and Goudsmit, pp. 201–211 in the *Zeeman Verhandelingen* (Martinus Nijhoff, The Hague, 1935).

   The allowed states for N noninteracting molecules in a box will have as wave functions a continued product of such one-particle wave functions (1) and the total energy of the N particles will be a sum of such one-particle energy expressions (2). Neglecting the symmetry restrictions on the wave function which would give rise to Einstein-Bose and Fermi-Dirac statistics, there will be one state of the system for each possible complete set of quantum numbers consisting of N individual sets $(l_\alpha, m_\alpha, n_\alpha)$, where $\alpha = 1$, $2, \ldots, N$ and the $3N$ quantum numbers $l_\alpha, m_\alpha, n_\alpha$ range independently over all positive integral values.

   It is important to know the total number of states of the system whose total energy is equal to or less than W for values of W large compared to the interval between energy levels. This is found by the method that has often been used in statistical mechanics. We introduce a $3N$ dimensional space. In this space the state of the system whose quantum numbers are $l_1m_1n_1, l_2m_2n_2, \ldots, l_Nm_Nn_N$ is associated with the point having this set of positive integers for Cartesian coordinates. There is thus one state per unit volume in this space. The number of states for which the energy is less than W is thus equal to 2^{-3N} times the volume of a $3N$ dimensional sphere of radius $(8\mu L^2 W/h^2)^{1/2}$.

   Denoting the number of states for the N particle system whose energy is less than W by $C_N(W)$, it is evident from dimensional considerations alone that $C_N(W)$ is proportional to $W^{3N/2}$ which fact is all that is needed in the derivation of the distribution law to be given below. From geometry it is known that the volume of an n dimensional sphere of radius r is

$$(\pi^{n/2}r^n)/\Gamma\left(\frac{n}{2} + 1\right) \tag{3}$$

and, therefore, the exact formula for $C_N(W)$ is

$$C_N(W) = \frac{1}{\Gamma\left(\dfrac{3N}{2} + 1\right)} \left[\frac{(2\pi\mu W)^{1/2}L}{h}\right]^{3N}. \tag{4}$$

   Now let us consider a system consisting of $(N + 1)$ noninteracting molecules in the box and ask for the probability that the extra molecule have an energy between w and $w + dw$ when all we know of the system is that the total energy is W. The probability, according to the statistical postulate, will be proportional to the number of states of the composite system for which w lies between $w + dw$ and hence the energy w_N of the other N molecules lies between $W - w$ and $W - w - dw$. Therefore, writing $P(w)dw$ for the probability that the energy of the extra molecule lie in this range, we have

$$P(w)dw \sim C_N' (W - w)C_1' (w)dw, \tag{5}$$

where $C_N'(w)$ is the derivative of $C_N(w)$ with respect to w. Using only that part of (4) which makes $C_N(W)$ proportional to $W^{3N/2}$, we have

$$P(w)dw \sim (1 - w/W)^{3N/2-1}w^{1/2}dw. \qquad (6)$$

Let us define an amount of energy called kT by the equation

$$W = 3NkT/2. \qquad (7)$$

Then, since N is very large, we can recognize that the first factor is equal to the exponential factor of the Boltzmann distribution, so

$$P(w)dw \sim e^{-w/kT}w^{1/2}dw, \qquad (8)$$

which is the usual Maxwellian distribution of velocities.

This mode of derivation brings out clearly that the $w^{1/2}$ factor arises from the fact that there are more states of the single molecule available in unit energy range at higher energies than at lower energies, whereas the exponential factor arises from the fact that if the single molecule gets more energy, there is necessarily less left for the other N molecules, and thus they are required to be in a range of energy where fewer states per unit energy range are available for them.

It is now a simple calculation to normalize (8) and find that

$$P(w)dw = (2/\pi^{1/2})e^{-w/kT} (w/kT)^{1/2}d(w/kT) \qquad (9)$$

and to calculate that the mean energy of a single molecule is

$$w_{Av} = 3kT/2. \qquad (10)$$

Physically, the N molecules in the composite system can be regarded as the perfect gas thermometer with which the single molecule is in thermal equilibrium.

Next, we can consider a slight generalization of the foregoing discussion, which leads to the Boltzmann distribution for systems having such a widely spaced set of allowed levels that they cannot be handled by means of a continuous $C(w)$ function.

We consider a composite system as before, which consists of N molecules in a box and in addition, the arbitrary quantized system whose allowed energy levels will be written $w_1, w_2, \ldots, w_\alpha, \ldots$ with the corresponding statistical weights (order of degeneracy) $g_1, g_2, \ldots, g_\alpha$. We suppose that the total energy of the composite system is known to lie between W and $W + \delta W$ and ask for the probability that the quantized part be found in the α the energy level, assuming that all states of the composite system with total energy between W and $W + \delta W$ are equally probable.

The probability of finding the quantized part in the αth energy level will, therefore, be portional to

$$P(W_\alpha) \sim C_N' (W - w_\alpha)g_\alpha\delta W,$$

which is also

$$P(w_\alpha) \sim (1 - w_\alpha/W)^{3N/2-1} g_\alpha \delta W.$$

As before, we may introduce the energy kT defined by (7) and recognize that the first factor is essentially equal to $e^{-w\alpha/kT}$ if N is a very large number. Therefore, we have

$$P(w_\alpha) \sim g_\alpha e^{-w_\alpha/kT}, \tag{11}$$

which is the familiar Boltzmann distribution law for systems having quantized energy levels. As before, we recognize the fact that the probability is proportional to g_α, the number of states of the quantized system of energy w_α, and the exponential factor representing the dependence on w_α of the density of states of the N molecules in a box which consitute the perfect gas thermometer-thermostat with which the quantized system is in equilibrium.

With the derivation of the Boltzmann law accomplished, one can proceed to develop the theory of the thermal properties of matter in the usual way by introducing the partition function

$$z(T) = \sum_\alpha g_\alpha e^{-w_\alpha/kT}, \tag{12}$$

from which the mean energy at temperature T is calculated by the formula

$$\overline{w}(T) = -k \frac{d \log z}{d(1/T)}. \tag{13}$$

The uses of the partition function are so well known that it will not necessary here to repeat that part of the development.

In deriving the distribution law for the distribution of translational energy of a single molecule in equilibrium with N molecules, the assumption was made that the energy of the $(N + 1)$ molecules is precisely known to be W. Looking back over the argument leading to (8), we see that the argument would have been essentially unaltered had we assumed that the composite system's energy had a value between W and $W + \delta W$.

Actually, when we have a gas at temperature T it will have a distribution-in-energy of its total energy, which we may derive by assuming the N molecules in a box to be part of a much larger composite system which includes N' more molecules where N' is very large compared to N. In that case we find that the distribution in energy of the N molecules is given by

$$P(w)dw \sim e^{-w/kT} w^{3N/2-1} dw, \tag{14}$$

where kT is defined by $W = 3N'kT/2$ which is supposed to be negligibly different from $3(N' + N)kT/2$ since $N' >> N$. From (14) we may calculate the mean energy of the system of N molecules in thermal equilibrium with the larger system. It comes out

$$\overline{w} = 3NkT/2, \tag{15}$$

which justifies the identification of T with the usual absolute temperature on the perfect gas scale. Similarly one may calculate the meansquare deviation from the mean of the energy of the N molecules

$$\Delta^2 = (w - w_{Av})^2{}_{Av} = 3N(kT)^2/2. \qquad (16)$$

Hence the fractional fluctuation in energy of the N molecules is

$$\Delta/w_{Av} = (3N/2)^{-1/2}, \qquad (17)$$

which is extremely small when N is large. This justifies the neglect of the distribution in energy of the N molecules in the argument leading to (8) by which the distribution in energy of one molecule was found.

The distribution law for the case of Fermi-Dirac statistics[2] is easily obtained as follows. We consider the system of N equivalent particles to which the Fermi statistics is to be applied, as a single quantized system in equilibrium with a larger perfect gas-thermometer thermostat. Then the distribution in energy of its quantized states is given by application of (11). In other words the relative probability of each of the independent states of the N equivalent particles governed by the Pauli exclusion principle is given by the Boltzmann factor, $e^{-w/kT}$.

The exclusion principle tells us that no two particles can be in the same quantum state and, therefore, if the allowed energy levels of a single particle are given by w_α, each allowed level of the system of N particles will be characterized by a set of quantum numbers N_α, one for each single particle state α, where N_α can have only the values 0 and 1 and $\Sigma_\alpha N_\alpha = N$. The total nergy is

$$W = \sum_\alpha w_\alpha N_\alpha.$$

To find the distribution-in-energy of the single particle in the case of Fermi-Dirac statistics, we have to calculate the mean value of N_β, the probability of occupation of the βth state by a particle. By the Boltzmann principle this is

$$(N_\beta)_{Av} = \frac{\Sigma_0 0 e^{-w/kT} + \Sigma_1 1 e^{-w/kT}}{\Sigma_{0+1} e^{-w/kT}},$$

where Σ_0 means the sum over all sets of N's consistent with $\Sigma N_\alpha = N$ and having $N_\beta = 0$ while Σ_1 means summation over all sets of N's having $\Sigma N_\alpha = N$ and having $N_\beta = 1$. If we write

$$W' = W - N_\beta w_\beta,$$

---

[2] A valuable account of Fermi-Dirac gas theory, together with its most important field of application, the Sommerfeld electron theory of metals, is given in the article by Sommerfeld and Bethe, *Handbuch der Physik*, Vol. 24/2 (Julius Springer, Berlin, 1933), p. 333.

this can be written

$$(N_\beta)_{Av} = \frac{e^{-w_\beta/kT}\Sigma_1 e^{-W/kT}}{\Sigma_0 e^{-w'kT} + e^{-w_\beta'kT}\Sigma_1 e^{-w'.kT}}$$

and, therefore,

$$(N_\beta)_{Av} = \frac{1}{A(T)e^{w_\beta/kT} + 1}, \qquad (18)$$

in which

$$A(T) = \Sigma_0 e^{-w'/kT}/\Sigma_1 e^{-w'/kT}. \qquad (19)$$

Here the parameter $A(T)$ is the ratio of the partition function for an N particle system from which the state w_β is excluded from the set of allowed single particle states to that for an $(N - 1)$ particle system from which this same state is excluded.

Evidently, if $A(T) \gg 1$, then the $+1$ in the denominator of (18) will be negligible and $(N_\beta)_{Av}$ will be approximately equal to $e^{-w_\beta/kT}$ times a factor that is independent of w_β. Therefore, in this limit the distribution law is not appreciably affected by the operation of the Pauli exclusion principle. On the other hand, if $A(T) \ll 1$ then for values of w_β small enough that $A(T)e^{w'\beta/kT} \ll 1$, we shall have $(N_\beta)_{Av} = 1$, that is, the low energy states are almost certainly occupied by one particle in each state which means a large departure from the classical distribution law.

The probability that a single particle have energy between w and $w + dw$ is, therefore, proportional to the product of the product of the number of states in this energy range, $c'(w)dw$ multiplied by the chance that a single particle state of energy w be occupied which is $[Ae^{w/kT} + 1]^{-1}$ so the distribution function is

$$P(w)dw \sim c'(w)\,[Ae^{w/kT} + 1]^{-1}dw, \qquad (20)$$

which is the familiar form from which the usual deductions of properties of the Fermi-Dirac gas may be made.

In conclusion, it is fitting to remark on the occasion of the seventieth birthday of one of the greatest teachers and productive workers that modern theoretical physics has known — Arnold Sommerfeld. Everyone of my generation grew up on atomic physics by way of his great *Atombau und Spektrallinien*, a large group have profited by the stimulation of his lectures on his American visits, and a fortunate few of us have derived boundless stimulation from the opportunity of working in his Institut für theoretische Physik in the former brighter days. All physicists join in whishing him a happy birthday and continued vigor with which to participate in the further developments of fundamental ideas to which he has contributed so extensively.

Physics in Industry[1]

E.U. CONDON

Associate Director
Research Laboratories
Westinghouse Electric and Manufacturing Company

To-day the nation's industries are working together as never before to do their part in the battle for freedom. It is fitting at such a time that we look over that part of civilized life for whose cultivation we as physicists are responsible, that we try to see how our science has been developing and in what direction we may cause it to grow in the future. We want to see clearly the job ahead so we can set about doing it earnestly and cheerfully; for we believe that, with effort, we can determine the future in peace as in war.

Although we can trace contributions to physics in America as far back as, say, Benjamin Franklin and his kite, the reasonably wide-spread organized research in physics that is now such an important feature of American academic and industrial life is a product of the last twenty-five years. Our professional group, the American Physical Society, is not yet fifty years old. For the first twenty years or so of its existence, very little was reported by its members which has affected the later development of our science in a basic way. The period was one of thoroughly sound beginnings marked by a few really outstanding contributions: Rowland discovered the magnetic effect of moving electric charge, and he perfected the diffraction grating for spectroscopy. Hall discovered the Hall effect, which is to-day a valuable tool in studying electrical conduction in metals and crystals. Michelson and Morley discovered the phenomena which laid the groundwork for the theory of relativity. Somewhat later Millikan measured the charge on the electron and the Planck constant.

It was a period in which very few of our physicists had the opportunity to contribute much to the advancement of science. Most of them were concerned with teaching, and the great industrial laboratories as we know them to-day simply did not exist.

A great change set in just after the first World War, although it is hard to see that that had much to do with it directly. Indirectly it did, for war acts as a stimulus to cooperative effort. The National Research Council was

---

[1] Essential substance of an address delivered at the opening of the Charles Benedict Stuart Laboratory of Applied Physics, Purdue University.

organized and soon after the war the Rockefeller Foundation established the National Research Council fellowships in physics, chemistry and mathematics. By this means some fifty young doctors of philosophy were each year enabled to continue their development as research men to such a degree of self-reliance that most of them could shoulder other burdens without losing their ability to carry on in research. I believe that these fellowships alone, supported at a cost that was absolutely trifling, have done more for the development of American research science than any other single thing. The fellowships reacted not only on the fellows but on the institutions and the faculties at the universities where the fellows worked.

In the 1920's research in fundamental physics expanded by leaps and bounds. The momentum carried by X-ray quanta was demonstrated by Arthur Compton. Davisson and Germer discovered electron diffraction; investigation of cosmic rays was inaugurated; atomic and molecular spectra were analyzed; and a group of young American theoretical physicists participated vigorously in the development of quantum mechanics.

It was also a period of rapid expansion of industrial physics in the electronic industries. From the birth of radio broadcasting in Pittsburgh in 1920 there was steady progress in the development of useful applications for phenomena associated with electric discharge in gases. The applied science of electronics was born to serve as a strong bond between physics and electrical engineering.

However, it was rather characteristic of that period that industry and physics were not closely associated. Academic physicists were a little inclined to be mistrustful of industry. In some vague way it was felt by many that the interests of engineering and of physics were best served by keeping them far apart after the engineering student had finished his sophomore course in general physics! Many a bright young man, turned physics instructor, felt he was casting pearls before swine when he had to expound the beauties of his subject to the sophomore engineers. Oddly enough, the engineering students could detect this air of condescension in their physics instructors and did not react to it with humility. I am often surprised at the vehemence with which leading engineers whom I meet will denounce the physicists who taught them in college. It was a most unhealthy situation. Happily it is now almost completely changed.

Then came the 1930's, largely a shameful decade in which people lived in insecurity, with much unemployment, little hope of professional advancement and in which there was a totally inadequate development of our natural resources in scientific talent and inventiveness. Like fools we watched while our present enemies bent every effort toward the mobilization of their universities and their industries to prepare the means that would crush us as a free people. In 1934 when Hitler started to rearm, many of our physicists were unemployed, and our young people could see little opportunity for a career in science. By 1936 the armament program had progressed so far in Germany as to require speed-up in the universities, such as we now have, in an attempt to supply the vast numbers of technical men which modern warfare requires. But we slept on, seeing in this not a marvelously coordinated effort to conquer us, but simply viewing it as a distant anti-intellectual attack on their universities. As late as 1940, even in 1941, there were those who argued against taking steps to resist these murderers of Lidice — because to do so would call for disturbing our way of life, and upsetting the calm, detached pursuit of knowledge in our universities!

On top of all this blindness to the international scene, we had to hear attacks on the scientific method as anti-social. Economic and social ills of society culminating in the depression were somehow blamed on science — as if all our troubles arose from too much of that calm analysis of carefully checked observations that is characteristic of scientific method! Somehow it just couldn't be admitted that there were some unsolved problems in applied social science. Somehow the fault lay with the physicists, for having taught us too well to understand the forms of matter and energy and how to use this knowledge to improve man's physical conditions of life.

Silly as such talk sounds now, it had quite a following in the 1930's. Support of scientific research was unduly curtailed except toward the very end of the decade. This situation might have been much worse if it were not for the fact that those physicists who had jobs, in spite of depression conditions, were having such a gloriously good time.

Heavy hydrogen was discovered, the neutron, the positron, artificial disintegration of the elements, artificial radioactivity, isotopic masses were measured with precision, new means for producing high voltage were developed, and finally there came the application of the new knowledge, while not yet five years old, to give medicine and biochemistry new tools of investigation that were soon to make vast changes in their approach to basic problems. With all this going on, is it any wonder that physicists did not find time even to answer the charge that they were responsible for the depression!

To be sure, some of them would have felt better to have had decent jobs; others would have liked, or at least their wives wished for, a little raise. And lots of fine young minds who would be most welcome in the profession to-day looked elsewhere for work, seeing how limited were the opportunities in science.

But the outlook was not entirely as dark as I have painted it. Toward the end of the 1930's there developed in America a new realization of the fruitfulness of bringing the newest results of physics into industry. Industrial research, already strongly established in the chemical field, began to expand in the direction of applied physics. The more progressive engineering schools established courses in applied physics. More physicists began to find their places in industry. Physicists began to see that their science, far from being restricted or polluted by an association with technical processes, could derive a new stimulus and a new significance that it could never possess so long as it was the private intellectual pursuit of a cloistered few.

For a time it looked as if there might be a separatist's movement whereby the industrial physicists would have professional societies of their own and would have little to do with the academic or "pure" physicists. Fortunately all such tendencies have been eliminated.

This war has completed the job of bringing all American physicists to a clear understanding of how much their science can contribute to the welfare of the people. At the moment their energies are devoted to applying their knowledge to winning the victory with a minimum of human suffering and material waste. Many university men are learning for the first time that there is a thrill and a deep satisfaction not only in discovering a new principle in science, but also in putting principles into application in new ways. When this war is over they will not forget this experience. Returning to the pure science laboratories in the universities they will have a sympathetic understanding of the problems of industry which will broaden and enrich the relationships of physics and engineering. I think they will even be more friendly and more tolerant of the sophomore engineers.

We must resolve not to neglect the cultivation of the basic science which we hope some day to apply. More and more, industry in America is recognizing the debt it owes to fundamental science — a debt it can hope to repay by fostering more basic research in its own research laboratories and by working in close cooperation with the universities.

I feel sure that those who are entrusted with furthering scientific research at colleges see this problem of applied physics in all its broad implications. They recognize, as we do in industry, that all physics is applied physics — so-called pure physics being simply that part whose application is to satisfy the curiosity of the physicists.

A Physicist's Peace

E.U. CONDON

Westinghouse Research Laboratories
Pittsburgh, Pennsylvania

This is being called a physicist's war. The physicists are being called upon to devote themselves, and they are devoting themselves, to the design and development of devices to help our army and navy win the fight for freedom.

There is no use contemplating the alternative to victory. One abhorrent glance at humanity degraded to slavery is enough. We must and will prevail in the conflict in which we are engaged. Of our ultimate military victory there is not the slightest doubt, in view of the comparative resources of the oponents. But let us be equally sure of victory in the peace — victory for the principles for which we fight in the world struggle today.

What, then, do we fight for? We fight for a world organization of society in which a maximum of human effort is available and effectively used for improvement of the physical, mental and spiritual well-being of all mankind. This calls right now for the destruction by force of those who have made this war for an opposite purpose — the utter degradation of all mankind to the service of the conquerors. The struggle will not be easy, but in it we shall learn many valuable lessons. We shall learn to work together, and we shall learn our strength when united in a worthy cause. We shall learn the joy of tremendous effort and sacrifice. We shall learn enough so that never again will it be necessary to go through a period of dull, stupid, enervating stagnation such as we experienced in the 1930's.

After the enemy's will to conquest has been broken, then our real battle begins. Great areas will have been devastated. Even today some of the battlefields of the first World War remain unreclaimed. Men will be battle-scarred and weary. Spiritual force will be at low ebb. We shall all feel like relaxing. But this we must not do!

In the first place, let us be clear about the fact that there will be an abundance of important work to do, more than ever before. Even in America — richest nation in the world — vast millions of our people are today undernourished, improperly housed and inadequately clothed. Think then of the enormous task that lies ahead in bringing to all mankind merely the material benefits which a small fraction of the people enjoys today. Let us not betray the heroes of this war in the peace. Let us make a pledge to continue the struggle for human betterment after the last shell has been fired — and with the same fierce earnestness that men are now displaying in defending their homes against the invaders. This Battle of the Peace shall be the most glorious adventure of the human race. Every man, woman and child in the world shall devote himself to it. The world's national and racial groups shall strive in keen and wholesome competition for the honor of making worthy advances and to help other less fortunate groups to go forward.

To win this battle completely we shall have to make many changes in our ways of life. We shall need to evolve a world political organization with power to maintain a society free from disturbance by aggressor groups. This is not nearly so difficult as many people suppose — given that we clearly understand our purpose and act accordingly. How much easier it would have been to stop Hitler in 1934 than now!

As citizens of a democracy we must mold our political organization to the form most suitable for the task ahead. The political solution resides, I am sure, in a close union of the allies now resisting the aggressors on a basis that provides for the gradual extension to all mankind of the liberal forms of democratic government of our own Constitution.

We shall learn to apply the rational methods of scientific investigation and experiment to the problems of political and economic management. We shall insist on a much larger public support of scientific research than ever before — not in the petty spirit of augmenting the private position of the scientists, but in recognition of the importance of research to the accomplishment of our goal. Instead of a NDRC we must have a vigorous and flourishing WPRC — World Progress Research Committee.

We shall face an enormous task in re-educating a whole generation of Germans, Italians and Japanese whose orientation at present deprives most of them of understanding the ideals for which we are fighting. This will have to be worked out by a mass program of occupational therapy in which these unfortunate individuals are given an opportunity to labor at reconstruction of the devastated areas, under conditions which will also open their hearts to an understanding of that for which we fought. This reconstruction, including the physical act of carrying back the loot they have taken from the invaded countries, will keep the Nazis busy for a long time — and definitely out of mischief if properly supervised.

Such a program for the peace will find physicists able to serve in a way that will entitle them to a place of honor. It is they who are charged with the duty of gaining as completely as possible a rational understanding of the physical forces of nature. Moreover, it is their additional duty to transmit this information to others who will put it to use in improving the physical well-being of all mankind.

The educational program we shall face is colossal. There has already been a hideous destruction by the Nazis of scientists, and of libraries and scientific equipment in Europe. There has everywhere been a terrible interruption in the training of scientists. And anyway, the number of persons who were trained in science before the war is now known to be totally inadequate for the work of the future.

It is entrusted to us to determine the future of mankind for a long time to come. Let us therefore work together with all men whose minds and hearts are ready: "With malice toward none, with charity for all, with firmness in the right as God gives us to see the right . . . to do all which may achieve and cherish a just and lasting peace among ourselves and with all nations."

Physics Gives Us — Nuclear Engineering

E.U. CONDON

Associate Research Director
Westinghouse Electric Corporation
and Vice-President
American Physical Society

Energy from the atom! No accomplishment in all human history has such stupendous or far-reaching consequences. It affects every field of human endeavor — political, military, social, industrial, and technical. To the engineer, who until now has been only mildly interested in atomic physics, it is a new field of engineering that he must actively study if he is not to be outmoded in his profession and citizenship.

To the familiar subjects of civil, mechanical, electrical, electronic, and chemical engineering, a new field has been given us by the physicists — nuclear engineering. This can be defined as the art of applying nuclear transmutations of matter to useful purposes. The subject of nuclear engineering, which has gradually been developing over the past fifteen years, is very much in everybody's mind because of its application to the making of a military weapon that ended the war eight days after it was first used.

It was considered in many official quarters that the war with Japan might have continued for another year. This might easily have meant the loss of another million American and British lives, probably the lives of even more Japanese, and a cost to us of upwards of 200 billion dollars. Instead, peace was restored at the cost of the lives of fewer Japanese and of none of the American lives that would have been lost and at a cost of only two billion dollors to ourselves. And moreover it has put us in possession of the means of assuring peace through world organization if the knowledge of the new weapon is used properly.

Before the atomic bomb, nuclear physics had provided a host of new ideas having peaceful uses — neutrons for cancer therapy, artificial radioactive materials for treatment of leukemia and of cancer, and for use in fundamental chemical studies both in biology and in chemical industry. So important had this field of work become just before the war that several large companies, among them Westinghouse, were considering the manufacture and sale of these radioactive materials. Although the war interrupted this activity and placed over all nuclear research a tight secrecy restriction, it enormously accelerated the research that ended so dramatically in three atomic bombs, one exploded experimentally and two dropped on Japan.

With the war ended we can now devote our energies to active cultivation of the applications of nuclear engineering to peaceful purposes — to better ways of producing neutrons and high-energy electrons for therapy and of artificial radioactive materials for all kinds of uses. Moreover we are standing on the threshold of the era in which atomic power will be developed, surely to be the most important engineering achievement of the next generation.

All sorts of prognostications are being voiced about the future of atomic power. Some say it will come only in the very distant future and may not then be practical; others are rashly predicting automotive power from U^{235} in a very few years. The wide variance in predictions comes about largely, of course, from the fact that most of the prophets have little but a crystal ball to guide them.

First let us get the main points of the story in terms of answers to some questions that occur at once to every technically trained man.

What is atomic energy? All energy used industrially comes either from the work done by falling water or from the combustion of fuels — coal and petroleum products principally. In combustion of coal the atoms of carbon combine with oxygen of the air to form carbon dioxide with release of energy. The characteristic thing is that the atoms involved in the combustion process are not changed intrinsically — the carbon atom from coal is still a carbon atom in the CO_2 of the flue gas. The energy used is that made available with the formation of CO_2 molecules from C from coal and O_2 molecules from the air. This is called chemical energy.

What is being called atomic energy is the energy associated with changes in the basic chemical nature of the atoms. In a chemical reaction, atoms of the same kind are present before and after, as in the familiar combustion process

$$C + O_2 \rightarrow CO_2$$

But in nuclear reactions, atoms are made to react in such a fundamental way that the product atoms are not the same as those we start with. For example, the nuclear

$$H^1 + Li^7 \rightarrow 2He^4$$

reaction is a process that actually occurs in the laboratory, an English discovery in 1932. Hydrogen reacts with lithium to give helium! This is atomic transmutation and quite outside the scope of the classical science of chemistry.

Physicists are accustomed to expressing the energy released in terms of electron volts per atom transformed, an electron-volt being the work done on an electron when it moves through a potential drop of one volt. An electron-volt is 1.60×10^{-19} watt-second and is therefore a definite amount of energy like foot-pound in ordinary work. Chemists usually express energies of reaction in calories per gram mole of material transformed. The heat of formation of CO_2 is about 94,000 cal per mole, that is for formation of 6.06×10^{23} molecules of carbon dioxide (for that is the number of molecules in a gram mole of any substance.) Because one calorie is 4.18 watt-seconds, it follows that the energy release in watt-seconds for forming one molecule of CO_2 is

$$\frac{94,000 \times 4.18}{6.06 \times 10^{23}} = 6.5 \times 10^{-19} \text{ watt-second} = 4.1 \text{ electron-volt}$$

On the other hand, laboratory measurements show that the energy released in the nuclear reaction by which hydrogen and lithium give helium is 17 million electron-volts per atom of lithium consumed — millions of times greater, weight for weight, than the chemical energy released in the burning of coal. The energy release in many such nuclear reactions is, generally speaking, of the order of millions of times the energy, weight for weight, released in chemical reactions.

Why is not atomic energy obtainable practically by "nuclearly burning" of hydrogen and lithium to form helium? The answer is furnished by comparison with coal. Coal (or any other chemical fuel) is valuable not only because of the energy release, but also because a self-maintaining fire can be made in which carbon and oxygen continue to burn. Of what use would coal be if a thousand dollars' worth of matches were used to burn a ton of coal? That was essentially the situation with all nuclear reactions prior to discovery of the phenomenon of uranium fission in 1939.

To make hydrogen atoms react with lithium it was necessary to ionize them and accelerate them in some kind of high-voltage apparatus. Of the many accelerated only a few struck lithium atoms in such a way as to react. The energy used to accelerate the others was wasted. The net result was that more energy is used in the experiment than that released. The overall output-input ratio was less than zero.

Hence from 1932 to 1939 we were inthe position of knowing that large energy releases were possible from many different nuclear reaction — but these could be produced only in laboratory apparatus that required more energy for their operation than was liberated by the nuclear reactions.

How did uranium fission change this picture? However, as soon as the word of discovery of uranium fission reached this country from Germany in January 1939, it was at once realized by physicists that the possibility of getting atomic power in useful form was within reach. But first let us say what uranium fission is. Uranium is the heaviest atom occurring in nature. The nucleus of uranium contains 92 protons surrounded by 92 electrons. One kind of uranium nucleus, U^{235} contains, in addition to the 92 protons,

143 neutrons, giving a total weight (i.e., atomic weight) of 235. *Another and predominating kind, U^{238}, contains 146 neutrons raising the weight to 238. When a neutron strikes a uranium nucleus in the right way, the nucleus breaks up by falling apart in two approximately equal fragments with the release of about 200 million electron volts per atom split. Great at this is it is no better, weight for weight, than the reaction that forms helium from hydrogen and lithium; in fact it is only about half as good from an energy release standpoint.

The essential thing about uranium fission is that the uranium atom falls apart in such a way as to produce two more or less equal fragments — *and to liberate several more free neutrons*. It is this neutron liberation that makes a self-maintaining process possible. The splitting requires a neutron to make it go — and the splitting process itself acts as a source of neutrons which can cause more uranium atoms to split. Here is the basis of a self-maintaining process, technically known as a chain reaction, such as is ordinary combustion.

Why, then, does not ordinary uranium explode or at least "burn nuclearly"? There are complications. Because several neutrons are released at every fission, a chain reaction is possible. But to make it an actuality, one of the several neutrons released must actually produce another fission to keep the process going. Otherwise the nuclear "fire" goes out.

If all the neutrons released produced more fissions the material would explode violently. But because neutrons move rather freely through matter (like X-rays) many are lost by escaping through the surface. Remedy: use a big enough lump to get a smaller surface-to-volume ratio. In other words unless the lump of fissionable material exceeds a certain critical size the chain reaction cannot proceed.

Another complication is that impurities in the uranium have a powerful effect on neutron absorption. This is very difficult to remedy for appreciable losses result from the presence of only one part per million of some materials, and it is no easy matter to manufacture anything of that purity on an industrial scale.

The worst complication of all was that uranium itself absorbs neutrons in other ways than those that produce fission. This phenomenon was both a blessing and a curse to the aims of the military project. It turns out that the over-all effect of this non-fission absorption of neutrons by uranium is sufficiently great to prevent the explosion of perfectly pure uranium even in so large a lump that escape of neutrons through the surface is negligible.

---

* For convenience, the approximate weight of an element is given as a superscript to the chemical symbol. The atomic number, when given, is a subscript preceding the symbol. Thus $_{92}U^{238}$ is uranium of atomic number 92 and atomic weight 238 approximately.

Neutrons given out in the fission process are "fast," i.e., have speeds corresponding to several million electron volts of kinetic energy. Such fast neutrons colliding with uranium atoms have a rather great chance of losing energy without being caught and without producing fission.

Neutrons of intermediate speed produced this way are unable to produce fission in U^{238}. They can do so only in U^{235}, which forms only 1/140 part of natural uranium.

Neutrons of a particularly low energy (about ten electron volts) are very likely to be captured by U^{238} to form U^{239}. This is very important! More on this later. This happens so readily that so many neutrons are used up this way that a chain reaction cannot be maintained in ordinary uranium.

An uncaptured neutron continually loses energy by colliding with atoms as it diffuses throughout any material until its average energy is that of the heat motion of the atoms of the material. Neutrons of certain extremely low energies are strongly captured by U^{235} to produce fission.

The clue to *possibly* making the chain reaction go with ordinary pure metallic uranium, which contains all kinds of uranium atoms but is predominantly U^{238} was to arrange the uranium in a lattice of small lumps so that many of the fast-moving neutrons would diffuse out of the uranium into some surrounding material. Here many of them would be slowed down before diffusing back into the uranium. The idea was that most neurtons would thus escape being caught by U^{238} until they had lost so much energy that capture by U^{238} was unlikely. Ultimately, though, they would return to the uranium lumps and be of sufficiently reduced speed to cause fission in U^{235}.

In the technical vocabulary of nuclear engineering this other material that keeps neutrons in custody and helps them lose energy until they are safe from capture by U^{238} is called the *moderator*. Evidently the moderator material must not absorb too many neutrons or the reaction will be stopped by this circumstance. Besides the quality of not absorbing neutrons, it is desirable to use material of low atomic weight. This is because the neutrons to be slowed collide elastically with the nuclei of the moderator material and so give up more energy at each impact if the two partners of the collision have nearly the same mass. The hydrogen content of ordinary water would be ideal from this viewpoint but absorbs too many neutrons. Heavy water is satisfactory from a neutron-absorption standpoint but previously had not been available in sufficient quantity. Metallic beryllium is a possibility but proved too expensive so that graphite was finally adopted, although not until processes were developed for manufacturing it to much higher standards of purity than is usual in ordinary industrial practice.

As this qualitative picture evolved prior to January 1942 the question of whether a chain reaction would go remained unanswered because of lack of exact quantitative knowledge of the various absorptions involved. But as knowledge accumulated, it became more and more probable that such a

Release of atomic energy depends on the phenomenon known as fission. A neutron moving with the right velocity strikes a uranium 235 atom (or plutonium atom), which breaks into two middle-size atoms whose total masses are very slightly less than the mass of the disintegrating atom. The difference appears as a whale of a lot of energy, according to the Einstein law of mass and energy equivalence.

lattice of uranium lumps and moderator — now called a pile — would go, i.e., a chain reaction continuously releasing atomic energy by fission of the U^{235} in it would be self-maintaining.

How can the pile be kept from blowing up? If a pile is so arranged that, on the average, more than one fission results from the neutrons produced by each fission, then clearly the number of neutrons present, and the amount of heat generated, increases by the compound-interest law. If a great multiplication happens rapidly — say in a small fraction of a second — then the phenomenon becomes an explosion. In short, we have an atomic bomb. Even if the reaction occurs slowly the pile would soon be destroyed by melting if the multiplication were allowed to proceed.

One way to control the pile is to provide passageways through it into which rods of material that strongly absorb neutrons can be placed. When these rods are in they absorb so many neutrons that the chain reaction is stopped. As these are slowly withdrawn a point is reached at which the reaction is just able to proceed. If pulled out farther the neutrons are able to multiply more rapidly and the pile operates at a higher power level. To stop the pile the absorbing rods are simply pushed back in farther. Cadmium and boron-containing steel are suitable materials for the control rods.

The language of the preceding paragraph implies that the time scale is slow enough for an operator to maintain control by manual operation of the rods or by use of a similar slow-acting control mechanism. That is in fact the case owing to another phenomenon in the fundamental physics of fission — delayed neutrons.

It was discovered in May, 1942, that most but *not all* neutrons emitted in the fission process come out instantly. The uranium nucleus in splitting apart spills out some neutrons immediately. But the atomic fragments formed are also in a highly unstable condition and some of them throw out additional neutrons after a short time delay, amounting on the average to half a minute. It is the delayed ones that set the time scale on which the neutron multiplication in the pile builds up and set it for such a long time

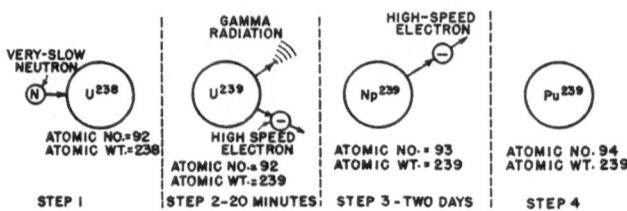

Plutonium is made by a four-step process in which a U^{238} atom absorbs a slow-moving electron. The product emits two electrons successively, resulting in a new — man-made — element of higher atomic number and mass than occurs in nature. It is plutonium, which like U^{235}, can break down.

that slow-acting controls are easily able to regulate the activity of the pile.

The first pile was built on the University of Chicago campus during the fall of 1942. It contained 12,400 pounds of uranium, a large part of which was supplied by Westinghouse, together with a graphite moderator. It was intended to be spherical in shape but as the critical dimensions proved to be smaller than the original calculations indicated, the sphere was left incomplete, giving the actual pile the shape of a large inverted doorknob.

It was first operated on December 2, 1942, at a power level of 1/2 watt and on December 12 the power level was stepped up to 200 watts but it was not allowed to go higher because of inadequate provision for shielding personnel from dangerous radiations. Further studies on piles were made by the construction of one in Tennessee designed for 100-kw level of operation. Later a pile using heavy water instead of graphite as moderator was built.

In summary, it should be remembered that although a pile is built with ordinary uranium, it is only the 0.7 percent of the metal that is U^{235} that is active. The U^{238} that forms most of the metal actually tends to stop the process. Only by ingenious lattice arrangement for slowing neutrons in a moderator is the pile able to operate in spite of the presence of the more prevalent U^{238}.

This means that, regarded as a fuel, only 1/140 of the total weight of uranium is being directly used; the rest is an inert material that remains largely untransformed by the pile.

How does the bomb chain reaction differ from that in the pile? The atomic bomb explodes, whereas the reaction in the pile proceeds in a slow way easily controlled by manual operation of absorbing rods. The big, fundamental distinction is that the bomb (one type) is made of essentially pure U^{235} and without the use of moderator. The chain reaction in the bomb is carried on by fast neutrons directly released by fission. As already remarked, this cannot happen with ordinary uranium because the U^{238} slows the neutrons to the point where they cannot produce fission in U^{235} and also absorbs many of them. With essentially pure U^{235} these competing absorption processes do not occur and the reaction is carried by the fast neutrons directly emitted from a U^{235} fission. These are utilized at once to produce fission in other U^{235} atoms. Here the main factors tending to stop the reaction are the loss of neutrons through the surface (which sets a minimum size to the bomb) and losses by absorption by impurities including any remaining U^{238}.

What is plutonium? This is a newly discovered chemical element not known to exist in nature but which is made from uranium by atomic transmutation. Plutonium is important because it, like U^{235}, is a material from which atomic bombs can be made.

That U^{238} can capture neutrons has already been mentioned as a phenomenon detrimental to the operation of a pile. When U^{238} captures a neutron it becomes U^{239} and emits gamma radiation as does radium. This U^{239} is not stable but emits high-speed electrons by a process of spontaneous radio-activity. The mean life of the U^{239} atoms is only about 20 minutes. By this activity they are transformed into atoms having essentially the same mass but one greater positive charge, 93, on the nucleus, and hence a new chemical element. It is called neptunium and written Np^{239}. Neptunium 239 is also spontaneously radioactive and emits another high-speed electron becoming thereby an atom having 94 positive charges on the nucleus but still essentially of mass 239. This process is slower; the mean life of the neptunium atoms is about two days. The resulting atom of charge 94 and mass 239 is another new element that does not occur in nature. It is called plutonium and written Pu^{239}.

Actually the purpose of piles in the military project was not to get atomic power but to produce the new element *plutonium*, which provides a second bomb material. It is, in short, a competitor to U^{235}. The process by which plutonium is formed — capture of neutrons by U^{238} — has already been mentioned as one that tends to stop the chain reaction in a pile. Nevertheless the uranium lumps in the pile are exposed to a dense atmosphere of neutrons, and so the means is at hand for changing a part of the U^{238} into Pu^{239}.

The several large piles but in operation, generated many hundreds of thousands of kilowatts as heat. This heat was, however, not utilized, as the main purpose of the operation was the production of plutonium for use in the atomic bomb. To utilize the heat would have required additional engineering to operate the pile at a high temperature and there was not time for that.

The pile when run at a higher power level also generates an enormous amount of radioactive material, far more potent than all radium ever mined. This greatly complicates the problem of operation of the large piles, requiring a high standard of reliable operation that must depend entirely on remote controls.

The plutonium is formed in the blocks of uranium in the pile. These have to be removed from the pile and the plutonium extracted by fairly simple chemical methods, because plutonium and uranium, being completely different elements, are dissimilar chemically. This process, however, is greatly complicated by the intense radioactivity of the materials.

How is U^{235} separated from ordinary uranium? The makers of the atomic bomb had plutonium at their disposal. An alternative material is U^{235}. It was felt desirable, in view of all the uncertainties involved, to develop several methods and provide production facilities for extracting in almost pure form the 0.7 percent of U^{235} contained in ordinary uranium. (The third isotope of uranium, U^{234}, is present in so minute proportion as to be wholly insignificant.)

Because of the almost complete identity of all physical and chemical properties of two isotopes of the same element — in this case U^{235} and U^{238} — this is an extraordinarily difficult problem. Several methods were tried, some of which were abandoned as not operative, or as requiring too great an effort, or as being too uncertain of success. These are mentioned in Smyth's report. Here we shall only deal with the three methods which were carried from the research stage into production plants. These are:

a — the mass-spectrographic method
b — the gaseous-diffusion method
c — the thermal-diffusion method.

In addition to these three methods a fourth, that of separation of gas in large high-speed centrifuges, was successfully carried to the pilot-plant stage. The centrifuges work on the same principle as the cream separator on the dairy farm, operating on the very slight difference in mass of the two uranium isotopes.

The *mass-spectrographic method* was developed by the physicists of the

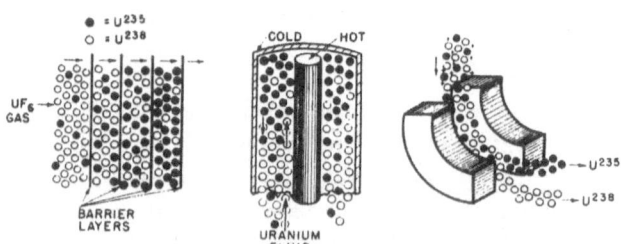

The small proportion of U^{235} can be separated from the predominating U^{238} in several ways, none of them easy. Three have been used on a large scale, and a fourth (mechanical centrifuge) successfully passed pilot-plant tests. In the multi-step barrier-layer method (left) the light masses pass through a little more readily than the heavy ones. In the thermal diffusion method (center) the lighter atoms tend to separate from the heavy ones under thermal action. By the mass-spectrometer method, the atoms are charged and passed through a magnetic field causing them to take separate paths.

Radiation Laboratory at the University of California during the year 1942. By January 1943 Westing-house was called in to design and manufacture the essential process equipment for the large production plant at Oak Ridge, Tennessee. In this method the poles of an electromagnet are enclosed in a large vacuum chamber. Uranium is introduced in the form of a volatile compound into an arc discharge which breaks the compound down and ionizes the uranium atoms.

A large potential difference between the ion source and the tank pulls these ions out through a slit in the source. Instead of moving straight across the tank, the ions are caused by the ordinary electrodynamic action of a magnetic field with a current, (the current being moving charged ions) to move in circular paths. The light ions move in a path of slightly smaller radius than the heavy ions (the ratio of the radii is as $\sqrt{238/235}$.) Therefore separate receiver boxes can be placed at the appropriate places to catch the material of each kind.

Naturally it is not as simple as this idealized description implies. The magnetic field must be exceedingly well regulated and the ripple in the high-voltage supplies must be exceedingly small; otherwise the ion beams will wander and either fail to be collected or get into the wrong receiver.

An interesting sidelight of this plant is that the many tons of conductor in the exciting coils of the electromagnets were made of silver — since the Federal Government had plenty of idle silver in its monetary reserves whereas copper was a critically short material during the war.

One of the most important features of the device is that the ion beam, in moving through the vacuum, ionizes some of the residual gas, providing free electrons that neutralize the space charge of the positive ion beam. This permits the use of beam currents; which although small, are nevertheless vastly greater than they could be if space charge were not neutralized. Without this feature the yield would be so low that this method in its present form would not be feasible.

The *gaseous-diffusion method* requires the use of the uranium in the form of a volatile compound, UF_6, the hexafluoride. When any mixed gas diffuses through a porous material, separation occurs because at a given temperature, light molecules move more rapidly than heavy molecules. However, there is only a very slight effect at a single passage through one sheet or "barrier" of the porous material. Therefore it is necessary to arrange for many successive fractionations, as in fractional distillation, at each of which only a small separating effect is obtained.

The impelling force causing the gas to flow through the barrier is of course a pressure drop that has to be made up by recompressing the gas in preparation for leakage through the next barrier. Because of the corrosive nature of the gas and of its great value after going partly through the plant, the design of these pumps presented many difficult engineering problems.

The problem of the barriers was itself of the utmost difficulty. This called for the mass production of many acres of barrier having microscopic passages of a kind that would resist corrosion and clogging up by the process gas.

The *thermal-diffusion method* is based on application of a curious and little understood physical phenomenon occurring in liquids as well as gases. In its barest essentials, thermal diffusion is this. Suppose we have either mixed gases or a mixture of two mutually soluble liquids and put them in a container that is hot in one part and cold in another — arranging matters so the material is not stirred up by convection currents. Then after a long time an equilibrium is established in which the composition of the mixture is not the same in the hot part as in the cold part. In other words a composition gradient accompanies temperature gradient.

This, too, is a small effect and is useful only when an arrangement is made for achieving many successive fractionations so as to build up a large resultant separation. In this method the process gas or liquid is placed in a vertical tube within which is placed another tube kept hot by any means while the outer tube is kept cool. Thus a radial temperature gradient exists in the process fluid. This provides a means for a separation, which is enhanced by the counter-current action due to convection as the hot fluid near the inner tube rises and the cold fluid near the outer tube descends.

By the spring of 1943 it was proved that this method could produce separated U^{235} with a plant whose initial cost in time and money was less than that of the other methods. The inner tubes were heated by steam and the chief drawback was the enormous consumption of steam so the thermal-diffusion plant, though relatively inexpensive to build, was rather expensive to operate.

As matters stand now separation of U^{235} from natural uranium is being done by production plants based on the three entirely different methods at the Manhattan District's reservation at Oak Ridge, Tennessee.

And now the bomb! Very little of this part of the story can be told as yet. Preliminary studies on this problem were made in 1941 and early 1942. At the end of the summer it was decided to concentrate all this work on a greatly expanded scale at a specially constructed laboratory at Los Alamos, New Mexico, about 40 miles northwest of Santa Fe. The first group of laboratory buildings, administrative buildings, homes for the personnel and barracks for the soldier guards were built during the winter of 1942–43 and the scientific staff began to arrive and start work in April 1943. What these people achieved, starting with empty buildings on a remote mesa with only an old Diesel-driven mine generator as the laboratory power supply thousands of miles from major industrial facilities and supplies, is an epic in the annals of science. The writer hand the privilege of assisting with initial arrangements during the first month but was unable to stay because of other war activities at East Pittsburgh. It is to be hoped that a fuller story than is contained in the Smyth report of the achievements of this group can be given to the public before long. The story of this group, continually growing in numbers, and communicating with outside suppliers only by devious channels, because of requirements of military security, will be most fascinating when properly told.

Although discussion of the bomb's details is not permitted, these essential points can be enumerated:

a — The active material is either Pu^{239} from the piles at Hanford, Washington, or U^{235} from the three different separation plants at Oak Ridge, Tennessee.

b — A bomb less than the critical size will not explode at all so it is not possible to experiment with little ones to learn how to make a big one.

c — Before firing, the active material must be kept separated into two or more lumps each of sub-critical size. The act of firing consists of assembling these rapidly into a mass that is above critical size for that shape.

d — This has to be done with great rapidity, using a firing mechanism, which was itself a difficult problem. The need for rapidity arises from the fact that if the parts come together slowly an explosive reaction begins before the parts are completely together. This would blow them apart again and stop the fission chain reaction with only an insignificant release of the atomic energy.

e — Even with the best design possible, the stopping of the reaction due to the bomb's blowing itself apart was expected to prevent the effective conversion by fission of all the material in the bomb. Some estimates placed this conversion efficiency as low as a few percent. What was actually attained at the Alamogordo, New Mexico, tests has not been disclosed to date.

f — The fission products are extremely radioactive and if all of them were to remain in a relatively small area (say a square mile) the radiations would be too intense to permit the existence of any living matter in the region for probably several weeks after the explosion.

g — To get maximum destructive effect from the blast the bomb is fired while at a considerable height above ground, which also favors the dispersal over a wide area of the radioactive products so that the contamination of the area is not thought to be an important attribute of the weapon.

What of the future? While no reputable scientist ever makes definite promises about anything that lies in the future, still it is possible to venture an opinion that the following significant developments are highly likely to be made within the coming decade:

a — More effective ways of producing U^{235} and Pu^{239} will be developed, permitting greater production at lower cost.

b — These materials in combination with ordinary uranium will make possible power-producing piles of smaller size than those thus far developed.

c — Piles will have important peacetime uses as special-purpose energy sources, and as sources of neutrons and radioactive materials for medical and other scientific work.

d — Piles will probably not be developed into small power units for automobiles or airplanes because of their overall weight including that of the material needed to shield the passengers from the dangerous radiations.

e — Also because of shielding difficulties, piles will probably not provide the driving power for railroad locomotives. However, it is reasonable to suppose that within a decade some ships may derive their power from piles.

Besides uranium it is known also that fission may be produced in thorium, which is much more abundant in nature than uranium and therefore may be the fuel in piles of the future. Whether release of atomic energy from other materials can be achieved is a question which can be decided only by future research. At present no means of doing this is in sight — but it should be remembered that the atomic bomb would have seemed fantastic to the best nuclear physicists in 1938.

Although atomic energy may seem strange and mysterious to the engineer, it will find its application in the power field as a source of heat. The fission chain reaction makes the pile get hot. Some heat exchanger fluid must go through the pile to get out the heat. The hot fluid will then be directly used as the working fluid in a standard heat engine; e.g., a steam turbine, possibly of special design. In other words the pile is a new kind of boiler and however mysterious it may seem now, it will not require a revolution in the well-known engineering practice by which heat is converted into mechanical effort and thence into electrical power. In the meantime the most urgent problem is that of international arrangements which will assure us that atomic power will only be used for peaceful purposes.

"... It was felt by those at Alamogordo that there was brought into being something big and something new that will prove to be immeasurably more important than the discovery of electricity or any of the other great discoveries which have so affected our existence." From the Smyth report.

Uranium Ores

Although uranium is contained in over one hundred minerals, only two — pitchblende and carnotite — are of great importance. It is estimated that uranium is present in the earth's crust in the proportion of about four parts per million. Early rough estimates were that the nuclear energy available in known world deposits of uranium is adequate to supply the total power needs of this country for 200 years (assuming utilization of U^{238} as well as U^{235}).

Pitchblende is found in metalliferous veins, notably Bohemia and Saxony. More recently deposits have been found in the Belgian Congo and the Great Bear Lake region of northern Canada. Most of the importations to this country during 1942 and 1943, the last years for which importation figures are available, were from Canada and the Belgian Congo.

Pitchblende of good quality contains as much as 80 percent of uranoso-uranic oxide (U_3O_8). It is a brown to black ore with pitch-like luster in the form of crystallized uraninite. Madame Curie was among the first to recognize this material as a source of radium.

Carnotite, the second main source of uranium, has been discovered in Arizona, Colorado, and Utah. It is found as a canary-yellow impregnation in sandstone. Production of this ore climbed steadily during the middle thirties from a low of 254 short tons in 1934 to a high of 6,256 in 1939. The actual pounds of uranium extracted from the ore produced in 1939 were 56,269. The actual extent of deposits has not been divulged.

Until recently, the only use for uranium was as a coloring agent for ceramics and glass. It was used in amber signal lenses and in glass of special coefficient of expansion for glass-to-metal contacts in radio tubes.

Chronological Highlights of the Atomic-Bomb Project

1905 — Einstein enunciates equivalence of mass and energy.

1912 — Rutherford initiates theory of nuclear atom.

1919 — Rutherford discovers transmutation of nitrogen by alpha particles from radium.

1928 — Quantum mechanics applied to understanding of radioactive disintegration by Gurney, Condon, and Gamow.

1932 — First transformation of lithium nuclei by artifically accelerated protons by Cockcroft and Walton of England. Discovery of heavy hydrogen by Urey. Discovery of the neutron by Chadwick.

1934–40 — Period of rapid development in many laboratories including discovery of artifical radioactivity by Irene Curie and F. Joliot; development of the cyclotron by E.O. Lawrence; development of electrostatic high-voltage atom smashers by van de Graaff (Massachusetts Institute of Technology), by Tuve (Carnegie Institute of Washington), by Herb (University of Wisconsin) and by the research physicists at Westinghouse. Extensive study of many nuclear reactions and parallel development of fundamental theory of nuclear structure. Much was learned about the special forces which bind together the constituent protons and neutrons in the nucleus.

1939 — Discovery of uranium fission by Hahn and Strassman in Germany. First reported in this country by Niels Bohr on January 16, 1939, and immediately confirmed and further studied in many laboratories in America. Possible military applications were recognized at once by a group of physicists, among whom were E. Fermi (Columbia), E. Wigner (Princeton) and E. Teller (George Washington).

1939, March — Pegram and Fermi (Columbia) made first approach to Navy Department to advise them of possibilities in fission.

1939, July — Einstein, Wigner, and Szilard enlisted aid of Alexander Sachs of New York in getting the facts on military possibilities before President Roosevelt. Roosevelt referred matter to "Advisory Committee on Uranium" headed by L.J. Briggs, director of the National Bureau of Standards. First meeting of this committee on October 21, 1939. Voluntary secrecy policy in this field began to be set up by physicists about this time.

1940, April 28 — Committee meeting to plan a larger research program. First definite reports of German activity on this subject having to do with military aims.

1940, Summer — Radiation Laboratory, University of California discovered possible use of plutonium for explosive chain reaction.

1940, Summer — Sachs active in urging more effort on this subject by contacts with President Roosevelt through his aide, General E.M. Watson. Uranium Committee under Briggs constituted as a part of the National Defense Research Committee by President Roosevelt on formation of latter body. Various small research contracts let to Columbia University, Princeton University for fundamental studies bearing on the problem.

1940–42 — Gaseous-diffusion method of separating uranium isotopes developed by a research group at Columbia University headed by Professors H.C. Urey and J.R. Dunning.

1940–44 — Investigation of thermal-diffusion method of isotope separation (on basis of research in Germany in 1938) by P.H. Abelson, first at the National Bureau of Standards and later at the Naval Research Laboratory. Pilot plant built at Philadelphia Navy Yard.

1941, Summer — Uranium Committee enlarged by addition of several new members. National Academy of Sciences Committee made an independent study of the situation. This study involved consideration of engineering problems as well as scientific problems.

1941 — Development of centifuge isotope separation initiated by Professor J.W. Beams, University of Virginia. First production models made by Westinghouse, and pilot-plant test carried out successfully by Dr. E.V. Murphree, Standard Oil Development Company.

1941, Fall — Previous cooperation and exchange of data with British scientists greatly extended especially by trip of Oliphant (Birmingham) to America and of Pegram and Urey to England.

1941, Fall — Preliminary studies of atomic bomb begun by Professor G. Breit of University of Wisconsin. Work continued in summer of 1942 under Professor J.R. Oppenheimer of University of California.

1941, December 16 — Top Policy Group consisting of Vice President Henry A. Wallace, Secretary of War Henry L. Stimson and Dr. V. Bush, recommended reorganization of program outside N.D.R.C. with greatly enlarged activity and Army jurisdiction. About this time the N.D.R.C. Uranium Committee became known as the O.S.R.D. S-1 Committee and was authorizing considerable expansion of the research programs at various universities.

1942, January — Major expansion of research activity authorized and started at Berkeley, Chicago, Columbia, Princeton, and several other places.

1942, April — Further cooperation with British developed during visit of F. Simon, H. Halban and W.A. Akers from England.

1942, May — Reorganization of O.S.R.D. S-1 Committee as a smaller group.

1942, June 13 — Bush and Conant send to Vice President Wallace, Secretary Stimson and General George C. Marshall detailed recommendations for major expansion of the program.

1942, June 18 — Colonel J.C. Marshall, Corps of Engineers directed to organize a new district to carry on the work.

1942, August 13 — The "Manhattan District" was officially established for this purpose.

1942, September 17 — Secretary of War placed Brigadier-General (now Major General) L.R. Groves in complete charge of all army activities on the atomic bomb which thereafter took over the previous O.S.R.D. activities.

1942 — Government sponsored research at Columbia University indicated that a nuclear chain reaction was of possible accomplishment.

1942, May — Discovery by Snell, Nedzel, and Ibser of delayed-fission neutrons.

1942 — Mass-spectrographic method of separating uranium isotopes developed by Radiation Laboratory, University of California.

1942, Fall — First pile built at University of Chicago (later moved to Argonne Laboratory, near Chicago) under direction of E. Fermi, W.H. Zinn, and H.L. Anderson. First operated on December 2.

1942, Fall — Construction begun at Los Alamos, New Mexico, of atomic-bomb laboratory, under direction of Oppenheimer.

1942, Fall — Design begun by Kellex Corporation of large-scale diffusion plant built at Oak Ridge by J.A. Jones Construction Company and operated by Carbide and Carbon Chemicals Corporation.

1943, Spring — One thousand-kw pile constructed at Oak Ridge, Tennessee for production of plutonium.

1943 — Plant at Handford, Washington for production of plutonium designed by E.I. duPont deNumours Company of the basis of research work at Metallurgical Laboratory, University of Chicago. Plant constructed and operated by DuPont.

1943, January — Westinghouse asked to design and build essential process equipment for large-scale mass-spectrograph separation plant at Oak Ridge designed and built by Stone and Webster. Plant operated by Tennessee Eastman Company.

1944, Summer — Large-scale, thermal-diffusion plant built at Oak Ridge, Tennessee.

1945, July 16 — First experimental bomb dropped on desert near Alamogordo, New Mexico.

1945, August 6 — First military atomic bomb dropped on Hiroshima, Japan.

1945, August 8 — Second atomic bomb dropped on Nagasaki.

1945, August 10 — Japan sues for peace.

Glossary of Important Terms in Nuclear Physics

Atom — Smallest unit of matter remaining unchanged in chemical reactions. All atoms are about 10^{-8} cm in diameter. They consist of a central positively charged *nucleus*, about 10^{-12} cm in diameter, surrounded by enough electrons to make the atom electrically neutral.

Neutron — A basic constituent particle of atomic nuclei having no electric charge and having a mass of about 1.67×10^{-24} gram.

Proton — A basic constituent particle of atomic nuclei having a positive charge numerically equal to that of the negatively charged electron, 1.60×10^{-19} coulomb and a mass about the same as that of the neutron. The proton itself is the nucleus of ordinary hydrogen atoms.

Electron — Smallest atomic particle. Unit of negative electricity.

Deuteron — This is the nucleus of heavy hydrogen atoms which occur in nature as about 1/5,000 of ordinary hydrogen. It is the simplest composite nucleus known, consisting of a combination of one proton and one neutron.

Alpha-particle — This is the nucleus of helium atoms and is a composite nucleus of two protons and two neutrons. The name originally referred to the alpha radiation from naturally radioactive substances like uranium and radium, later recognized to be fast-moving nuclei of ordinary helium gas.

Atomic number — An integer characteristic of each chemical element which tells how many protons there are in the atomic nucleus and also how many electrons there are in the atom, outside the nucleus. Usually denoted by Z. Examples: hydrogen, Z = 1; helium, Z = 2; neon, Z = 10; uranium, Z = 92.

Isotope — A particular variety of atom or nucleus characterized by a particular atomic weight as well as a particular atomic number. Example: all uranium atoms have a charge Z = 92, those of the light isotope have an atomic weight of about 235 while those of the heavy isotope have an atomic weight of about 238. There is also a very rare isotope having an atomic weight of 234.

Neptunium — A new chemical element not known to occur in nature having Z = 93 and an atomic weight of 239. This is formed by radioactive decay of U^{239} which emits a β-particle (high energy electron) to become Np^{239}.

Plutonium — A new chemical element not known to occur in nature, having $Z = 94$ and an atomic weight of 239, formed by radioactive emission of a β-particle from Np^{239}.

Moderator — A substance (carbon, heavy water, or beryllium) used as a means of slowing down neutrons by means of elastic impacts of the neutrons with the atoms of the moderator.

Heavy water — A kind of water whose molecules consist of the heavy hydrogen isotope, deuterium, in combination with oxygen, written D_2O instead of H_2O.

Chain reaction — Any reaction, chemical or nuclear, in which the process continues by virtue of the action of one of the products to cause the reaction to continue. Example: uranium fission is caused by a neutron and the fission process releases more neutrons which can cause more fissions.

Pile — Any arrangement involving lumps of fissionable matter, e.g., uranium, together with moderator, so arranged as to utilize the neutrons well enough to result in a chain reaction.

Foundations of Nuclear Physics

E.U. CONDON

Director
National Bureau of Standards
Washington, D.C.

Because of his historic contributions to many fields of science and technology, NUCLEONICS features this introduction to nuclear physics by Dr. E.U. Condon — key figure in the United States wartime development of nuclear energy, member of the National Academy of Sciences, and president of the American Physical Society in 1946

With the establishment of a journal devoted to the subject, there can be no doubt that nuclear technology has come of age. On this occasion, it seems appropriate to review the history of nuclear physics and something of the way in which the science back of nuclear engineering started. So that it need never appear on these pages again, perhaps it is well to record, too, the inevitable pun that "nuclear" physics is really "unclear" physics.

About 50 years ago, the electrical nature of matter and the atomic nature of electricity were first recognized. Men then began to speculate on how this electricity was arranged in the atoms. In other words, they tried to interpret the known properties of atoms in terms of a structural hypothesis.

Positive Particles

The most decisive step in this direction came in 1912 when the experiments of Sir Ernest Rutherford showed that the positive electricity in the atom must be concentrated in a single particle whose diameter could not be greater than 10^{-12} cm. He studied the distribution in angle of the scattering of alpha particles from radium by thin foils of various substances. The alpha particles are known to be high-speed, positive helium ions. Before Rutherford's experiments, it was often supposed (on no evidence) that the positive electricity was more or less continuously distributed over the whole region occupied by the atom, which is a sphere about 10^{-8} cm in diameter.

The charge on an electron is now known to be close to 4.8×10^{-10} esu. Therefore, since 1 esu = 300 v, the electric potential in volts, V, at a distance r from such a charge (where r is measured in 10^{-8} cm) is

$$V = \frac{4.8 \times 10^{-10} \times 300}{r \times 10^{-8}} = \frac{14.4}{r}$$

In this article, electric, charges will be expressed in terms of the electronic charge as the unit. The amount of positive and negative charge in any atom in this unit is known as the atomic number for that element, usually denoted by Z. Thus Z equals 2 for helium and 79 for gold.

If the positive electric charge in a gold atom were spread out over a sphere about 10^{-8} cm in radius, the greatest electric potential at any point within the space, even in the absence of the negative electricity, would be less than 10,000 volts.

It is known from experiments on their deflection in electric and magnetic fields that the alpha particles from radium have energies corresponding to an acceleration between two and three *million* volts. A particle of this much energy would go right through the gold atom and be only slightly deflected. Yet Rutherford found experimentally that a very few of the alpha particles were scattered by gold through very large angles, even up to 180 deg, that is, a complete reversal of direction. From this he concluded that the positive electricity in the atom was not spread throughout the volume of the atom but must be confined to a much smaller region in order that its electric potential be great enough to produce the observed large-angle scattering of alpha particles.

He found that the actual scattering was quantitatively accounted for, in great detail and for many different elements, by supposing the alpha particles to be scattered by their motion according to mechanical laws under the inverse-square-law electrostatic force between the positively charged alpha particle and the very small positively charged and massive core of the atom of the scattering material. This very small central core of the atom. Rutherford called the *nucleus*.

Atom Electrically Neutral

His conclusions have stood the test of time through a vast and ramified development. The atom of the chemical element whose atomic number is Z consists of a heavy central nucleus about 10^{-12} cm in diameter, carrying a positive electric charge of Z units, and surrounded by Z electrons, each carrying a negative electric charge of one unit. Thus, the whole atom is electrically neutral. The mass of such an atom is denoted by A on a scale in which the most abundant kind of oxygen atom is arbitrarily assigned the exact value 16. In terms of this unit of atomic weight, each electron has a weight of about $1/1840$, so that nearly all of the mass of the atom is in the positively charged nucleus.

The simplest atom of this kind is the hydrogen atom for which $Z = 1$, and $A = 1$ (approximately). Its central nucleus is the smallest known and has a special name. It is called a *proton*. The exact values of A are always close to an integer. The significance of the exact values is considered later.

It is the number Z which determines the chemical properties of an atom. It was soon discovered that the atoms of most chemical elements are not all of the same weight. Atoms of the same Z, but differing weights A, are said to be *isotopes*. The simplest example is provided by hydrogen itself. Its heavy isotope was not discovered until 1932 by Urey, Brickwedded, and Murphy. They found that about one part in 5,000 of ordinary hydrogen consists of atoms for which $A = 2$; that is, they are about twice as heavy as the atoms of ordinary hydrogen. Likewise it was found that neon is a mixture of three isotopes for which $A = 20, 21$, and 22 approximately. It is now known that uranium is a mixture of three isotopes for which $A = 234$, 235, and 238.

For the purposes of classical chemistry, all isotopes of an element have the same properties. Chemistry in the usual sense depends only on Z, not on A. Very careful study shows extremely slight dependence of chemical properties on the weight, but these effects are so small that they do not mar the truth of the statement.

For nuclear physics and nuclear engineering, A is every bit as important a property as Z. We are naturally led to consider what kinds of atoms occur in nature as specified by the pair of numbers (A, Z), the first giving the atomic weight to the nearest integer, the second giving the nuclear charge which is exactly an integer. In Fig. 1, a graph is made of the values of $(A-Z)$ as ordinates and Z as abscissas. A black dot shows the values of (A, Z) for which stable atoms occur in nature.

Such a graph shows the general tendency is for A to be about equal to $2Z$ for elements occurring in nature (except for light hydrogen), and for A to be greater than $2Z$ for elements with Z greater than about 20. Evidently there must be something more in the nucleus than the Z protons needed to give it its charge, for a nucleus consisting only of Z protons would have an atomic weight equal to Z or about half the actual atomic weights occurring in nature.

Curiosity prompts the asking of these questions: (a) What else is in the nucleus to make up the extra weight not accounted for by the protons? (b) Why are there only certain elements occurring in nature? For example, why do we not find a hydrogen of mass $A = 10$ or a uranium of mass $A = 100$? (c) What holds the particles together in an atomic nucleus? (d) What special properties do the very heavy elements have that make them break up spontaneously by throwing off helium nuclei, otherwise known as alpha particles? (e) Since some naturally radioactive elements throw off high-speed electrons with energies so great that we must suppose that they came from inside the nucleus, does this not show that there are electrons in the nucleus?

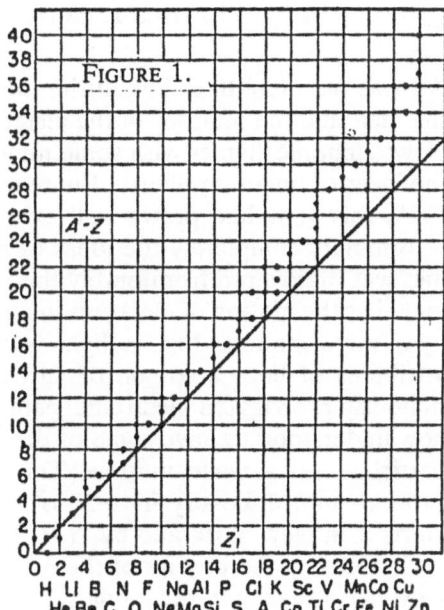

FIGURE 1.

At the start of this century, theoretical study of the continuous spectrum of radiation from hot bodies led Planck to the radical hypothesis that light did not behave entirely like a continuous wave motion, but had some of the attributes of a stream of particles. These individual particles of light were called *light quanta*, and the associated theoretical developments were known as the *quantum theory*.

Soon after Rutherford was led to the idea of the nuclear atom by his studies of the scattering of alpha particles by thin foils, Niels Bohr began to study the orbital motion of the electrons surrounding the nucleus. He found that by an appropriate application of the ideas of quantum theory to these orbital motions he had the clue to the interpretation of the structure of the line spectra of atoms. Thousands of wave-lengths of light emitted by atoms, wave-lengths which can be measured to six-decimal-place accuracy, fitted nicely into the highly detailed theoretical scheme which developed from Bohr's ideas.

This development of the understanding of the motions of the electrons around the nucleus occupied the principal attention of physicists and chemists in the period from 1914 to 1930. A complete interpretation was given not only to the vast amount of data of spectroscopy, but also to a basic understanding of the main facts of the periodic system, of the different types of chemical bond, of the electronic structure of molecules, of the electron theory of metals and of the basic mechanisms of chemical reactions. Those were fruitful years.

Quantum Mechanics Discovered

All was not smooth sailing during this period. Physicists started out with the idea that the motions of the electrons could be understood by classical Newtonian mechanics with some minor patchwork changes occasioned by the ideas of quantum theory. But matters proved to be not that simple. By 1925, this view had pretty well exhausted itself and it was necessary to reformulate completely the ideas of atomic mechanics. Quantum mechanics was discovered from purely theoretical considerations, leading to the idea that the motion of electrons in atomic systems was somehow governed by a wave motion. Just as light was recognized as having some of the attributes of both a stream of particles (that is, light quanta) and continuously propagated electromagnetic waves, so a stream of electrons was now understood to have some of the attributes of both a simple stream of particles and a wave motion. This idea was brilliantly verified in the experimental discovery in 1927 of electron diffraction by Davisson and Germer in New York and by G.P. Thomson in England.

Concerning the physics of the nucleus, it may be said that the first and most decisive step in the modern period came in 1928 with the application of the quantum theory to the interpretation of natural radioactive disintegration of the heavy elements by the emission of alpha particles. This was done quite simultaneously and independently by Gamow, a Russian, working in Göttingen, and by Gurney and the writer, working at Princeton.

Consider the atomic species U^{238}. The atoms of uranium now remaining on earth are what is left from an original larger stock formed, somehow, geological epochs ago. The average life of a uranium atom is 6.75×10^9 years or 2.14×10^{17} sec.

Uranium Disintegration Rate

There is no way of knowing when a particular uranium atom will disintegrate. Some have done so long ago and are no longer with us. Others will not do so for billions of years. There is nothing we can do which affects this rate. Radium disintegrates much more rapidly. The average life of its atoms is only 2,440 years, so there would not be any appreciable amount of this on earth if it were not being continuously generated as a disintegration product of uranium. The problem is to understand what peculiar sort of mechanism is responsible for the very wide differences in the actual life of different uranium atoms, and why the average life of radium Atoms is so much shorter than that of uranium.

The theory of Gurney-Condon-Gamow showed that this strange behavior was one of the most remarkable consequences of the new wave mechanics, which had originally been developed to explain the motions of electrons in the outer part of the atom. It is a fundamental characteristic of wave motion that it can go a little way into regions where it ought not to go according to oversimplifed theories. For example, in geometrical optics, if a ray of light is traveling through a block of glass and makes too great an angle with the normal to a glass-air interface, then when we calculate the direction which the ray should have on emerging into the air, it comes out imaginary. Physically the light is totally reflected, and there is no transmitted ray of light. But experiment shows that a little light does penetrate a short distance into the air, its intensity dying off exponentially to negligible values in a distance equal to one or two wavelengths of light. The same sort of thing happens when we put microwaves into a wave guide that is too small to give real transmission to the waves in question. Simple theory says the propagation of the wave is imaginary and this means that the wave is propagated a little distance, dying off exponentially in such a wave guide. Practical attenuators for microwaves are built on this principle, so there cannot be too much "long-haired science" about this idea.

This also applies in quantum mechanics to the motion of alpha particles in the nucleus of the uranium atom. According to classical mechanics, a particle cannot go into regions where its potential energy (if it were there) would be greater than its total energy. This would require its kinetic energy to be negative; hence, its momentum would have to be imaginary. But in wave mechanics, an imaginary momentum is associated with an imaginary propagation which means an exponentially damped wave rather than an oscillatory one.

If we now consider the motion of a particular alpha particle in relation to the other constituents of a uranium nucleus, we have to suppose that, when it is inside, it is held by special attractive nuclear forces that are responsible for the general stability of all nuclei. This is a spherical region about 10^{-12} cm in radius. There is reason to believe the alpha particles move back and forth in this region with a velocity of the order of 10^8 cm per sec, averaging about 10^{20} oscillations per sec. If an alpha particle tries to leave, it finds itself drawn back in by the nuclear attractive forces. Said otherwise, it would have to go into a region of space where its potential energy would be greater than its total energy. The region where this is so is not very thick, however. Once the alpha particle is definitely outside the nucleus, the strong electrostatic repulsive forces between the like charges of the alpha particle and the rest of the nucleus greatly reduce the potential energy. This is shown schematically in Fig. 2.

If the alpha particle were governed by classical mechanics, it could not penetrate from the inner region, where its potential energy is less than its total energy, to the outer region where this is true. To do so, it would have to pass through a region where its momentum would be imaginary. In quantum mechanics, it has a slight probability of doing this because its

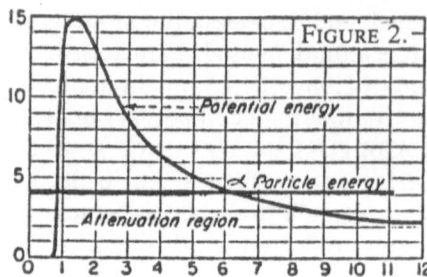

FIGURE 2.

behavior is governed by a wave that can penetrate this region even though it is exponentially attenuated in doing so.

The alpha particle oscillates up to the edge of the nucleus about 10^{20} times per sec. The chance of disintegration in a second (reciprocal of the mean life) is about 5×10^{-18}. Therefore the attenuation introduced by the potential wall around the nucleus must be about 10^{-38} or 380 decibels.

This theory had a striking initial success in that it was possible to make quite good quantitative estimates of the attenuation from reasonable hypotheses about the unknown law of force between the rest of the nucleus and the alpha particle. Moreover it made clear a previously puzzling empirical correlation, known as the Geiger-Nuttall relation, according to which those elements decaying rapidly are the ones that send off alpha particles of highest energy. Referring again to Fig. 2, we see that, if the alpha particle energy is higher, the attenuation region will be smaller and the rate of disintegration greater.

Developed late in 1928, this theory provided a great stimulus to the study of nuclear physics which up until that time was hardly more than a potential branch of physics. Aside from the way it served to clarify ideas about the nature of natural radioactive disintegration, it also soon served as a great stimulus to experimental work.

As early as 1919, Rutherford had succeeded in disintegrating nitrogen atoms by bombarding them with alpha particles. But the total yield of disintegrations was extremely small and the associated phenomena, therefore, were difficult to observe. This is due to the fact that when the nucleus of an atom is bombarded with another charged particle, the two repel each other. The force of mutual repulsion tends to bring their relative velocity to zero before the two colliding particles have got very close together. After that, the mutual repulsion causes them to fly apart again.

High Barrier Potential

It is natural to suppose that two such colliding particles must really come within a distance of about 10^{-12} cm in order for any real disintegration to take place. Looking at Fig. 2, we see that an alpha particle arriving from the outside would have to have an energy of at least 15 Mev in order to surmount the repulsive potential energy barrier representing the interaction of the alpha particle with the rest of the nucleus. Of course, this barrier is high because it is considered for uranium. For an element of $Z = 10$ instead of $Z = 92$, the height of the barrier is only about one ninth as great and lower voltages would be necessary.

But even considering the collision of the lightest elements, it appeared that it would be necessary to accelerate particles with a voltage in excess of 1,000,000 in order to get them to go over the barrier and penetrate the nucleus to produce artifical nuclear disintegration. In those days (before 1930), the cyclotron had not been invented nor were any large Van de Graaff electrostatic generators available. Yet it seemed that it was necessary to have voltages above 1,000,000 if a beam of protons was to be accelerated to sufficiently high energies to produce artifical disintegration. This view was completely altered when the concept of barrier leakage or the tunnel effect was applied to it. For, according to quantum mechanics, if a particle can leak out, it can also leak in, and therefore it is not necessary to have a voltage supply capable of accelerating the ions with enough energy to go over the top. This consideration encouraged Cockcroft and Walton to try to disintegrate lithium artificially by bombarding it with protons at unexpectedly low voltages.

This reaction can be written like a chemical reaction

$$_1H^1 + {}_3Li^7 \rightarrow {}_2He^4 + {}_2He^4$$

which means that when a proton (nucleus of hydrogen) collides with a lithium nucleus in the proper way the total material rearranges itself into two helium nuclei that fly apart. So well does barrier leakage work in this case that this reaction has actually been observed to occur with a bombarding voltage as low as 13 kilovolts.

High-voltage Accelerators

This result gave an immense impetus to the construction of high-voltage accelerators to obtain beams of hydrogen ions, deuterons, and helium ions of as great energy as possible. Soon afterward the discovery and development of the cyclotron made possible the rapid extension of investigation of nuclear reactions to a wide variety of substances. As one goes to heavier elements than lithium, higher voltages are needed because the potential barriers are higher but, in all cases, their effective height is considerably reduced by the potential barrier leakage.

The next great steps in the development of nuclear physics came in 1932 with the discovery of heavy hydrogen by Urey, Brickwedde, and Murphy, and the discovery of the *neutron* by Chadwick. The neutron is a particle (also governed by the laws of wave mechanics) which has a mass approximately equal to that of the proton, but has (hence its name) no electrical charge. These were found when beryllium and several other light elements were bombarded by alpha particles. Because they have no charge, they are quite penetrating and at first were confused with gamma rays. However, their behavior in many ways was soon found to be very different from that of gamma rays and all the observations were in accord with the idea that they were neutral particles of mass essentially the same as that of the proton.

Nucleus Constituents

Here at last was the answer to the question of what contributes the extra mass in complex atomic nuclei. As we have seen, the mass of most nuclei, A, is equal to or greater than twice the mass Z that comes from the protons it contains. It was now natural to assume that the nucleus was made up of two kinds of particles, protons and neutrons. The nucleus of mass A and charge Z consists of Z protons and $(A-E)$ neutrons. That is the reason a plot of $A-Z$ against Z was chosen for Fig. 1. In terms of present views, the ordinate is the number of neutrons in the nucleus and the abscissa is the number of protons.

From this point of view, the nucleus of the atom of heavy hydrogen is especially interesting because it is the simplest possible compound nucleus. It consists of the simple binary compound of one proton and one nucleus. A full study of its properties is the most direct way to get information about the fundamental nuclear forces that bind a neutron and a proton. Such study is more fruitful than that of heavier nuclei because of the greater mathematical complexity in analyzing the wave mechanics of the behavior of a nucleus containing more than two particles.

If a nucleus is to possess stability, it must be because some new kinds of specific forces come into play between neutrons and protons, between protons and protons, and between neutrons and neutrons, at distances of the order of 10^{-12} cm or less. Such forces overcome the repulsive effects due to the electrostatic interaction of the like charges on the protons. Naturally the study of these forces and their full description must be the foundation stone of basic nuclear physics in the same way that the inverse square law is the foundation stone of electrostatics. Attractive forces lead to negative potential energy; that is, the potential energy of two close particles is negative if it is reckoned from zero when the particles are at infinite separation. Thus it is necessary to do a positive amount of work in separating them against their mutual attraction. In the same way, the energy of a water molecule is negative compared to the energy of its constituents, two hydrogen atoms and one oxygen atom.

Energy Information

Information about the relative energies of different atomic nuclei can be obtained by studying the yields and thresholds of various nuclear reactions. For example, in the case of bombardment of lithium by protons, not only can one measure the energy of the protons which strike the lithium by measuring the voltage applied to the tube in which they are accelerated, but one can also measure the energy of motion of the two helium atoms formed. In ways like this, one can find that the two helium atoms have together more kinetic energy than the original kinetic energy of the bombarding proton.

From this, it follows that the neutrons and protons in two helium atoms must be in a state of lower total energy than when they are combined as in one lithium nucleus and a free proton. Determination of the energy changes in nuclear reactions in this way proceeds by identically the same kind of reasoning used in thermochemistry to get at the heats of formation of various chemical compounds by measurements based on heats of reaction of various reactions involving the compounds. To be sure, the experimental technique is different from that of classical calorimetry, but the basic idea is the same.

In the early 1930's, important developments of precision mass spectrometers were made. This was largely done by Bainbridge in this country. There are many kinds of mass spectrometers for special purposes. They are all devices in which a beam of positive ions is formed and made to pass through regions occupied by various electric and magnetic fields. The ions of different masses travel different paths. Sometimes the mass spectrometer is used as an analytical device to find out what material is present in the source by observing the masses of the ions produced. But the form of interest here is one that is adapted to the precise measurement of the magnitude of the individual atomic masses. In this way, it has been possible to determine many atomic masses to six-figure accuracy.

It has already been pointed out that the masses of the atoms are not accurately given by the integer A which we have associated with them on the scale that makes the O^{16} isotope have the arbitrarily assigned precise value of 16. But now we come to an even more extraordinary point. When the mass of the deuterium atom is measured precisely it is found to be 2.014725. Therefore, the mass of two deuterium atoms will be twice that number, or 4.029450. Now two deuterium atoms have just the makings of one helium atom, namely two outer electrons, two protons and two neutrons. Yet when the mass of the helium atom is precisely determined in the mass spectrograph it is found to be only 4.00386; in other words, in the process of forming helium from two deuterium atoms, there is a loss of 0.0259 units of mass which seem to have just disappeared. When the accurate masses are used, many example of the same sort of thing occur.

To give another example, consider the formation of two helium atoms by the bombardment of lithium with a proton. The mass balance is

	INITIALLY		FINALLY
Li	7.01816	He	4.00386
H	1.00813	He	4.00386
	8.02629		8.00772

giving a loss of mass of 0.01857 units.

The mass is not really lost in cases like this. As long ago as 1905, Einstein was led by the theory of relativity to recognize that mass and energy are inseparably bound. They are two aspects of the same thing. It is not correct to say, as is sometimes done, that mass can be converted into energy or energy into mass. Mass *is* energy, and energy *is* mass. Wherever energy is, there is mass, but the experimental techniques by which a given quantity of mass-energy is recognized may sometimes be more like those involved in a mass determination and in other cases those in an energy determination.

Einstein showed that the amount of energy in ergs, E, associated with a mass M (grams) is given by

$$E = Mc^2$$

where $c = 3 \times 10^{10}$ cm/sec, the velocity of light. In atomic physics, one commonly measures mass on the scale in which unity is one sixteenth the mass of the O^{16} atom, which is 1.6604×10^{-24} gm. Energy is usually measured in terms of the electron volt, that is, the energy acquired by any particle possesing one atomic unit charge, like the electron or proton, when freely accelerated through a potential difference of one volt. Since the charge on the electron is 4.8×10^{-10} esu and there are 300 volts in one esu of potential, it follows that one electron volt is equal to 1.6×10^{-12} ergs. Therefore, the energy equivalent of mass in these units is expressed by the relation

$$\text{one atomic mass unit} = 932 \text{ Mev.}$$

We are dealing in the examples cited with the "disappearance" of mass of the order of one or two hundredths of a mass unit — therefore a quantity of the order of several Mev. This corresponds in rough magnitude with the observed energies of the particles released in nuclear processes. Exact data on the energies gives exact check with the mass discrepancies, and for the past dozen years the Einstein relation has proved accurately valid on dozens of different nuclear reactions.

In this way, the measurement of precise masses of the atoms enables one to calculate in advance the energy release that accompanies any particular atomic process. It is possible, therefore, to estimate that the fission of the uranium atom will be associated with a release of about 200 Mev of energy per uranium atom split, since from the masses involved it is known that about one fourth of a mass unit disappears. Naturally this makes the precise measurement of atomic masses of the utmost importance in all matters concerned with atomic energy.

One of the questions which often preplexes the beginning student of nucleonics is this: Is the neutron a combination of a proton and an electron? There are some textbooks that say it is. Others will vouchsafe the opinion that a proton is a combination of a neutron with a positive electron. (The positive electron has not been mentioned in this article because, while such things exist, they seem only to have a transient existence and are not a permanent constituent of ordinary matter.)

Neutron Composition

The modern point of view on this is not entirely satisfactorily developed, but the ideas are quite different from those usually connoted by such a statement that a neutron "consists of" a proton and an electron in combination. Measurement of the mass of the neutron indicates a value of 1.00895 as compared with the mass of the hydrogen atom of 1.00813. The neutron thus has more energy than a hydrogen atom and could, so far as energy is concerned, spontaneously split into a proton and an electron, rushing apart with enough energy to account for the mass difference. Possibly free neutrons are unstable and radioactive in this way, but so far there is no experimental evidence for it.

The nearest thing we have to experimental evidence on changes of this kind is the study of both the natural and artificial beta particle emitting radioactive materials. A radioactive material which is beta-active is one that emits from its nucleus a high-speed electron (or positron). All of the artificially radioactive materials are beta-active substances. For these we do not have a simple barrier leakage theory like that applying to the emission of alpha particles from the heavy, naturally active alpha emitters such as uranium or radium. From the modern point of view, there are no electrons in the nucleus, *as such*, apart from what may be there on the hypothesis that a neutron is made of a proton and an electron in a peculiarly tight state of binding. But the hypothesis of this peculiarly tight state of binding is not very consistent with the fact that a neutron has more mass than a proton and an electron separately.

Energy Spectrum

The most remarkable thing about the beta-emitting radioactive materials is that, on disintegration, the different disintegrating atoms send out electrons of different energies. There is observed a continuous spectrum of emitted electron energies. This is true in spite of the fact that all of the atoms seem to be alike before and after disintegration. Therefore, there must be a definite energy difference between them.

In the case of the beta-active emitters, it is supposed that the true energy difference between parent and daughter substances corresponds to that carried away by the most energetic electrons emitted. The fact that in some disintegrations the electrons are emitted with a lesser amount of energy is formally accounted for by supposing that, really, there are two particles emitted: In addition to the emitted electron, there is supposed to be emitted a *neutrino* (Italian for little neutron, as in "bambino"). The emitted electron and the neutrino share the total energy between them in such a way that statistically, in some disintegrations, the electron gets all the energy. In most cases, however, it shares the energy in varying degrees with the other particle. The neutrino is postulated to have zero mass, or a mass considerably less than that of the electron and, having no charge, it completely escapes detection. Naturally this is not considered to be a very satisfactory way out, the neutrino having been invented for the purpose of balancing the books on the missing energy and having been postulated to have properties which make it almost, if not impossible, to detect. One of the most important fundamental discoveries that could be made in nuclear physics today would be some real evidence of the existence of the neutrino, or of its non-existence. Unsatisfactory as the neutrino idea may seem, it is the best that physics has had to offer for well over a decade.

Regarding the neutrino idea, however, one has to suppose that a neutron is capable of spontaneously changing into a proton with simultaneous emission of a proton and a neutrino:

$$N \mid \rightarrow P \mid + e + n$$

N is written for the neutron and P for the proton in this equation. But for the positron-emitting radioactive materials we have the opposite case in which a proton presumably changes itself spontaneously into a neutron and a positron and a neutrino,

$$P \rightarrow N + p + n$$

where p is the positron. It is necessary to suppose the emission of a neutrino occurs here for the same reason as before, namely to account for the continuous distribution in energy of the emitted positrons.

It must be confessed that the situation here is not satisfactorily understood, but it is clear that, because of the occurrence in nature of both positron- and electron-emitting radioactive materials, it is no more correct to suppose that the proton is a positron and a neutron that it is to suppose that a neutron is an electron and a proton.

Science and the National Welfare

E.U. CONDON

Director
National Bureau of Standards
Washington, D.C.

Society is at this moment at the threshold of an undreamed-of mastery of our material environment, for science, which provides that mastery, is in its Golden Age.

In particular, achievements in nuclear physics promise incredible advances in the years ahead. Energy from atomic power plants has been much talked about, but even more important are the tools provided by nuclear physics for research in other fields. Radioactive isotopes, for example, will permit us to explore the structures and constitution of molecular aggregates, for such isotopes can be introduced into a system as scientific detectives. They will behave as the usual atoms of the particular element behave, but they can be traced and studied by means of the radiation they emit. Tracer studies of this kind will unravel secrets in biology, physiology, medicine, chemistry, and metallurgy.

The combined effect of tracer studies, of a variety of sources of radiation, of various sources of high-intensity, highly-accelerated sub-atomic particles, and fundamental knowledge of the nucleus means that spectacular advances in many fields are at hand. The problem of curing fatal diseases will be successfully attacked; fundamental biological and physiological processes will be understood; new types of therapy will be developed in medicine; better control of intricate chemical manufacturing processes will be feasible; new products, like petroleum fuels and metals with unusual properties, will be possible; and even new forms of plant life can be created. The speed with which these possibilities are realized depends primarily on how much effort we put into such activities. For there is no question that the impetus of the new knowledge in nuclear physics, in conjunction with steady advances in other fields of science during the last 50 years, means a general efflorescence of the physical and life sciences.

But if we are to profit from this happy situation, there are major problems to be solved, and their solution will not wait. From one point of view life today is a race — a race between knowledge in the physical

An address delivered before the American Council of Commercial Laboratories at the Statler Hotel, Washington, D.C., December 4, 1947.

sciences, which gives material mastery, and general ignorance, which retards or rejects mastery of our environment. Rejection means no more and no less than destruction of civilization as we know and cherish it.

The problems confronting us, approaching them from the standpoint of the sciences, exist on several planes and two in particular: the specific problems of science as science, and the question of these sciences in relation to the other activities of man.

Problems of Science

The problems arising within the sciences themselves are extremely practical ones, and, on the whole, they are not complex. Several axioms are at once apparent. First, science is universal. Second, science is unlimited in its material. Third, the rate of scientific progress depends on the amount of effort put into science. These axioms are important: they mean that no individual and no nation has a monopoly in science, that science affords an inexhaustible mine of valuable knowledge and discoveries, and that we must be willing to support science appreciably if we expect to gain heavily and to maintain leadership.

The Steelman Report

A comprehensive and cogent analysis of the problems of science is to be found in John R. Steelman's report to the President, *Science and public policy*. Taking into account the three major groups engaged in research and development activities — the universities, the industrial laboratories, and the Federal research agencies — Dr. Steelman points out that each of these groups is "especially adapted to the performance of a particular type of research and each can make a unique contribution to our total research and development effort," with university emphasis on basic research, industry on development, and government laboratories engaged in both.

As a "basis for our progress against poverty and disease" and as the basis of national security, the Steelman report analyzes the present scope of our scientific effort, the deficiencies now present, and the needs in terms of a broad program. The main recommendations of the report are 8 in number, and I would like to discuss them briefly.

(1) It is recommended that expenditures for research and development be expanded as rapidly as facilities and trained manpower can be provided. A suggested goal is that, by 1957, 1% of the national income should be expended in research and development in university, industry, and government laboratories.

The report shows that a little over $1,100,000,000 is being spent this year for research and development, excluding the social sciences. With a national income of $200,000,000,000, this is an expenditure of little more than .5%. Only about $110,000,000, or less than 10% of the total, is spent for basic research. Almost half — that is, $460,000,000 — enters into the development of military weapons and needs, not including the amount spent for atomic bomb development now considered to be a civilian activity.

(2) It is recommended that heavier emphasis be placed in the future on basic and medical research. More specifically, it is recommended that the total research and development budget be doubled, coincidentally quadrupling basic research activity and tripling research on health and medicine.

(3) It is recommended that support for basic research be provided by the Federal Government at a progressively increasing rate, reaching an annual rate of $250,000,000 by 1957. The present rate of total expenditures for basic research is $110,000,000, while quadrupling would require $440,000,000. This proposal, therefore, leaves ample scope for large-scale and expanding support of basic research by private groups and state governments.

(4) It is recommended that a National Science Foundation be established with a Director appointed by, and responsible to, the President to administer the program of grants in support of basic research. It is also recommended that the Director have a board of advisers, half of whom should come from government laboratories in order to provide for proper correlation of the work with that of the government laboratories.

(5) It is recommended that a program of Federal scholarship aid to university students be developed in order to provide for the proper training of the increased number of scientists needed and that this program be a part of a general program of assistance to university students in all fields of interest.

(6) It is recommended that suitable Federal assistance be given to colleges and universities in developing their scientific research facilities, and that this should be administered as part of a broad program of aid to universities in all fields.

(7) It is recommended that the work of the several Federal research establishments be better coordinated by the establishment of an Interdepartmental Science Committee, by a coordination of all scientific research programs through the Bureau of the Budget, and by the assignment of a number of the White House staff to devote himself to problems of liaison at the top policy level of the Federal Government.

(8) Lastly, it is recommended that aid to the reconstruction of European scientific research be made part of our European Recovery Program. This recognizes, first, that science is universal in that its truths are part of the universe accessible to all investigators; second, that we gain as much by original discoveries made elsewhere as by those which we make; and, third, that the progress of other nations in science and technology is necessary if they are to become self-sufficient again.

The program outlined in the Steelman report is splendidly conceived, and every point is vital if we are to live up to the responsibilities with which we are confronted by our good fortune in natural resources and freedom from war devastation.

One of the great obstacles in the way of a major program of expenditures on basic research is the difficulty of explaining to an appropriations committee — and even to management in private business — precisely what the program will accomplish with that degree of definiteness expected and demanded in other fields. It is necessary to entrust funds for research programs on faith, on the competence of the leaders of such programs, and the trust must be maintained for a sustained period of time. It is characteristic of most fundamental research that several years are required for the completion of any work of importance and that the end result may be difficult to evaluate by anyone except specialists. What, for example, is the cash value of Einstein's discovery of the relation, $E = mc^2$? No doubt it is an astronomically large value now. But what was its worth at the time of its formulation, and who was qualified to make the evaluation? The point simply is this: pure knowledge cannot be evaluated in cold cash, and pure knowledge is independent of such evaluations.

Unfortunately, appreciation of this fact is not as widespread as it should be, which suggests the story of two partners who had long operated a chemical manufacturing business. They finally decided to employ a research chemist. Along about 11 A.M. of the first day of his employment, one partner said to the other, "Shall we go see whether that research chap has discovered anything?" "No," replied his partner, "It's a little too soon. Let's wait until after lunch."

Zones of Danger and Weakness

One of the dangers facing us in the present situation is overconfidence. The United States has led the world in technological progressiveness and in the techniques of mass production. We are, without question, the most powerful nation in the world. In these very facts lies the essential danger, for overconfidence is a product of precisely this set of circumstances. Illustrations of pride preceding fall fill the pages of history, and civilization after civilization has perished in this fashion. We need glance backward no farther than the recent war to see a once scientifically sophisticated power lose leadership and initiative — Germany. For many years, during the latter half of the 19th century and the early 20th, science in Germany was in a position of international prominence, and yet we now know how misguided and superficial were their efforts in the direction of atomic energy. I believe that two factors were at play here: First, the Nazi leaders eliminated the truly first-rate scientific leaders and installed second-rate party-men in positions of scientific leadership. Second, there are obvious evidences of overconfidence on the part of the scientists as well as the nation in their scientific ability and achievement. Thus, after the revelation of our work in atomic energy, we had the spectacle of, first, the German refusal to believe that accomplishment, and second, childish attempts to pretend that they had not wanted to develop an atomic bomb but that they really had progressed in atomic research and that their researches were to be devoted to peacetime uses. The rationalizations would be merely amusing were they not also sardonic.

Again, we have the spectacle of England's dilemma in this century. Prior to the 20th Century, the English had led the world in technology, one of the consequences of their early industrialization. This leadership had lulled the British into accepting this pre-eminence almost as a law of nature, and progress in modernization of facilities and in mass-production technique was not pursued vigorously. The result was that England fell behind Germany and the United States. A reluctance to accept scientific advances, in the face of obsolescence, is thus dangerous.

The obvious lessons of the past, as far as science is concerned, indicate that competent leadership must be fostered in science (remember that for every thousand scientists adequate to contribute in a rather routine way there is only one with great and inspiring creative ability), and we must never take for granted future achievements on the basis of past performances. This thought leads to another danger confronting us: as a nation we have been outstanding in applying science; we have not been outstanding in basic scientific discoveries or theory. If we are to attain our goals, it is imperative that basic research be supported on a large scale.

In atomic energy, for example, we were essentially dependent upon the work of European scientists for our basic knowledge, and European scientists in this country contributed heavily to our success, in particular Fermi and Szilard. Again, during the first half of the war, we were dependent on British research and development in radar for our own program, and it was not until the latter portion of the war that we contributed in a basic way to this field. Then our contributions, particularly in microwaves, were significant.

Research in Rubber

Still another field, vital to our economy, in which we have been dependent on European research is rubber, representing in the recent conflict a vast Federal investment second only to atomic energy and radar. The need for synthetic rubber during the war, as a result of the unavailability of natural rubber, is well known. What is not so well known is that the synthetic rubbers we used were developed largely by the Germans. The four types of synthetic rubber which we produced during the war were GR-S, Neoprene, Butyl, and the Nitrile rubbers. Of these, only Neoprene is purely American, a development of the Du Pont Company. Butyl is partially an American development, for its constitutes a radical improvement of the German material, polyisobutylene; yet it was based on this German work. Fundamental patents were taken by the Germans on the remaining two types — the Nitriles (under the German name Buna-N) and GR-S (under the German name Buna-S) — in the early 1930s. Of all these rubbers, GR-S is the most important: more than 80% of our total production was of this type because it is not only cheaper but best for tires.

Now that natural rubbers are again available, the problem of what to do with the synthetic industry, which involved a Federal investment of more than $700,000,000, is acute. This industry will be called on for only limited production, primarily to insure plant potentialities in the event of any future emergency and to provide the synthetic product for certain applications. The magnitude of the investment, the size and scope of the plants, and the relations between the synthetic and natural commodity are major commerical problems. For this very reason, the need for continued research and development is obscured.

The National Bureau of Standards has long been active in the research and development phases as they pertain to both synthetic rubbers and natural rubbers. From the standpoint of the national economy and security, it is necessary that a major and coordinated program of research and development be maintained in this field. Basic research is necessary if new types of synthetic rubbers are to be developed; developmental research is needed to develop desirable characteristics in the rubbers now available, to determine their properties. Much also remains to be done in measurements and instrumentation associated with the synthetic rubbers.

In the future, this country must have a vigorous program of rubber research to maintain "a technologically advanced and rapidly expandible domestic rubber-producing industry" as part of our national policy outlined in a the Crawford Act (Public Law 24, 80th Congress). The cost of such a program would involve an annual expenditure of about $4,000,000, which is less than 1% of the amount spent for the 1,000,000 tons of rubber that this country consumes annually. Industry should expend a corresponding amount for the development of new rubbers, in addition to its expenditures for research on end-products.

The cost of such a program is actually relatively small in terms of the value of the commodity and in terms of its national importance. Merely to maintain the present synthetic plants in a stand-by condition involves an annual expenditure of over $8,000,000, and these plants may well be obsolete at the time of another emergency. Therefore, a Federal expenditure of half this, to insure our future in this field, is, from any practical point of view, trifling.

Research in Optical Glass

A comprehensive and broad program of research in the field of synthetic rubber is a matter of national wisdom, and similar programs are needed in other fields, many of them not of such vital concern on the surface. For example, a national program of basic research on optical glass is a primary desideratum, and yet the thought of the importance of optical glass is not likely to occur to those not engaged directly in military problems, because the annual requirements of this country for precision optical instruments for civilian purposes during a period of peace are almost negligible when compared with the demands made upon our industry by our military agencies during war.

Here is a field in which we were long dependent on European developments. Prior to World War I, all optical glass used in this country was imported from abroad. It was during this period, under the sponsorship of the Navy, that the Bureau started experiments on the production of optical glass and succeeded also in fulfilling military requirements during that conflict; this this was possible only because the United States did not enter the war until the fighting in Europe had been going on for over two years. In the years between World War I and World War II, experimental work was supported at the Bureau by the Navy Department as a hedge against any future emergency, and the foresight of the Navy Department was amply rewarded in the recent conflict, for not only were satisfactory types of glass available as a result of prior experimentation, but actual production in this emergency period was necessary by the Bureau, attaining a peak of 236,000 pounds in 1943. Moreover, the Bureau was able to train industrial engineers and technicians so that their plants could enter into the production of this specialized kind of glass, and assistance was rendered to other branches of the military establishment.

If we are to be again prepared for future eventualities, a program of research and experimentation must be maintained. Stockpiling of optical glasses is not a solution, for stockpiles tend to maintain the status quo, saddling the military services with obsolete instruments and making the introduction of better glasses and instruments difficult. As a general rule, with valid exceptions only in the case of basic raw materials, stockpiling is futile, for it tends to hinder progress.

The only sensible solution is a progressive research program involving the development of new types of optical glass, analysis of the chemistry and physics of such glasses, the development of new and more efficient methods of making and processing optical glass, the investigation of new optical materials for such systems as the ultraviolet and infrared, studies of polished surfaces, and the development of control methods in production of highly precise optical components.

Research in Buildings and Structures

Finally, let me mention a field somewhat removed from pure science and related more to applied science and engineering — building technology.

The need for research in this field needs not stressing in this critical period of housing shortages, but it is significant to note the technical reasons behind our apparent backwardness in this field. In almost every field where American science and industry have teamed together to produce spectacular results, production has involved a centralized operation — for example, the production of automobiles, tires, typewriters, and so on. In the building industry, however, no single firm has specialized in the production of a building as such, and practically every material and product known enters into a completed structure. In each of the fields supplying components for a building, research has been done, depending on various conditions too many to outline here, and varying tremendously in extent and scope. No one, on the other hand, has attacked the problem from an integrated point of view, with the single exception, to my knowledge, of the work of the Bureau of Standards in building materials and structures.

Even here, as a result of the extremely limited funds granted for this purpose, the attack has been on a relatively small scale. Recently, all of the sections engaged in this type of work at the Bureau have been unified into a consolidated Building Technology Division, and an accelerated and coordinated program is under way. Groups are engaged simultaneously in investigations of the properties of materials: structural strength; fire resistance; acoustics and sound insulation; heating, ventilating, and air-conditioning; durability and the exclusion of moisture; building and electrical equipment; and other projects.

Unified scientific research in other fields of industry has been responsible for productive results, and it is reasonable to assume that the effect of this approach, applied generally throughout the $10,000,000,000 construction industry, will achieve similar results.

Science and Man's Other Activities

Even these few illustrations indicate that science does not function in a vacuum, divorced from everyday life. It is a pre-eminently practical thing, dealing with crucial problems affecting industry, business, the nation, and the world. It costs money, and it demands the efforts not only of scientists but of every segment of our population. Too often science is pictured as an "ivory tower" affair with no, or little, relation to reality. On the contrary, it is concerned immediately with the nature of the universe. It is the cause of our industrial economy, it operates within the full context of social existence, and its deals with practical problems as much as, if not more than, with theoretical ones. One of the discouraging attitudes widely prevalent in the contemporary world is the high regard placed upon what is called "practical" and the low esteem granted the "theoretical." In point of fact, the two differ only in time, relative to application; and pure, fundamental knowledge precedes applied knowledge.

The operations and progress of science can therefore be understood fully only in terms of the framework of our general society and in relation to the other activities of man. This context is particularly significant when we consider that science has now placed in our hands tools that are equally potent for good or evil. I have been talking, for the most part about the good, but actually the potential evil is more important, because of what value is this growing potential of good if science is used to destroy the civilization from which it has sprung?

It is fashionable to cry down the so-called pessimist who suggests this dangerous possibility, partly because no one loves a pessimist, partly because man is largely a hopeful creature with a belief that, at worst, he will muddle through, and largely because the dangers are difficult to group and appraise as a consequence of the staggering difference in kind and degree of present dangers in the form of scientific warfare. It is sufficient to say for my purposes that science has presented us with several weapons, each of which, unleashed, can mean almost total, if not total, destruction.

The question, then, is how to prevent such a situation. The answer is not to be found in the physical sciences. It is to be found in other realms of man's activity — in economics, in sociology, and in political science. Man's conduct in the physical sciences is rational; in these other fields it is largely arbitrary.

Research in the "Humane" Sciences

It is often said that man's social irrationality is a consequence of the fact that economics, sociology, and political science are not sciences but merely individual judgments and personal opinions. Now this is palpably untrue even at present, for much is known about cause and effect in these fields, and such statements are made only because habit, custom, tradition, and heritage tend to make us cling to whatever we know rather than to re-examine the data, coolly and critically. So far, no readily demonstrable experiments exist in what I shall call the "humane" sciences as exist in the physical sciences.

Admittedly, these "humane" sciences are younger than the physical sciences. Moreover, the variables to be accounted for are vastly greater than those we deal with in the physical sciences. But these are not adequate reasons for belittling the "humane" sciences and denying them support. On the contrary, these are compelling reasons for supporting them, and the present state of civilization demands that this be done. As a matter of fact, since the physical sciences have outstripped man's capacity for using them wisely, sanely, religiously, it is of the utmost urgency that we attempt to forge ahead in the "humane" sciences lest all be lost.

This is the time for intensified activity in these fields, not only because of the urgency of our need but because now the physical sciences have two tremendous tools to contribute to the "humane" sciences, tools that will permit "scientific" analysis of data having a large number of variables.

The first of these tools is statistics, which provides the theoretical, mathematical basis for analysis, the mathematical techniques for handling data, and the criteria for evaluating results. Mathematical statistics is now a substantial and well-developed discipline, and it does, in fact, offer these tools. Automatic electronic computing machines, on which many laboratories and companies are at work, constitute the second tool shortly to be available to the "humane" sciences. These machines will permit the handling and analysis of data, rapidly and comprehensively. Until the present, one of the major problems in fields where vast amounts of data are obtained has been the handling and classifying of the data. Literally thousands of man-days are needed in even relatively simple problems. This means that research is expensive, and the "humane" sciences have not usually been able to afford such luxuries. As an example of the labor involved in handling data of this type, consider a relatively simple problem. At the present time, a typical census problem involving 100,000 pairs of 5-digit numbers, representing statistical data, takes approximately 12 working days, exclusive of card handling and data punching. An electronic digital machine will handle the same sequence in 10 minutes at the most.

The Steelman report does not consider research in the economic, social, and political sciences. The study of the physical sciences in itself was a major effort, requiring 5 volumes of summary findings. It is to be hoped, however, that a similar analysis of the "humane" sciences will be made in the near future and that a program for these sciences will be mapped out and implemented.

Research in the "Mental" Sciences

Just as there is a disparity in the evaluation of research between the physical and the humane sciences, so too there appears to be an analogous disparity in the attitude of most people toward research between the medical and the "mental" sciences. Like the physical sciences, the medical sciences produce what are called "tangible" results — for example, new drugs, new clinical techniques, and so on. Like the "humane" sciences, the "mental" sciences do not appear to produce materialistic results and have suffered similarly in the support granted them for research. This, too, is a situation that needs remedy. Psychology, psychiatry, and psychoanalysis are disciplines pertinent in the solution of current problems. Aside from the statistical fact that 3 out of every 7 beds in the hospitals of the United States are occupied by the mentally ill — a vast drain in terms of lost manpower and cost — and that untold numbers of borderline cases permeate the entire social structure, we need to know more about the workings of the mind. For there is little doubt but that nonevident factors affect human behavior profoundly, factors like frustrations and fears.

These factors affect every activity of man, his personal, social, political, and even scientific life. From the standpoint of science we can say not only that science affects individuals and nations but that these individuals and nations affect science. Even from this restricted approach, then, what has happened or happens to men's minds and spirits is of interest if we have scientific objectives in view. We have seen how entire nations have apparently succumbed to a schizophrenia that has led to the espousing of mad, undemocratic, bestial beliefs. We have seen at least one nation despoil its scientists as a result of such an aberration.

Compartmentalization in the sciences and in other fields in inimical to a coordinated attack on the problems of man. This compartmentalization is actually breaking down in the sciences. The distinction between chemistry and physics, for example, has almost vanished. Competent research in the social sciences now depends on mastery of mathematics and on the utilization of the electronic tools. The complexity of modern life depends on specialization for progress in particular fields but, for over-all progress and for a solution to the dilemma of unbalances, integration and coordination are essential. In short, education of a comprehensive nature, embracing many fields, is needed for the survival of our civilization.

The sciences, like those other truth-seeking activities of man, require a free environment, an environment, above all, free from fear, petty arbitrariness, and tyranny. The pursuit of the sciences is fundamentally nothing more or less than the pursuit of truths. In the last analysis, all of man's activities are subservient to what happens to his spirit — to his spiritual welfare, "For what shall it profit a man, if he shall gain the whole world, and lose his own soul?"

Electronics and the Future

E.U. CONDON

Member AIEE

The vast influx of reports on new and not easily understood developments in the various fields of electronics easily can be quite baffling to the electrical engineer. An understanding, such as is provided here, of recent expansion in electronics research is helpful to give the correct perspective with which to view the electrical industry.

Electronics has not been given, so far as I know, an officially agreed-upon definition. Generally speaking, electronics is a part of science, it is an art, and it is an industrial practice. *It is the science, art, and industry concerned with electrical phenomena involving electrically charged atomic particles, outside of solid and liquid bodies.* With this understanding we have a fairly clear separation of the field from the classical branches of electrical engineering and at the same time a definition that is broad enough not to restrict us to those particular phenomena in which only electrons are involved.

Electronics is thus a very broad subject. It embraces all phenomena connected with the passage of electric currents through gases and high vacua. Such phenomena are utilized for a wide variety of purposes: the generation of high frequency electric power, its amplification, the control and rectification of electric power at all frequencies, production of electric current from light sources, and production of light from electric currents (including X-rays and gamma rays as special forms of light). Phenomena associated with the propagation of radio waves in the ionosphere of course also are included, as is the study of cosmic radiations from interstellar space, and special means for producing beams of atomic particles of high energy for the study of nuclear physics.

Electronics may be subdivided into 11 major topics. The first four of these are branches of our subject which are already well launched as a result of the work of the war years. These are

---

Essentially full text of an address presented at the National Electronics Conference, Chicago, Ill., October 3, 1946.

E.U. Condon is the director of the National Bureau of Standards, Washington, D.C.

1. The general field of radar, loran, and air and sea radio navigation aids including associated microwave techniques.
2. The broad field of television, now ready for development as a popular service after long delays caused by the war.
3. The broad field of industive and dielectric heating for which there seem to be limitless possibilities of special application in the metllurgical and plastic industries.
4. The field of electronic instrumentation and control, including the general concept of the servomechanism.

These four main branches of our subject are all relatively new developments, but there already exist large industrial activities built on them, so that they are becoming mature and ready to join the main classical branches of electronics — namely, the art of radio communication and the art of rectification and control of large blocks of power by gaseous conduction devices such as the thyratron and the ignitron. Because these four branches are discussed in such great detail so often, I shall endeavor to point out seven additional fecets to modern electronics.

Electronic Computing Devices

Only those who are skilled in mathematical physics realize how very limited is man's ability to obtain analytical solutions to important problems in applied mathematics. Great as has been the progress of the past century, the time has come when many problems great importance, especially in hydrodynamics, aerodynamics, and meteorology can be handled only by methods based on elaborate arithmetical computation. For many years now it has been the practice to introduce artificial and unjustified approximations into the setting up of many such problems in order to reduce them to a form tractable by our limited analytical attainments. As a result one often is confronted with this unsatisfactory situation: A theoretical calculation is made and the results are compared with experimental data and discrepancies are found. The question now is, are these discrepancies due to an improper physical formulation of the problem, or due to the inadequacy of the mathematical methods used in making specific calculations from a correct physical formulation? Such a situation is clearly intolerable.

However, calculations required for problems of current interest are so elaborate as to require many man-years of work by skilled computers. Aside from the cost involved, such a situation slows down progress because one has to wait too long to get an answer by old-fashioned hand-operated computing machines.

A good deal of progress was made during the war in opening up the field of application of electronic methods to rapid calculations. The first major step in this direction is not, strictly speaking, electronic but electromechanical. It is the sequence controlled calculator built by Professor Aiken at Harvard from standard International Business Machine units. In addition, several relay computers have been or are being built at the Bell Telephone Laboratories and the International Business Machine Corporation.

Until recently, this work at the Bell Telephone Laboratories has been under the technical direction of Doctor G.R. Stibitz and S.B. Williams. Such machines have been built for the Naval Research Laboratory in Washington, for the Aberdeen Proving Ground and other Army research centers, and for the Watson laboratory of Columbia University under the direction of Doctor W.P. Eckert.

The first major development in the electronic field was the *ENIAC* (electronic numerical integrator and computer) built under the direction of J.W. Mauchly and J.P. Eckert at the Moore School of Engineering, University of Pennsylvania, for the Army Ordnance Department. The next major step in the field was the development of the *EDVAC* (electronic discrete variable computer), now under construction at the Moore School in Philadelphia. Additional projects in this field include the joint project of Professor J. von Neumann of the Institute of Advanced Studies in Princeton and Doctor V.K. Zworykin of the Radio Corporation of America laboratories in Princeton. Also worthy of mention is the research program of the National Bureau of Standards in this field being organized under the joint direction of Doctor John H. Curtiss and Harry Diamond. This project includes an immediate program of production of machines for the Bureau of the Census and for use by the Navy in a proposed National Mathematical Computation Center together with a long range program of fundamental development of components for such machines.

The *ENIAC* is a rather remarkable electronic device involving a large assemblage of 40 panels, employing approximately 18,000 vacuum tubes and 1,500 relays. Numbers are transmitted through the machine as a sequence of pulses for the various digits, the individual pulses lasting 2 microseconds separated by approximately 10 microseconds. The basic arithmetic operations are carried out in terms of synchronized performance of various calculating units, each elementary operation being assigned a time of 200 microseconds. This basic time unit is to be made shorter in the newer machines under development. Some idea of the speed of operation of the machine is given by remarking that it can multiply one 10-digit number by another 10-digit number in about three milliseconds.

Evidently such a machine is only useful if it is made to carry out a whole sequence of arithmetical operations without intervention of human beings with their enormously slow reaction times of the order of 200,000 microseconds.

Probably the part of this field which today provides the greatest challenge to inventive ingenuity is the development of memory units. In a long and involved calculation one needs to be able to store in the machine some numbers which are produced in intermediate stages of the calculation and then be able to produce them as input to the machine at later stages of the calculation. Space does not permit a more exact formulation of the need.

Mass Spectrometers for Chemical Analysis and Control

By a mass spectrometer is meant any high vacuum device in which chemical substances in the vapor phase are ionized by electron bombardment and the products of ionization subsequently are passed through electric and magnetic fields in such a way as to separate out the ions by mass. The mass spectrometer emerged from the pure physics laboratory some 30 years ago after initial development by J.J. Thomson and F.G. Aston in England. Especially valuable contributions to the development were made in Chicago by Professor A.J. Dempster of the University Chicago. Later important developments were made at Princeton by Professor H.D. Smyth and Professor W. Bleakney, and at the University of Minnesota by Professor J.T. Tate and his students, especially Professor A.O. Nier.

Of what use is the mass spectrometer? in fundamental physics it makes possible the determination of the existence of different atomic isotpes and their relative abundance, and thereby becomes the detection device that is essential to all use of stable atomic isotopes for chemical researches by the tracer technique. Special forms of the mass spectrograph have been developed, particularly by Professor K.T. Bainbridge of Harvard, for the precise measurement of atomic masses, thereby giving important data on the energy of binding of atomic nucleuses. Other special forms adapted to

FIGURE 1. Gas-filled tubes that control the d-c power supplies of the ENIAC (electronic numerical integrator and computer)

production of high ion current were developed under Professor E.O. Lawrence at the University of California for use in separation of the uranium isotopes. This development matured into the construction of one of the multi-million-dollar isotope separation plants at Oak Ridge, Tenn. This is a special use which probably is already obsolete since other methods of isotope separation have already proved to be superior.

The form of mass spectrometer which, it seems to me, is most likely to be of continuing research and industrial significance is that adapted to chemical analysis of complex gaseous mixtures. Such instruments have been developed by Professor Nier of Minnesota, by Doctor J.A. Hipple of the Westinghouse Electric Corporation, and by Doctor H.W. Washburn of the Consolidated Engineering Corporation in Pasadena, Calif. They are already finding important application in the hydrocarbon industries — that is, in the manufacture of automotive fuels and of synthetic rubber. Although they have not as yet been adapted directly to plant process control, so far as I know, I predict that this will become one of the important developments of the future. Great advances in this field have been made by the study of infrared absorption spectra and through the use of the mass spectrometer. Important contributions to the development of standard methods of analysis are being made at the National Bureau of Standards through a co-operative venture with the American Petroleum Institute, directed by Doctor F.D. Rossini, involving the development of infrared methods by Doctor C.J. Humphreys and applications of the mass spectrometer by the group headed by Doctor F.L. Mohler.

I am sure that in this direction lies a great opportunity for further co-operation between electronics men and chemists, leading to further development of instruments suitable for commercial service and their special adaptation to particular problems of research and plant control.

Physics of Radio Propagation

It is important to remember that were it not for the ionized regions of the upper atmosphere, radio communication would be a purely local, almost line-of-sight affair. The ability of communication systems to extend over great distances is now known to be due to the reflecting properties of the ionized regions which we call the ionosphere. If the ionosphere were a simple mirror located at some definite height and reflecting strongly radio waves of all frequencies, the situation would be quite simple. However, the degree of ionization in the upper atomsphere is shifting continually with changing conditions of radiation from the sun in ways that are far from fully understood. The exact behavior of the ionized regions depends on the earth's magnetic field as well as the degree of ionization over the path of communication. The causative factors which determine the earth's magnetic field are themselves by no means well understood in spite of many years of diligent study.

Yet it is hardly possible to overestimate the importance of complete and accurate understanding of the phenomena of the ionosphere. Not only is ordinary radio communication service over long distances completely dependent on such knowledge, but also the proper operation of aid to long distance aerial and sea navigation, and eventually the means for guiding long-range missiles which some people think we some day may have to direct against our fellow men while they do the same to us in some future war.

The whole study of radio propagation includes also a proper analysis of the factors affecting radio wave propagation in the lower atmosphere or troposphere. This region, where ionization effects are negligible, apparently does not affect the propagation of radio waves of broadcase frequency or the usual short waves. But meteorological conditions do affect the propagation of ultrashort waves and especially microwaves. This cuts both ways, at the same time giving the weather men a new tool for study of weather conditions, and making it necessary for the radio man to consider the way in which meteorological factors affect his operations.

In this field great advances were made during the war. Stations for observation of the ionosphere were operated in the Pacific theater by the Army and Navy and also the Department of Terrestrial Magnetism of the Carnegie Institution of Washington, D.C., which had pioneered in ionospheric studies since the basic research of Breit and Tuve in 1924 when radio pulses first were reflected from the ionosphere and the results interpreted as giving a measure of the height of the reflecting layers as in modern radar. Similar programs also were conducted by the British throughout the British Empire and by the Russians. Of course, the Germans and Japanese also had extensive programs of ionospheric observation, but their results were not available to us until after the war.

All available data were channeled into the National Bureau of Standards where it was handled by the staff of what was known as the interservice radio propagation laboratory, operated co-operatively under the auspices of the Joint Communication Board of the Joint Chiefs of Staff. After the close of the war, arrangements were made for the continuance of this work on a peacetime basis. This plan has called for organization of the central radio propagation laboratory as a division of the National Bureau of Standards, under policy guidance of an executive council representing government and private industry interests in radio propagation.

The central radio propagation laboratory now has a staff of approximately 200 and has taken over the operation, or has contracted for the operation, of a chain of ionospheric observing stations located at Fairbanks and Adak in Alaska, at Guam, at Manila, at Leyte, at Palmyra, in Hawaii, at Louisiana, at Trinidad in the West Indies, and at Stanford University in California, in addition to its home station at Sterling, Va. This laboratory has a broad program of fundamental research which includes co-operative arrangements with the Harvard College Observatory at Climax, Colo., the United States Naval Observatory in Washington, D. C., the Mount Wilson Observatory near Pasadena, Calif., and the McMath-Hulbert Observatory in Michigan.

In addition this unit of the National Bureau of Standards maintains close liaison with similar groups working in Britain and in Russia. Co-operative work with the British and the Russians in this field is thoroughly satisfactory. The output of all this work, in addition to fundamental research in the physics of radio propagation, also leads to the preparation of predictions of conditions affecting radio propagation, which are available to the public in the form of a special series of publications from the National Bureau of Standards. This is a continuation of a prediction service that was available, on a restricted basis, to the Armed Forces during the war, but which is now generally available for use. Predictions of ionospheric conditions are made three months in advance in a monthly publication. In addition, special warnings of disturbed conditions are sent out over the National Bureau of Standards broadcasting station, *WWV*, and are provided to newspapers through special reports distributed by Science Service.

This very vital work is clearly of the utmost importance to the future development of electronics, and by its nature is dependent for success on whole-hearted international co-operation. So far there is every indication that the service established by the United States Government will co-operate effectively with other national services to the end that all of us may quickly gain a better understanding of all the factors affecting radio propagation.

Device for Accelerating Atomic Ions

Nearly all that we know about the structure of the atomic nucleus was learned from experiments conducted by bombarding various targets with streams of high energy protons, deuterons, electrons, or helium ions, obtained by multiple acceleration as in the cyclotron, by means of the electrostatic generator of Van de Graaff, or by various circuits for rectification and multiplication of voltage provided from the 60-cycle supply.

Prior to the war there were a number of transformer-rectifier sets which operated up to a million or a million and a half volts. These could provide steady voltage and currents of tens of milliamperes and have been developed into machines for production of X-rays of great penetrating power used during the war for industrial radiography and also finding increasing application in hospitals for cancer therapy. Before and during the war there were further developments of the Van de Graaff type of generator carried out by Professor John Trump of Massachusetts Institute of Technology and also by Professor R.G. Herb of the University of Wisconsin, which resulted in machines capable of giving up to four million volts, producing an ion beam of great homogeneity in energy of the several particles.

The Cyclotron. Development of the cyclotron as initiated by Professor E.O. Lawrence of the University of California, had gone to the point where there were a number of such instruments available in American universities and in Europe and Japan, capable of producing beams of protons, deuterons, and alpha particles of energies of the order of ten million electron volts. At the time when the war interrupted such research there were 20 or 30 such instruments in use throughout the world, providing the essential means for production of artificial radioactive materials for chemical tracer studies and for basic studies of the physics of nuclear transformations.

Back in 1940, Lawrence had projected a much larger cyclotron to be built with a magnet having pole pieces 184 inches in diameter, which would produce a beam of particles having energies of the order of 100 million volts. This development was interrupted by the atomic bomb project when it was decided to devote this large electromagnet to the development of the mass spectrograph method of separating uranium isotopes already mentioned.

The giant cyclotron requires more than a mere scaling up of the smaller ones. The operation of an ordinary cyclotron depends on the fact that the period of revolution of a charged particle is independent of its energy: at higher energies it moves faster but moves in a larger circle, the speed and radius being proportional so that the period of revolution remains the same. Thus the alternating electric field which provides the acceleration means for the particles remains the same. Thus the alternating electric field which provides the acceleration means for the particles remains in phase with the revolving ions as they wind their way out from the center to the rim of the cyclotron chamber. But this result is only valid at energies sufficiently low so that the variation of mass with velocity is negligible. For the proton the energy equivalent of the rest mass is about 920 million electron volts. Hence, when a proton has an energy of 9.2 million electron volts, its mass has been increased by 10 per cent and a corresponding discrepancy results in the condition of synchronism of the particles with the alternating electric field which accelerates them.

One way to minimize this difficulty is to operate with the greatest possible voltage amplitude on the accelerating electrodes of the cyclotron. If the particle gets a large energy increment each time it crosses the gap, its whole trip will involve very few revolutions and there will be little opportunity for the alternating electric field to get out of synchronism with the revolving ions. However, this is not a very neat solution since high voltage operation of the radio frequency oscillators brings many special insulation problems and requires power losses that go up with the square of the desired voltage amplitude.

The Betatron. Another important development which came along just before the war was the practical development of the betatron by Professor D.W. Kerst of the University of Illinois. This now has been brought to the point where practical means exist for the production of a beam of electrons having energies of the order of 100 million volts, and still larger machines of this type are projected. In the betatron, the electrons are accelerated by the electric field associated with a changing magnetic field. For many years the general possibility of doing this was common knowledge among physicists, many of whom devoted a good deal of attention to the idea, but it remained for Kerst to get the idea of a means for utilizing this principle in a practical device.

Considering that the rest energy equivalent of the mass of an electron is only half a million electron volts, it is plain that the betatron provides us with electrons whose energy is so great that their mass is some 200 times the mass of ordinary slow electrons such as we encounter in commercial electronic devices.

The Synchrotron. During 1945 some new ideas were introduced into this field which give us a broader conception of its possibilities than hitherto was realized. Corresponding to the truly international character of scientific advance, the important ideas seem to have occurred quite independently in Russia to V. Veksler and in America to E.M. McMillan of the University of California. McMillan calls the device a synchrotron. Its application to the acceleration of positive ions is being called a frequency-modulated cyclotron. Besides we have the betatron, and Pollock of the General Electric Company has shown how the betatron and synchrotron principles can be combined in a single machine for electron acceleration.

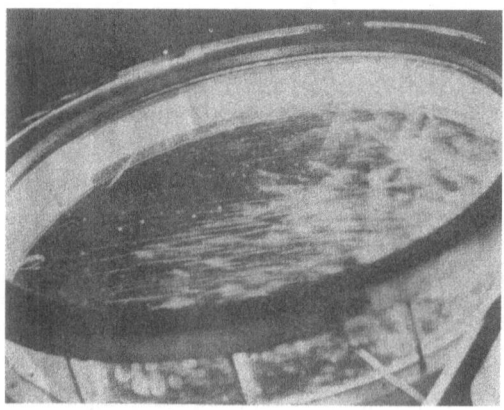

FIGURE 2. Vapor trails produced by cosmic rays passing through a cloud chamber

Briefly it may be stated that McMillan has shown that, in a device of the general type of a cyclotron or a betatron in which the particles move approximately in circles or spirals and are acted on by an alternating electric field, the most effective mode of operation is to work with the ions which are in nearly equilibrium orbits, that is, those which have such an energy that the orbital frequency is the same as that of the alternating electric field, and whose phase is such that they cross the accelerating gap just as the electric field is zero and changing in the sense that earlier arrival of the particle would result in acceleration. Such equilibrium orbits are stable in phase in the sense that a particle lagging behind tends to be accelerated and one running ahead tends to be decelerated. The position of the orbit in space may be stabilized by arranging matters so that the magnetic field falls off slightly with motion out from the center, so that the lines of magnetic force are concave toward the center.

Particles in such equilibrium orbits are moving then in a manner which is stable in phase and in space. To accelerate the particles it is necessary to change the energy of such equilibrium orbits, which can be done either by changing the magnetic field, as in the betatron, or the frequency, as in the frequency-modulated cyclotron.

It is not possible here to go into a more detailed discussion of these important new ideas. Suffice it to say that they have been tested on the small 37-inch cyclotron at Berkeley and found to work when the frequency of the oscillating circuit is varied by a mechanical rotating condenser. As a result the large 184-inch cyclotron is being built in accordance with synchrotron principles, which will result in the production of a beam of particles of several hundred million volts. Construction has advanced to the point where a beam is expected before the close of 1946.* The synchrotron principle of operation does not require such high voltages as in the usual cyclotron. Because the particles have phase stability, they may be accelerated quite gradually even though the total acceleration produces a considerable variation in mass, say of the order of 30 to 50 per cent. This permits the great simplification of using low power oscillating electric fields to produce the acceleration.

Besides the giant cyclotron at Berkeley, Calif., several other large frequency-modulated cyclotrons are now under construction, notably at Harvard University, at Columbia University, and at the University of Rochester. In consequence there soon will be quite good facilities for researches on the nuclear transformations produced by bombarding atoms with ions having energies of the order of 300 million electron volts, that is, about 20 times the maximum available in the cyclotron laboratories of prewar days, and 50 per cent more than the great amount of energy released in a single act of uranium fission.

---

* The machine is now in operation at Berkeley, Calif.

Who knows what will come of such research? Here we are really on the frontiers of knowledge without the slightest idea of what kind of results are to be expected. All we know is that from past experience we can expect adventures in this new region to be at least as exciting as the exploration of lower voltages was in the two or three decades just closed.

At the National Bureau of Standards we have plans for the construction of a 100-million-volt betatron, and probably of a similar acceleration device for positive ions, in order that the bureau may continue to discharge its basic responsibility for the establishment of measurement techniques and of safety standards for workers in this field.

So far I have dealt with only devices of the broad cyclotron class in which the particles are brought repeatedly under the influence of the same accelerating electrodes by the deflecting action of a magnetic field. But considerable progress also is being made in the application of the war-born microwave and cavity resonator techniques to the development of linear accelerators whereby particles can be accelerated to high energies by traversing alternating fields phased in such a way that the field is always in a direction to accelerate the particle at the place where it happens to be. Noteworthy among major projects of this class is the one being developed by Professor L.W. Alvarez of the University of California. In this it is expected to attain an acceleration of 1 million electron volts per foot of length on a tube 40 feet in length, the ions being injected at an energy of 4 million volts obtained by the aid of a pressuretype Van de Graaff machine based on the design of Professor Herb of the University of Wisconsin.

Uranium Chain Reactions for Power

Two types of atomic bombs have been developed, both based on the use of atomic fission induced by fast neutrons, one employing plutonium, an element which does not exist in nature, the other employing essentially pure U-235 which has to be separated at great expense from the U-239 of nearly identical physical and chemical properties with which it occurs in nature.

The element plutonium, which has atomic number 94 and atomic weight 239, is produced in a uranium pile, the quantity production in this country being carried out in a great plant built for the purpose at Hanford, Wash. It first was demonstrated, on the University of Chicago campus less than four years ago, that a suitable arrangement of lumps of ordinary pure metallic uranium interspersed with pure graphite could be made to undergo a controllable slow neutron chain reaction. In this equipment a U-235 atom undergoes fission on absorption of a slow neutron. In so doing it falls apart into two new radioactive atoms of medium atomic weight which move apart rapidly but are brought to rest quickly in the solid, their energy appearing simply as heat to warm the uranium. The splitting U-235 atom also spills out several more neutrons. These diffuse in the graphite where they are slowed down and diffuse back into the uranium to cause the fission of more U-235 atoms. Some of the neutrons are lost by absorption by impurities, others escape through the walls; so that in the steady state, although several are released in each fission, only one produces another fission. Of those which do not produce fission, a certain number are captured by the U-238 atoms in such a way as to form U-239. The isotope of uranium is radioactive and soon disintegrates into Np-239, an isotope of neptunium, the element next above uranium in the periodic table. This in turn disintegrates spontaneously with the emission of a beta particle converting into Pu-239, an isotope of plutonium. Although itself radioactive, Pu-239 is comparatively stable and can be separated from the uranium lumps from time to time for use in atomic bombs.

These operations produce a vast quantity of heat per unit of plutonium formed, but in the war project this heat was not utilized. In order to do so it would have been necessary to solve the engineering problems associated with high temperature operation, as heat has no value as a power source unless it is supplied at a high temperature. In this field the most important research projects before us consist in the development of the most economical forms of this equipment in order that the heat may be utilized to drive heat engines for conversion to electric power.

With the organization of a Federal Atomic Energy Commission as provided by legislation passed by the last Congress, we may look forward to a vigorous research and engineering development of the power generating possibilities of the uranium chain reactions. In accordance with the policies formulated in that legislation as recommended by President Truman, the interim management of the project by the Army's Manhattan District already has started developments looking toward the study of the means of using atomic power for peacetime uses.

It would be wrong to raise false hopes that an economic millenium is at hand. We know that the cost of basic production of power in our economy is a small fraction of the total productive effort required, so that even if power suddenly became free at the primary generating points, it would not make such a vast change in the economic picture as some imaginative writers have supposed. Moreover, at present our knowledge is so uncertain that it is very difficult to predict what costs will apply to power from atomic fission. Naturally much will depend at first on arbitrary questions of accounting, such as the question whether the research costs are simply to be absorbed as governmental expense, or charged to the operations resulting from such research.

Although it is too early to speak very definitely of the probable future prospects for atomic power, there is every indication that power from this source will have an important effect on the economy of ship propulsion, and also on the power supply of communities which have neither water power nor a convenient coal supply.

In any case this topic is one of the utmost importance to electronics. Fundamentally it belongs to electronics under the broad definition here adopted. In the narrower sense this field will require a vast amount of electronic instrumentation, because of the necessity of equiping these plants with completely automatic self-regulating control devices due to the fact that the great amounts of radioactivity generated in their operation make it impossible for workers to come near them. If we get an atomic power industry of the future, it will be largely electronic in character.

Radioactive Tracer Studies and Therapy

Before the war a considerable number of artificial radioactive substances were available in minute amounts, and chemists were beginning to develop the technique of using them in fundamental studies of the mechanism of chemical reactions, especially in biochemistry. Also they were finding application for these substances in a limited way in therapy of malignant for these substances in a limited way in therapy of malignant tumors and leukemia, using their radiations in combination with special chemical properties to give a wider range of possibilities than is afforded by X-rays or natural radium or radon.

During the war a considerable number more of such radioactive elements was discovered, so that today there are approximately 450 such radioactive isotopes known, including at least one for each of the chemical elements from atomic numbers 1 to 96. Moreover, the large densities of neutrons available in the experimental uranium pile at Oak Ridge, Tenn., and the large production piles at Hanford, Wash., as well as the large amount of radioactive fission products, have made available quantities of radioactive materials of many kinds that are hundreds and thousands of times greater than were obtainable before.

The Army's Manhattan District recently has made arrangements to conform to the policy of giving wide distribution of these radioactive materials to research workers. Announcement of details concerning availability of these materials for civilian research purposes finally was made in *Science* on June 14, 1946, and the materials are beginning to flow out to workers in various biological research centers.

Some of the accomplishments made with tracer chemistry are quite marvelous. For example, the entire chemistry of plutonium necessary for the design of elaborate separation plants was learned from studies made with an unweighably small amount of the material, its behavior toward various reagents being studied by following the radioactivity.

Probably C-14 is the most important of these elements from the standpoint of future possibilities for chemistry. This is produced by the action by the action of neutrons on some nitrogen containing compound, the neutron entering the N-14 nucleus and knocking out a proton and thereby transforming it to C-14. Because the whole field of biochemistry and medicine is founded on the chemistry of carbon, there is no limit to the scope of possible application of this material in learning about animal and plant metabolism in health and in disease. Likewise the H-3 isotope, tritium, is of very general importance.

Other radioactive isotopes which are certain to find many important applications are the 14.3-day P-32, the 180-day Ca-45, the 47-day Fe-59, the 250-day Zn-65, the 53-day Sr-89, and the 8.0-day I-131, where the times are the characteristic half-lives of the element in question.

To this rapidly developing art, electronics is making important contributions in steadily improving the convenience and reliability of the equipment used for measuring the radioactivity of these tracer elements.

Cosmic Radiation

The study of cosmic radiation began some 20 years ago when the researches of Millikan and his associates, principally Bowen, showed clearly that the agent active in causing ionization in the air was absorbed by the atmosphere and by water and was clearly coming from outer space, or at least the outermost layers of the atmosphere.

In the ensuing years this cosmic radiation has been the subject of a steadily increasing and fruitful range of studies. It has been found that the radiations consist of charged particles coming in from outer space, with energies of the order of several billion electron volts, that is, ten times the maximum energy in the most ambitious of the acceleration devices described in an earlier section. In arriving at this conclusion studies are made of the dependence of the cosmic ray intensities on the magnetic latitude, the entire earth behaving like a giant mass spectrograph to help us determine the energy of the incoming particles.

Among the most important discoveries of recent years from cosmic ray research is that there are in the cosmic radiation, as determined from the tracks they produce in a cloud chamber in a magnetic field, a new type of basic particle, called mesons, whose mass is about 200 times that of the electron, and therefore about one-tenth that of the proton. These particles apparently have but a transient existence and their role in the scheme of things is far from being well understood. Since the mass of the electron is equivalent to 0.5 million electron volts, it follows that the rest mass of the meson is equivalent to about 100 million electron volts. Therefore it is natural to suppose, although not certain, that particles accelerated to well above 100 million volts in the new giant synchro-cyclotrons now under construction will be able to give us a controlled yield of mesons with the aid of which we can learn where they fit into the scheme of things.

That is why physicists are so anxious to get accelerating machines which extend to this range of voltage: from cosmic ray studies they have clear indications in advance that important things are to be learned there.

Is There a Science of Instrumentation?

E.U. Condon

Director
National Bureau of Standards
Washington, D.C.

Instrumentation, as the word is used here, means the development and application of measuring devices which respond quantitatively to some physical property of a situation and give an output which depends on this property. The word instrument is often used in a broader sense, as when forceps are referred to as surgical instruments, but in this discussion it refers to measuring devices.

Nearly all of the progress of modern science and industry is directly traceable to the development and use of a wide variety of measuring instruments. With monotonous regularity people who discuss this subject quote Lord Kelvin's saying to the effect that one's knowledge of a subject is of a poor kind indeed unless one knows how to measure quantitatively the factors involved. Quoted often enough so that it now has the status of a eliché, this saying is nonetheless true.

With all this appreciation and understanding of the importance of measurement, the problem of devising suitable measuring instruments for use with different quantities and in different situations has, up until recent times, been left pretty much to individual scientific specialists to work out as best they can. But in recent years, especially in the past decade, it has come to be more and more widely recognized that the problems met in designing various kinds of instruments have a great deal in common. Because of the importance of these common elements there is a useful body of general doctrine and data which can be termed the science of instrumentation. Recognition of this fact has found expression, for example, in the founding and growth of the Instrument Society of America. The society is devoted to this over-all point of view on the design of measuring instruments — not on such an abstract philosophical basis that no particular results are attained, but rather in practical terms whereby the designer of an instrument for a particular purpose can derive maximum benefit from experience gained from other instruments which have been built, perhaps, for quite unrelated purposes.

This trend deserves to be more widely known and cultivated among scientists as a whole. Too often it happens that a man confronted with a measurement problem requiring instrumentation is such a specialist in the phenomena being studied that he is unacquainted with the mechanical, electrical, and optical design principles which enter into good design of his instrument and is impatient with them. He then works out a "gadget" to fill his needs as best he can and in many cases it depends greatly on the happy ingenuity of his machinist or glassblower. What the science of instrumentation is trying to do is to codify general principles and design data referring to instruments as a whole so that their design can itself be put on a scientific basis worthy of the scientific problems it is to help solve.

Instruments may be classified first as to whether the objects whose property is to be measured are discrete or continuous. For example, an instrument that measures the capacitance of a large number of small condensers, one by one, is measuring a discrete property, and one that indicates changing hydrostatic pressure is measuring a continuous property. Evidently all instruments of the discrete class have a great many problems in common with regard to the mechanism for feeding in the objects to be measured, independently of what quantity is measured, and the same is true of all instruments responsive to a continuous variable.

Of the instruments for discrete measurement, perhaps the simplest is that which merely counts the total number of events or objects to which it responds. In nuclear physics it has been necessary in recent years to develop electronic circuits capable of counting objects at very rapid rates. In these the object or event to made to give rise to an electrical pulse fed to a scaling circuit and eventually to a mechanical register. Such techniques will undoubtedly find application in many other fields, such as in counting blood cells in a blood sample and in automatic digital computing machines.

Second, an instrument can be classified as to the nature of the element in it which is sensitive to the quantity to be measured and the nature of the response made. Such a sensitive element is known generically as a *transducer*, since it changes a physical quantity of one kind into another. For example, temperature measuring devices are made which depend on (a) the differential expansion of two solids, producing elastic deformation of them, (b) differential expansion of a solid vessel and a contained liquid, producing relative motion, (c) change in electrical resistance producing an electrical output signal and many others.

At first sight it may seem that the sensitive elements used for different quantities are so various as not to have much in common. If this were so there would be no general basis to the science of instrumentation. But it is not so. Every sensitive element gives some kind of output quantity Q (such as a linear motion to be observed visually or an electrical voltage to be dealt with appropriately in another part of the instrument). Let x be the physical quantity to be measured. It is a general characteristic of all responsive elements that they possess properties that need to be considered in design of an instrument independently of the physical nature of x and Q. Among these are (a) *sensitivity*, the quantity $Q'(x)$, giving the rate of change of output with change of input under steady equilibrium conditions, (b) *unsteadiness*, the measure of the fluctuation in output $\overline{(Q - Q_0)^2}$ given by the transducer under a steady value of x due to looseness and erratic disturbances inherent in the situation, (c) *slugishness*, the measure of the time rate at which Q assumes its new equilibrium value Q_1 from the value Q_0 when x abruptly changes from x_0 to x_1, and (d) *hysteresis*, which is a shortcoming of some transducers whereby Q is not dependent on x alone but on the past history of all values earlier assumed by x, and (e) *permanence*, the quality whereby $Q(x)$ retains its constant functional form over a long period of time whether the instrument is in use or not, thereby maintaining the instrument's calibration.

Evidently the design principles involved in specifying these characteristics of a transducer are quite general and will have similar limiting effects on the performance of the instrument no matter what the nature of Q and x. And as soon as we are past the transducer, evidently, we are dealing with the physical quantity Q instead of x. Hence from here on the design principles involved do not depend in any way on the physical nature of the quantity to be measured. In passing it may be remarked that sluggishness is not always an undesired element in a transducer, for sometimes the quantity which it is really sought to measure is not really variation of x at each instant, $x(t)$, but some sort of moving time average like

$$\bar{x}(t) = k \int_{-\infty}^{0} e^{-k(t-\tau)} x(\tau) d\tau$$

which is based essentially on the values of x in the past for a time of the order of $1/k$.

The next essential part of the over-all instrument may involve some sort of further transformation or amplification of the transducer output Q, as when the optical image of the a linear movement is projected with enlargement on a screen, or when an electrical voltage is passed through a transformer. Evidently the design data for such devices will depend solely on Q, not at all on x, and therefore will be the same no matter in what branch of science or technology the instrument is to be applied.

The last part of the instrument involves the transformation of Q into some sort of indicating, recording, or control mechanism, depending on the use to be made of the data the instrument provides. Where indication is desired for individual readings or continuous observation, the coupling to the observer's sensory equipment is usually made through the sense of sight. Usually the indicating device involves the motion of a pointer relative to a scale, the position being noted visually by the observer. This led Eddington to regard all science as the systematization of relations betwwen pointer readings obtained in various ways. This favored position of the sense of sight arises from its great sensitivity and resolving power. But one could also devise instruments in which, for example, the output, dependent on the quantity to be measured, controlled the pitch of a sound heard by the observer.

The transforming and amplifying elements of an instrument, as well as the indicating elements, need to be judged according to the same kind of qualities as were listed for transducers. In the over-all analysis there is nothing to be gained, for example, in having very little sluggishness in the transducer if the indicating element has much sluggishness.

A particularly simple kind of indication is that of the "go — no go" variety, which simply gives a signal if x is greater or less than preassigned limits or lies outside a preassigned range.

A recording mechanism is an element that gives a material record of the variation in x observed by the instrument. This is usually in the form either of a continuous curve drawn on a strip of paper or of a discrete series of digital numbers stamped on a strip. But it need not take either form and might be a magnetized tape with the output coded on in some way. Such a form would be applicable where some computing needs to be done on the output data by an automatic computer arranged to accept data on a magnetized tape as its input. The design considerations involved in a recording mechanism may be quite independent of the nature of the quantity x that is being measured.

Finally, a control mechanism is an element that takes the output, which may also be given presentation by an indicator or recorded in a recorder, and uses this to actuate some device in the original situation affecting to have the control mechanism provide a situation which leads x to assume a constant value. But it is only slightly more complicated to require that x follow some preassigned functional variation with time. Here all sorts of new problems present themselves, such as the tendency of the system to "hunt", that is, for the correction action brought into play by the control to produce greater effects than needed to bring x to the control value $f(t)$.

In more complex cases there may be more elaborate combinations of the basic instrument elements, as when several transducers respond to several quantities $x_1, x_2 \ldots$ and the signals so obtained are combined mathematically in some analogue or digital computer to produce one or several output quantities $z_1, z_2 \ldots$ which may actuate a control mechanism. Also it may happen that the output does not actuate a control mechanism related to the system being measured but actuates another device which maintains a prescribed relationship to that system. Thus, a gun director's input is optical or radar data about position of a target, but its output does not act directly on the target to control that position, but on a gun with the object of aiming it so as to launch a projectile which will hit the target.

Enough has been said to bring out the fact that in modern complex instruments a large part of the instrument is independent of the nature of the quantity to be measured. It is this fact that is giving rise to the recognition that there is a science of instrumentation concerned with these common elements and their systematic study and improvement. It seems quite likely that the instrument scientist will in the future play a role of steadily increasing importance in the development of all the sciences. It is a profession that should lend itself to specialization and workers in all fields should be relieved as much as possible from the distraction of having to devote too much attention to the detailed design of instruments for their particular needs.

In accord with these general ideas, the National Bureau of Standards is planning a program that will give just this kind of over-all study to the basic problems of instrumentation in all the physical sciences. Up to now, corresponding to the older and still common practice, instruments have not been considered from the general viewpoint, but separate and uncoordinated attention has been given to mechanical, optical, electrical, and electronic instruments in different parts of the organization. The program is being planned in cooperation with the Office of Naval Research and with research groups in the Army and the Air Force.

The objectives of this program are three: (1) systematic analysis of available methods and devices in terms of their precision and reliability; (2) studies of materials, components, and elements which are now known to impose serious limitations on instrumentation; and (3) development of specific instruments not now available.

Under the first heading, as part of process of reviewing the literature, some particular limited field in which specific instrumentation problems are of interest will be studied. Careful analysis at this stage of how related measurement problems were solved in the past, of available related instrumentation, of its precision, and of the materials involved, will establish a basis for actual development work and provide critical data on the status of instrumentation in this field. Over a period of years, this approach should not only yield the concrete instruments needed — a need determined in this case by the cooperating agencies — but also accumulate information in a systematic manner, permitting a logical approach to subsequent problems.

Present demands for instrumentation improvement are essentially two-fold: higher sensitivity and faster response. These are necessary because the time intervals with which present day work is concerned are extremely small, only minute displacements may be allowed, and the signals are low in power. In general, sensitivity can be increased if the time during which phenomenon is under observation is increased, but this often cannot be done because the existence of the phenomenon may be of short duration — as in, for example, nuclear and cosmic physics and in the study of biological cells. At the same time, noise level becomes a serious limiting factor where signals are weak.

In spite of these incompatible objectives, instrumentation is succeeding in extending sensitivity and signal response, but in the process large quantities of complex data are the result, creating the problem of processing such information. Here other instruments, whose functions are the orderly and integrated presentation of data, are needed. Recording devices are typical of this class of instruments, but more sophisticated types are necessary now for the analysis of large amounts of data, the performance of predetermined operations, and the presentation of the summarized findings, discarding unwanted background. Fortunately, machines like electronic digital computers, originally planned for solving mathematical equations, afford a means of attaining this end. Specific problems to be considered in the program of the National Bureau of Standards include transducers, electron optical field mapping, data transmission and reduction, improved microscopes, and so on. Work on some of these is already in progress. Coincidentally, the bureau's program in electronic digital computers is continuing: two machines are under construction in its laboratories, in Washington and Los Angeles, while another five are being built by contract with private laboratories and industries.

One aspect of instrumentation, already mentioned, is the problem of material and components. Related to this is the existence of principles of measurement whose application cannot be realized until developments in another field of science are achieved. Thus, electron optical field mapping represents an application of principles in optics to electronics, yet the realization of this important research tool had to wait until the science of electronics had advanced to its present state. An illustration of the dependence of instruments on materials is the wire strain gage. The principle was known as long age as Ohm's law. Yet it was not until Simmons, Ruge, and De Forest developed methods of bonding the wire to paper and ways had been found to measure and record small changes in current that the principle was applied successfully. It is for these reasons that the study of materials and components, as well as the study of past methods proposed for measurement problems, is an integral part of the bureau's program.

The need for cross-channelling information in the sciences has grown acute in recent years. When it comes to instrumentation, this need has two aspects: basic developments in one field may provide the basis for instrumentation in another field and instruments developed in one field may have applications in others. As an illustration of the first aspect, scientists in the bureau's radio propagation laboratory recently developed and constructed an atomic clock which has remarkable precision. The heart of the development is the application of the knowledge of resonance absorption in atomic physics and of microwave radiofrequencies to the problem of time. Again, the bureau has developed a standard of length based on the green line of mercury 198. This development depended on knowledge of atomic and nuclear physics, while the realization of the instrument and associated apparatus is a combination of knowledge in electronics and optics. An analogous story could be told about the recent measurement at the bureau of the magnetic moment of the proton. Here there is, at least in part, with excellent possibilities for full development in the future, the basis for atomic standards of the three basic measurements — time, length, and mass.

Indicative of the importance of cross-channelling information concerning instruments is the bureau's experience with diamonds as tools for measurement. Many years ago the bureau was interested in diamonds for ruling lines on precision scales, and devices and methods for cutting diamonds were successfully developed. The same group of scientists devised a diamond indentation method of hardness testing, which has been standardized and is used in the Tukon tester.

In the course of this work, scientists engaged in the study of aircraft cylinder wear, who had been using conventional measurement methods, recognized the possibility of determining cylinder wear by the change in length of a series of standard indentations, using diamond indentation. Modifications were needed, and a special instrument was developed capable of measuring wear to a precision of 0.00002 inch and of detecting even smaller amounts. Auxiliary developments continued in such fields as the cutting of diamonds and the drilling of small diamond dies for fine wire production.

These examples not only indicate the need for wide dissemination of information among the fields of science but reinforce the concept that instrumentation is a science in its own right. There is no reason why the practitioner of this science, conceived of as the logical approach to measurement problems, cannot provide measurement devices of a given type for a variety of fields. Certainly this would save the experimentalist much time and effort. Prior to this century, an analysis of the experimentalist's activity might have shown that the bulk of his time was spent in getting ideas and in analyzing the data of his subsequent experiments while a minimum of time was spent in the construction of instruments. In the present period, too often the scientific situation is such that the bulk of his time has to be spent in devising and constructing his instruments. From the point of view of research, this is unfortunate; and it is here that the science of instrumentation can make significant contributions not only to instrumentation as such but to the progress of original research.

Effect of Oscillations of the Case on the Rate of a Watch

E.U. CONDON and P.E. CONDON

National Bureau of Standards
Washington, D.C.

Eighty years ago Lord Kelvin[1] presented an interesting paper on the effect of the mode of suspension on the rate of a clock or chronometer. This paper reported on, and gave the correct machanical interpretation of, the effects observed. The problem constitutes an interesting example of the theory of coupled oscillations, which readily lends itself to use in a junior course in mechanics as a laboratory experiment. For that reason it seemed to us that a modern presentation of the subject might be of interest.

It happens that many watches have the property that when hung up their period of oscillation as a pendulum is nearly in resonance with the frequency of oscillation of the balance wheel. In consequence a watch hung up in such a way as to be free to swing is set in forced oscillation as a pendulum. Sometimes on looking at a wall case full of watches will be seen to be swinging away merrily on their hooks.

The rate of a watch that is swinging in this way is considerably changed, and therefore watches should not be so suspended when they are being kept in the shop for rate adjustment. We have made observations in the extreme case in which a well-regulated watch would gain or lose 10 or 15 min/day when its case was undergoing such forced oscillations.

We found, when a watch is hung up by its stem and free to swing as a pendulum, that not only is the rate greatly affected but also it varies considerably from day to day. This is probably due to accidental variations in the particular way in which the watch is hung on the nail. Table I gives data so obtained on a group of six watches. The first three lines give values of the rate in seconds per day[2] taken over 24-hr periods on three successive days when the watches were hung on nails and free to swing. The second three lines give the rates observed for the watches on the next three days

---

[1] Kelvin, *Popular lectures and addresses* (Macmillan, 1894), vol. 2, p. 360.
[2] The sign + means that the watch is *losing*; − means that it is *gaining*. This sign convention arises from the definition of the *rate* of a time-piece as being the rate of change of the *correction*, where the correction is defined as the amount that must be added algebraically to the reading of the watch to give the correct time.

when the watches were in the same position but clamped so they could not swing.

The problem is one of self-excited oscillations. The details will depend on the manner of action of the escapement, which is designed to excite torsional oscillations of the balance wheel with respect to the case.

In making an experimental study we found it convenient to make a small holder for the watch so that it could be suspended by a stiff wire as a torsion pendulum, with the plane of the watch horizontal. A threaded brass rod about 10 in. long was attached to the holder so as to be horizontal. This carried some brass nuts by which a fine tuning adjustment of the moment of inertia of the watch and support could be made.

First the length of the supporting wire was adjusting to produce resonance between the external oscillation of the case and the internal oscillation of the balance wheel when the nuts for fine adjustment were in the center of their range. Then a determination was made of the rate of the watch for each of a series of positions of the nuts. Later, with the watch not running, the period of free oscillation was determined for several positions of the nuts so that one had a calibration of the period of the torsion pendulum in its dependence on the position of the nuts.

In this way the data shown in Fig. 1 were obtained. Ordinates show the fractional change in rate of the watch; abscissas, the fractional departure of the oscillation frequency of the mount from equality with the undisturbed balance wheel frequency of the watch. It will be seen that the largest effect is obtained near resonance, as one would expect. Moreover, when the case

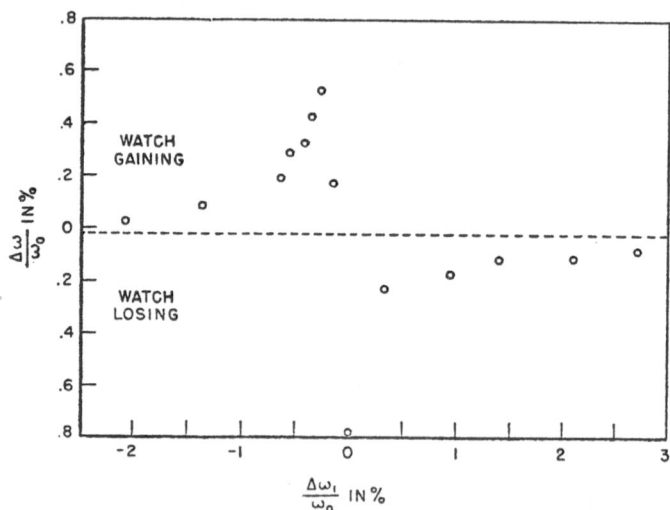

FIGURE 1. Fractional change in rate of watch as a function of the fractional departure of the oscillation frequency of the mount from synchronism with the undisturbed balance wheel.

frequency is below resonance the watch runs fast, and when it is above resonance the watch runs slow. The maximum effect was of the order of some minutes per day on a watch which when mounted rigidly was correct within a few seconds per day. This treatment seems not to produce any ill effects in the watch, which kept good time when rigidly mounted again at the conclusion of the experiments.

The theory of these effects is most conveniently considered first by studying the undamped oscillations of the free system. Let x be an angular coordinate measuring the departure of the case from its equilibrium position and let y similarly specify the orientation of the balance wheel with respect to fixed axes. Let I_0 be the moment of inertia of the balance wheel, and I that of the case and works and other parts of the torsion pendulum (with the balance wheel not turning relative to fixed axes). Likewise let k_0 be the torque per unit angle of the hair spring, and k that of the torsion wire by which the watch is supported.

The equations of motion of the free oscillations are then

$$I\ddot{x} + kx + k_0(x - y) = 0,$$
$$I_0\ddot{y} + k_0(y - x) = 0. \tag{1}$$

For convenience, write $\omega_0^2 = k_0/I_0$ and $\omega_1^2 = (k + k_0)/I$. Here ω_0 is the angular velocity corresponding to the proper vibrations of the balance wheel ($\omega_0/2\pi = 2.5$c/sec in most watches), and ω_1 is the angular velocity for the torsional oscillations of the case with the balance wheel not turning with respect to fixed axes. The equations of motion become

$$\ddot{x} + \omega_1^2 x - \lambda\omega_0^2 y = 0,$$
$$\ddot{y} + \omega_0^2(y - x) = 0, \tag{2}$$

where $\lambda \equiv I_0/I$. Assume $x = Ae^{i\omega t}$ and $y = Be^{i\omega t}$; then ω must have a value that determines the same value of A/B from each equation of motion:

$$A/B = 1 - \omega^2/\omega_0^2 = \lambda\omega_0^2/(\omega_1^2 - \omega^2). \tag{3}$$

TABLE 1. Rates (sec/day) of swinging and clamped watches.

Watch	A	B	C	D	E	F
Swinging	−403.4	−102.5	+668.5	− 99.2	−276.6	−329.0
	+948.9	− 93.5	+195.1	−111.9	−255.7	−160.8
	− 12.6	−196.7	+120.6	− 27.2	−245.3	−113.4
Clamped	− 2.9	− 3.5	− 45.0	+ 5.3	− 0.1	+ 1.0
	− 3.0	− 3.8	− 42.3	+ 6.1	− 2.4	+ 0.2
	− 2.0	− 4.4	− 47.5	+ 14.9	− 0.9	+ 0.5

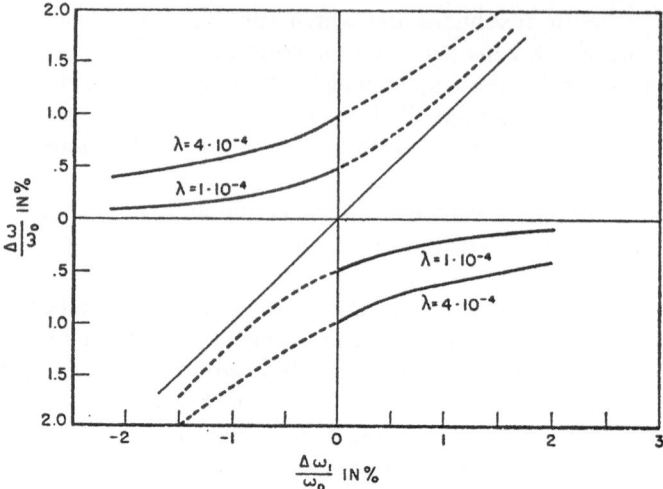

FIGURE 2. Roots of Eq. (4). The solid lines correspond to observed values.

In practice $\lambda \ll 1$. The interesting phenomena occur when the system is tuned so that ω_1 is nearly equal to ω_0. It is convenient to choose ω_0 as the unit of frequency. Solving Eq. (3) for ω^2 one finds

$$\omega^2 = \tfrac{1}{2}(1 - \omega_1^2) \pm \tfrac{1}{2}[(1 - \omega_1^2)^2 + 4\lambda]^{1/2} \qquad (4)$$

In Fig. 2 the two roots are plotted as a function of ω_1 for two different values of λ, namely, $\lambda = 10^{-4}$ and $\lambda = 4 \times 10^{-4}$. At each value of ω_1 there are two roots, one corresponding to $\omega > 1$, watch gaining, and one corresponding to $\omega < 1$, watch losing. The first of these roots, from Eq. (3), corresponds to motion in which the case and balance wheel swing in opposite phase, while for $\omega < 1$ they swing in the same phase.

But the experiments show behavior corresponding to only one of the two modes at each setting of the nuts. The mode actually observed to be excited in the experiments corresponds to the full parts of the curves in Fig. 2; the dotted parts were not realized in practice. Just why one of the modes is excited on the other side of resonance, is not dealt with in Kelvin's paper.

In seems quite likely that this depends on two factors, both of which favor the observed behavior. The mode excited is that one in which there is lesser amplitude of motion of the case compared to the balance wheel. Because of the large moment of inertia of the case relative to that of the balance wheel, the driving mechanism which exerts equal and opposite torques about the axes corresponding to the coordinates x and y, will be more effective to exciting such a mode. Likewise, the damping, hitherto neglected, due to air resistance, imperfect elasticity of the support and so on, will be mostly in the x coordinate. This has the effect of introducing a small term $+ a\dot{x}$ in the left-hand member of the first of Eqs. (2). The equation determining ω now becomes (with ω_0 as unit)

$$(\omega_1^2 - \omega^2 + ia\omega)(1 - \omega^2) = \lambda. \tag{5}$$

Since $\alpha \ll 1$, we may assume that the root will be of the form $\omega + ia\theta$ to the first power in α, where ω is the root of the equation with $\alpha = 0$. Substituting in Eq. (5), one finds

$$\theta = \tfrac{1}{2}(\omega^2 - 1)^2/[(\omega^2 - 1)^2 + \lambda]. \tag{6}$$

This shows that the decrement of the motion ($e^{i\omega t}$ becomes $e^{i\omega t} \cdot e^{-a\theta t}$) becomes very small for that root in which ω^2 is close to unity. At each value of ω_1 one root is such that $(\omega^2 - 1)^2 < \lambda$ and the other is such that $(\omega^2 - 1)^2 > \lambda$. hence there will always be a considerable difference in damping of the two modes, and the mode observed is the one of low damping.

The case is which there is no restoring force in the supporting wire was also commented on by Kelvin. From Eq. (3) with $\omega_1 = 0$ we see that $\omega^2 - 1 = \lambda$, so that the watch always gains when suspended in this way. On the other hand, at resonance, when $\omega_1^2 = 1$, we see that $(\omega^2 - 1) = \lambda$, so that the effect there is greater in the ratio $\lambda^{1/2}$ than when there is no stiffness in the suspension.

It is a pleasure to acknowledge the cordial aid given by Mr. H.A. Bowman and Miss M.L. Scott of the Metrology Division of the Bureau in connection with the experiments.

The Development of
American Physics[*]

E.U. CONDON

National Bureau of Standards,
Washington, D.C.

I had an informal understanding with our Secretary that my address was to cover the growth of physics in the last fifty years, But, now that the program is out, I see that the title has been changed to read, The Development of American Physics. I happen to be one of those who never went along with the idea of a distinctly racial or regional attitude toward our science, even when put forth by such eminent authorities as Hitler or Goebbels. I think it would be provincial of me consciously to stress the American side of the story, although this will undoubtedly happen because we are all products of our environment.

It would be easy indeed to fill up my time with a lot of specific instances of the way in which experimental physics has moved out of the horse-and-buggy age — of the age of simple little apparatus put together by one man with "love and string and sealing wax" — and of how it has moved into a gigantic affair of multi-million dollar electronuclear machines requiring great teams of trained men in various specialties for their design, construction and operation.

It would also be easy indeed to fill up the time with a lot of specific instances of the way in which, during the past half-century, new branches of applied physics have given rise to whole new industries, such as the widespread use of electric power, the telephone, the radio, television, X-rays and radiation therapy in general, the use of radioactive materials as tracers in chemistry, the deeper understanding of the solid state, and its associated technical improvements in metals, magnetic materials, photoelectric emitters, semiconductors, insulating materials, and so on and so on — using up the entire forty-five minutes allotted to me with the mere listing of important topics.

It would, moreover, be easy indeed to fill up the time with dates and specific items about the growth of theoretical physics. Fifty years ago even Max Planck had not heard of Planck's constant, nor had Lorentz heard of

---

[*] Address delivered at semi-centennial meeting of the American Physical Society, Harvard University, Cambridge, Massachusetts, June 17, 1949.

the Lorentz contraction, nor had Einstein heard of the theory of relativity. For that matter, forty years ago Bohr had not heard of the Bohr atom, nor had von Laue heard of the diffraction of X-rays by crystals. Thirty years ago Hess and Millikan were just beginning to recognize that cosmic rays come from outer space, and as for Dirac, he was just a boy in school.

Twenty years ago we were all just beginning seriously to grapple with the ideas and experimental consequences of quantum mechanics. Only a short time before, Goudsmit and Uhlenbeck had introduced the idea of electron spin and Pauli had enunciated the Pauli exclusion principle which was the last major step in clarifying our basic ideas of the analysis of atomic spectra. Twenty years ago was when we first began to realize not only that the wave nature of matter would manifest itself in diffraction of electrons and other atomic particles, but that it also made it possible for particles to leak through potential barriers even though they did not have energy enough to go over the top — a process which if discovered today might be called *infiltration*. Twenty years ago Chadwick had not discovered the neutron, nor had Urey discovered deuterium. Nor had Irene Curie and Frederic Joliot discovered artificial radioactivity, nor had Pauli and Fermi balanced the books in beta-decay processes by inventing the neutrino, nor had Yukawa thought of the mesotron. And yet all of these things were to happen within a very few years, so that twenty years ago we stood on the very threshold of the recent very great advances in our knowledge of nuclear physics.

Ten years ago we were pretty well equipped with general ideas about nuclear physics and by that time we had developed some fine equipment which was available for experimental work. Besides the sledge-hammer boys with their Van de Graaff generators and cyclotrons, there were also the watch-maker boys like Rabi at Columbia and Bloch at Stanford who were learning precise and interesting things about nuclei by the gentlest of methods. Ten years ago we had just learned of the discovery of uranium fission, when Bohr brought the news to a physics conference in Washington from the Berlin laboratories of Otto Hahn in January 1939.

Ten years ago we were just beginning to recognize that — if certain as yet not accurately measured parameters had suitable values — it might be possible to produce a slow neutron chain reaction using uranium fission to release atomic power, or to produce a fast neutron chain reaction leading to a military weapon of quite unprecedented destructiveness. This was the time of our first encounter with top policy in government and politics. Probably things will never again be the same for us in this respect. This, I think, is a good thing for it has helped to give many of us a better insight than we had before into the complexities and duties of responsible citizenship in a democracy.

It would be easy to fill up more than the allotted time — some of which must already have gone by — with the development of this theme. But there is not time enough for that either.

Since our whole subject is clearly too big to be dealt with by the method of enumeration and brief description of even a few of the main lines of specific development, how then, shall we deal with it? I would like to try to touch on a few very broad considerations affecting the significance and role of physics in the world of modern thought, and the changes which have come about during the period in question as to the methods of theoretical physics, and the expanding development of our ideas of what constitutes progress toward *explanation* in physical science.

The past half-century has seen enormous change with regard to these matters although they are more apt to escape out attention because such ideas and trends are more subtle than the specific, concrete phenomena which we treat in articles for the *Physical Review*. It seems to me that the greatest change which has occurred in the past half-century in this broad light, has come in the recognition generally that the framework of ideas of classical Newtonian mechanics is inadequate as a basis for the description of all physical phenomena. The modern scientific period in physical science is only about three centuries old and, as we know, owes its strength and main initial impetus to the formulation of Newtonian mechanics, and the enormous success of Newtonian mechanics in explaining the motions of the moon and the planets. The Newtonian concept was then rapidly extended to dynamical systems of all kinds, including fluid and elastic systems.

Its successes were so great that many men jumped to the conclusion that all physical phenomena must be ultimately describable in such terms. Thus, it was natural in the middle of the nineteenth century that Lord Kelvin and others should seek to find an explanation of the second law of thermodynamics in terms of dynamical models involving hidden cyclic variables — not with the statistical concept as we know it from Boltzmann and Willard Gibbs. Likewise Maxwell sought a completely dynamical model of the electromagnetic field in which neighboring elements of space were regarded as being somehow in a state of strain and as acting on each other by dynamical connections. We have come so far away from that point of view in this century that I dare say that many of the younger physicists of today are hardly aware of the emphasis which Maxwell placed on the mechanical model of the medium which propagates electromagnetic effects.

Historically the first great addition to purely dynamical idea was made in the development of statistical mechanics by Gibbs and by Boltzmann. Here the notion of a statistical ensemble of possible states of a system is introduced in order to arrive at a quantitative model for the observable property, entropy. So far as I can make out from the older literature the fact that this step revealed the inadequacy of Newtonian mechanics for dealing with thermodynamics was not recognized very fully at the time. This came about, I suppose, because there still seemed to be the possibility of giving a full and complete description of complicated molecular systems by ordinary mechanics, in principle at least. Therefore, there was a certain tendency to regard the statistical part as a temporary expedient, a kind of scaffolding which could be removed when the construction was far enough along. However, that is not correct, for entropy is a physically observable attribute of matter which requires the statistical method for its description.

In the first few years of the century we find the modern attitude in electrodynamics gradually forming. More and more there is observed a tendency to study the properties of the field equations and the relation of special solutions to specific observational situations. In this process we find physicists more and more concerned solely with getting a suitable self-consistent set of equations that describe and correlate observable effects, and spending less and less time considering what sort of dynamical constitution the hypothetical aether must have in order to be governed by such equations.

Prior to this century Maxwell made the first application of what has since become a very powerful method in theoretical physics. The known phenomena, first studied by Ampere, concerning the magnetic fields due to galvanic currents, were represented by the equations

$$\text{curl}H = 4\pi i,$$

where H is the magnetic field strength and i is the current density in electromagnetic units. Maxwell recognized that this equation could not be complete because div curl $H \equiv 0$, identically. On the other hand, the equation div $i = 0$ only holds exactly when the currents are steady. Therefore, Maxwell argued, the quantity i must be replaced by some more general expression which reduces to this in case the currents are steady.

A simple exploration of known relations showed that the simplest physical quantity which reduces to i under steady conditions, and whose divergence is zero under all conditions is

$$i + \dot{E}/4\pi i$$

where $\dot{E}$ is the time rate of change of the electric field and c is the ratio of electromagnetic and electrostatic units. He therefore boldly guessed that this was the proper term to replace i under nonsteady conditions. By this purely formal consideration he was able to make an extension of incomplete field equations in a way which led to the prediction of electromagnetic waves and a description of their properties, hence to radio, and the whole electromagnetic theory of light.

The next great step of this kind was taken by Einstein in 1905 when he formulated the principle of relativity. He found that the transformations between the reference systems of different observers in differing states of relative motion with constant relative velocity which were applicable in electrodynamics were different from those which were applicable in classical mechanics. It also appeared that the experimental data, of an optical kind, favored the validity of the electromagnetic scheme of transformation rather than that indicated by classical mechanics. Therefore, he had to seek a way to modify the Newtonian equations so that they would be governed by the Lorentz transformations but would nevertheless reduce to ordinary classical mechanics when all velocities were small compared to that of light.

From that time on, the method was fully established as a basic tool for construction of new theories in physics. The basic principle involved may be stated something like this: If we have really discovered the appropriate kind of mathematical framework for a given situation, then the actual law of nature will usually be found to be the simplest of the various mathematical possibilities consistent with the known requirements.

Thus, in Einstein's general theory of relativity, it is recognized that gravitational fields are describable by the introduction of a general Reimannian curvature tensor for the four-dimensional space-time. But this consideration alone is not specific enough to be definite. The simplest additional mathematical condition is to require that the contracted Reimann-Christoffel curvature tensor shall vanish. And, it seems, this is the law which gives the correct description of gravitational phenomena.

In the modern period of the last twenty-five years, this method has been extraordinarily fruitful. Schrödinger discovered the specific equations of wave mechanics by the simplest heuristic considerations of this sort once he was on the trail of having a wave mechanics. Perhaps its most striking and powerful achievement was its use by Dirac when he set up the simplest appropriate relativistic wave equation for the electron, and showed that this automatically brought with it the properties of spin and the prediction of the existence of the position. The whole theory of mesons and meson fields is a construction of this kind. The main trouble here seems to be that with this mathematical tool the younger generation of theoretical physicists is so facile at constructing new theories that we have an embarrassment of riches.

Closely related to this general idea, but somewhat more special, has been the growth of the broad idea of invariance in physical theories. This got into physics first in general terms leading to the development of vector and tensor analysis, designed to give expression to the experimental fact of the isotropy of space. This is expressed by requiring that all equations be of such a kind that their form is invariant when referred to a second set of coordinates oriented in an arbitrarily different way relative to the first frame.

This is where the mathematical theory of groups makes its entry into physics. The different possible frames of reference which are observed to be physically equivalent are connected to each other by transformations. These tranformations form the elements of a group. All quantities and relations entering into any physical theory must transform like one of the representations of the group. This principle serves in a broad way of limit, select and control the mathematical forms which may be used in building theories. First we recognized invariance with respect to the group of

Euclidean rotations of axes, then with respect to the group of Euclidean rotations of axes, then with respect to the larger group which includes Lorentz transformations. Then we had to deal with the more special symmetry groups applicable to the description of symmetric molecules or to crystalline structures. Finally, in the case of systems consisting of many identical, and therefore indistinguishable particles, our equations must possess invariance with respect to interchanges or general permutations of such particles.

However, we must be on guard against assuming that invariance properties which hold in one field of physics must necessarily hold in every other branch of physics. Invariance is a broad generalization from experiment for a certain class of phenomena, and this does not exclude the possibility that the invariance does not obtain with regard to some other phenomena. Thus, rocksalt is isotropic with regard to refraction of ordinary light, but far from isotropic with regard to diffraction of x-rays. Again, I would be greatly surprised if the basic equations of short-range nuclear forces should turn out not to be Lorentz-invariant, but I do not think we have any experimental evidence today that definitely requires that they do have this property. This whole broad discipline of invariance under a group of transformations is one of the fruitful ideas that is essentially characteristic of the modern theoretical physics of this century.

In passing we may note that some broad ideas which in earlier times were thought to be of great significance seem nowadays to be less fruitful than formerly it was expected they would be. In this category I would put the whole class of variational principles. While people still like to show how their field equations can be derived from a variation principle, that is, the stationary property of some field integral, this general connection does not seem to have been really fruitful in guiding us to new results in physics.

So much for some general characteristics of the modern point of view in the construction of physical theories. I would like to close by commenting briefly on the general question: What is the intellectual goal of physics? More specifically, what do we mean when we *explain* something in physics? This used to mean that we had successfully correlated the observed phenomena with the behavior of some mechanical model behaving according to Newtonian principles. And for a long time, extending well into this century, many of the older physicists insisted that phenomena were not *explained* unless they had been explained in this sense. But we have seen that the construction of theories in modern physics has gone far beyond that limitation, and so we must ask for a fuller description of what we mean by explanation.

It seems to me that in this direction, too, our half-century has seen a great deal of progress, or at least change toward a more sophisticated outlook. In this the American philosopher John Dewey has played a great role, as has also our past-president and one of our Harvard hosts, Percy Bridgman.

Dewey points out clearly that the growth of rational thought processes may be considered as a response to the biological necessity of adapation to the environment. Its ultimate function, he says, is that of "prospective control of the conditions of the environment." It follows then, continuing to quote, that: "The function of intelligence is therefore not that of copying the objects of the environment, but rather of taking account of the way in which more effective and more profitable relations with these objects may be established in the future."

Bridgman's critique of the concepts and procedures of physics is built on an insistence that maximum emphasis be placed on the idea that the concept must be identical with the fully determined by the physical procedures which we actually employ in observing the quantity which expresses the concept. His critique also illustrates one of the characteristic modern attributes of theoretical physics in its strong insistence on elimination of any kind of dependence on non-essential or adventitious elements in a theory.

What then remains? Once eleven years ago I attempted to answer this question for myself and came up with the following statement which I still think is a good working basis for considering the subject:

The object of physics is so to organize past experience with relation to matter and energy, and so to direct the acquisition of new experience, that we may advance toward a goal in which it may ultimately be possible to predict the outcome of any proposed experiment which is capable of being carried out — and to make the prediction with the expenditure of less human effort than it would take actually to carry out the proposed experiment.

This statement you will see is well-hedged. I do not say that we can ever reach the goal, but certainly the experience of the past fifty years has shown a great deal of advance toward it. And you will observe that I restrict the field of predictions to experiments which are capable of being carried out. And finally that the whole activity is clearly tied in with the pragmatic function of developing intelligence in giving us a more economical way to know about our physical environment than by making the trial each time.

It is an exciting and challenging game. It is lots of fun being a physicist, and I am sure the physicists are going to have a lot of fun in the next half-century if they are allowed to work at physics.

Present Program of the National Bureau of Standards

E.U. CONDON

Dr. Condon, who was born at Alamagordo, New Mexico (or the heart of today's rocket-experiment country), took his Ph.D. in physics at the University of California in 1926, after which he studied for a year in Göttingen an Munich. Later he taught at Princeton, Minnesota, and Columbia, and was Associate Director or Westinghouse Research Laboratories from 1937 until 1945, when he became director of the National Bureau of Standards. On October 1, Dr. Condon will take up his new post as Director of Research at Corning Glass Works.

The fields of activity of the National Bureau of Standards are physics, chemistry, mathematics, and engineering. One cannot, of course, sharply define the division between these disciplines and, as a matter of fact, as the fields of biophysics and biochemistry show, it is becoming increasingly difficult to distinguish between border regions even when the biological and physical sciences are juxaposed. Thus the National Bureau of Standards is also engaged in activities touching upon the biological and medical sciences. The work in X-rays and natural and artificial radioactivity reveals how inevitably necessary this is while there is considerable activity relating to instrumentation for the biological and medical sciences. For example, Newman, Swerdlow, and Borysko recently developed a new method for microsectioning of biological tissue, and Broida has developed a heart pump and an artificial heart valve. At the same time progress is being made in the important field of synthetic blood plasma.

Within the physical sciences, the work of the bureau consists of both pure and applied research, development and engineering, testing, calibration, and specifications, and, finally, general scientific services — largely of an advisory kind.

A major aspect of NBS activity is concerned with national defense. A large number of important military problems are under study in nearly every section of the NBS laboratories, but the larger activities of military importance are (*a*) the research on factors affecting ionospheric and tropospheric radio propagation and the operation of a prediction service concerning ionospheric conditions, (*b*) the research and development responsibility in the field of proximity fuses for nonrotating projectiles, (*c*) work on chemical properties of atomic energy materials and on problems related to protection of personnel from their radiations, and (*d*) continued

The essentially sheer, 2,000-foot drop from summit to base makes the Cheyenne Mountain site ideally suited for simulating communication between aircraft and ground. Permanent installations on the Colorado and Kansas plains as far away as 300 miles will receive the transmitted signals.

development of special guided missiles for the Navy following on the pioneer lines opened up by the work on the Bat.

Arrangements for carrying on the radio work were made by agreement and support of the Joint Chiefs of Staff, who decided that the military needs for research in radio propagation should be made the responsibility of a civilian organization at NBS, which would also take care of civilian interests in the same field. As a result of this decision the Radio Section, which had been part of the Electricity Divison, became a division, and it is in fact one of the largest in the bureau. Besides radio propagation work, this division also handles the standards work on electrical measurements at radio frequencies. This has become a very demanding task in view of the rapidity of progress in recent years in making use of an ever-widening range of frequencies, now extending up to some 100,000 megacycles/ second.

The radio division, called the Central Radio Propagation Laboratory, has a staff of some 400 technical people. Besides the laboratories here, it operates a dozen ionospheric observing stations, some with NBS personnel, and some by contract. These are located in the Pacific islands, in Alaska, Greeland, and Canal Zone, as well as at several places in the continental United States. Data from these and other stations operated by other governments with which we cooperate pour into Washington for immediate use in the prediction service and analysis, as part of a research program to improve our fundamental understanding ionospheric processes. This part of the work involves close cooperation with the solar astrophysicists at both the Mount Wilson Observatory and the High Altitude observatories at Boulder and Climax, Colorado. In addition, the radio division operates the well-known standard frequency transmitter WWV from Beltsville, Maryland, and a 400-acre field station for radio

For intensive investigation of radio propagation phenomena, NBS has recently installed powerful radio transmitters on the summitt of Cheyenne Mountain, Colorado. (See figure on preceding page.)

experiments at Sterling, Virginia, about thirty miles west of Washington. We are now building transmitters and receiving stations for an elaborate observing program to determine tropospheric factors affecting the propagation of radio waves at the range 100 Mc-1000 Mc, and perhaps higher. The transmitters will be located atop Cheyenne Mountain, just west of Colorado Springs, and the receivers on the plains several hundred miles to the east.

Architects are at work on plans for new modern radio laboratories to be built at Boulder. Present prospects are that construction will start in the spring of 1952 and that the laboratory will be ready for occupancy in the fall of 1953. The citizens of Boulder, which is the seat of the University of Colorado, presented the federal government with a splendid 220-acre tract for the NBS Boulder Laboratories.

The acquisition of a large site in this part of the country has already proved valuable for other than radio work. Construction of a special-

To develop efficient antenna systems for ionosphere sounding equipment, NBS maintains a model antenna range at Sterling, Virginia. At the vertex of the 50-foot "A" frame is a target transmitter, and on the ground directly below the transmitter is the antenna under test.

purpose laboratory for certain work that NBS is doing for the Atomic Energy Commission is now going on. This building will be completed and occupied on a high-priority schedule before the construction of the radio laboratory even begins.

Within the past year there has been a great increas in all kinds of military work, some of it in wholly new lines of activity and some in the direction of stepping up the effort in fields in which we were already engaged. The electronic ordnance work, for example, must be greatly increased in order to take care of the peculiar requirements of fuses for guided missiles. Although a splendid new laboratory building for this work was provided on our grounds by the Army Ordnance Department just at the end of the war, these facilities are now so inadequate that the total laboratory space devoted to this work must be enlarged by a factor of four or five. The bulk of this expansion will take place elsewhere, at a site not yet determined, although some of it must take place in Washington because of the urgent time schedules we are trying to meet for the armed services.

The work we have been called upon to do for the Atomic Energy Commission has also tended to increase rather than decrease during the past several years, and it carries over into a number of distinct fields that cannot be publicly discussed. The work in radiation physics has been strengthened by construction of a laboratory building to house a 50-million-volt betatron and a 180-million-volt synchrontron.

Need for better facilities for the Navy guided missile work than the several crowded and unsatisfactory Quonset huts now used as laboratories here has been recognized this year by the Congress. For reasons of economy, both of money and of time, there has been assigned to NBS a large unused naval hospital at Corona, California, about forty miles east of Los Angeles. This has recently undergone the necessary remodeling and was activated last summer as the Corona Laboratories of the National Bureau of Standards. Here we have more than 100 acres and a large group of good buildings, which will take care of the guided missile work and other related development projects.

In the period under review several major organization changes have occurred. One of the most important of these has been the development of the Applied Mathematics Division under the leadership of John H. Curtiss. This was started in a small way to continue the work of the Mathematical Tables Project formerly located in New York City and also for the purpose of introducing modern statistical techniques for design of experiments and analysis of laboratory data into all the NBS research and testing work.

Hardly had this been done when we were approached by Admiral H.G. Bowen, then chief of the newly established Office of Naval Research. He was keenly aware of the military needs for a central computing facility of a higher sort, which would be capable of rendering a consulting service on problem formulation, as well as being able to handle large blocks of mechanized computations. He suggested the appropriateness of meeting this need through an NBS-operated facility.

At the same time extensive interest was developing in automatic digital electronic computers. The Bureau of the Census, and soon thereafter several other government agencies, including the Office of the Air Comptroller and the Army Map Service, asked for NBS assistance in developing and procuring such computers. All these things have worked together to make the work in applied mathematics much more extensive than was envisaged when it first started. Besides the mathematical work on computers in the Applied Mathematics Division, there is a section in the Electronics Divison under S.N. Alexander which has done a splendid job on development of new engineering designs and techniques in this field. This work has resulted in many specific contributions to the art and to construction of the SEAC (National Bureau of Standards Eastern Automatic Computer), which has been operating for many months on a twenty-four-hour-a-day, seven-day-week schedule on military mathematical problems and in testing new circuits and components to be used in future more elaborate computers.

The mathematics program has also involved the establishment of a western unit known as the NBS Institute for Numerical Analysis, located

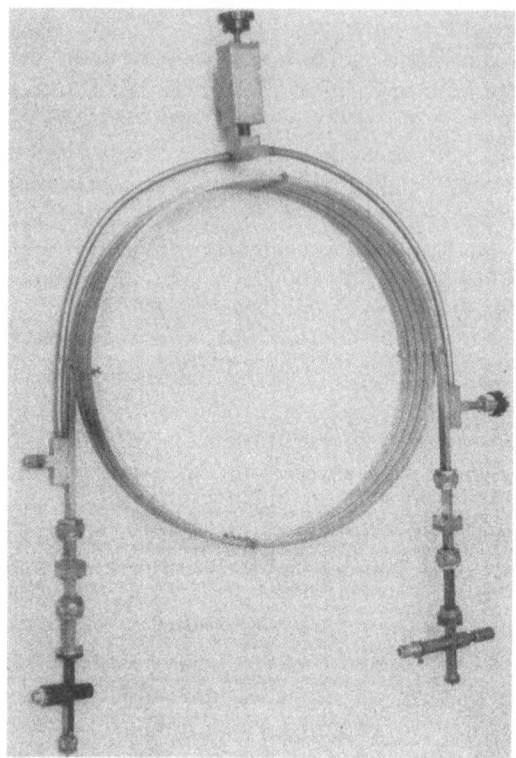

This 30-foot copper-tube absorption cell, containing ammonia gas at 10 μ to 15 μ pressure, is the heart of an improved atomic clock, which shows great promise as a new, atomic standard of frequency and time. The constant frequency for controlling the clock is derived from the absorption line of the gas when a 23,870.4-Mc signal is passed through the cell. The ammonia clock is one of four different atomic clocks now under development at the National Bureau of Standards.

on the campus of the University of California at Los Angeles. This unit carries on fundamental research on methods of numerical solution of mathematical problems and renders a computing service to government agencies and government contractors in the Pacific Coast area.

It would be impossible to mention all the specific things that have been and are now going on. Under present conditions, development problems, especially those for the military, tend to demand most of our attention, but it would be wrong to leave the impression that there has not also been a series of interesting advances in basic science, particularly in matters affecting precise measurement and determination of basic physical constants. In the postwar period, we have been able greatly to improve our spectroscopic facilities. William F. Meggers turns out a steady stream of analyzed spectra and has done a great deal to perfect the use of Hg^{198}, artificially made from gold in the nuclear reactors, as an ultimate standard for precise measurement of length by interferometry. Standard lamps have been supplied to the International Bureau of Weights and Measures near Paris and the National Physical Laboratory in London and are now available for general scientific use.

The NBS new Atomic Beam Frequency and Time Standard, in which a sharp spectroscopic line in the frequency spectrum of cesium is used to stabilize a quartz oscillator. The device forms an integral part of a servo system that will supply correction signals to a frequency and time generator that has the highest potential accuracy of any type atomic clock, perhaps as high as one part in 10,000,000,000.

The attainable precision is such that the old platinum meter bars that have served as material standards for preservation of the meter are now scientifically obsolete. As soon as the new technique can be made more widely known and more generally available, it will be proper to propose an international legal agreement for use of the better method.

An interesting achievement was what is popularly known as the "atomic clock," in which the resonant absorption of microwaves in ammonia gas provides the frequency control element of a time-measuring device that can be made precise and completely independent of any vagaries in the period of the earth's axial rotation. Other atomic time-keeping possibilities involving control by resonant frequencies arising from the hyperfine structure of the atoms in a molecular beam of cesium atoms are also under active study.

In the field of fundamental atomic constants, a most useful contribution was the work of H.A. Thomas, R.L. Driscoll, and J.A. Hipple, in

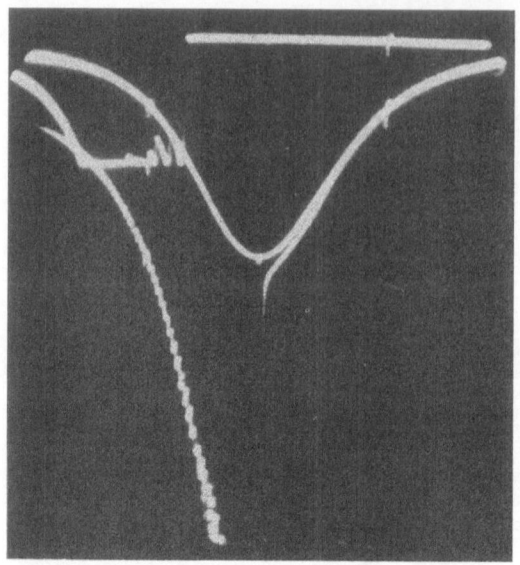

Oscillogram of engine "knock" (jagged portion of curve, upper right) obtained by NBS in studied of pressure changes within a specially constructed high-compression engine. The sinusoidal curve shows pressure changes during a normal nonburning cycle. The second curve, in which knock occurs, follows the first curve through most of the first half of the cycle; it then rises sharply as the compressed gas ignites.

So-called Grasshopper in upright position, ready for operation. This air-launched automatic weather station, developed by NBS for the Navy Bureau of Ships, will automatically set itself up and transmit weather data by radio. After the station has parachuted to earth, explosive charges are used to disengage the parachute, raise the station to an upright operating position, and erect a telescoping antenna to a height of some 20 feet.

determining the gryromagnetic ratio of the proton in absolute units. This makes possible the use of proton magnetic resonance methods for the convenient determination of magnetic fields in many other related experiments, and in combination with work in other laboratories it is improving knowledge of several basic atomic constants. Another contribution of importance is the development of the "omegatron" by Hipple and his colleagues. This is a miniature precision-built cyclotron, the object of which is to determine accurately the resonance frequency of positive ions circulating in a uniform magnetic field. With it a totally new determination of the faraday has been made, which seems to have cleared up the puzzle of the longstanding discrepancy between the values obtained electrochemically from the silver and the iodine voltameters. The present position strongly favors the iodine value.

Another field in which we have been able recently to strengthen our basic work is in fundamental research in cryogenics. Under construction is a small addition to our low-temperature laboratory where there will soon be installed a large helium liqufier built for us by Arthur D. Little, Inc. Besides problems of fundamental thermometry and thermodynamics, here the principle fields of interest are research on the puzzling phenomena of superconductivity, and the equally puzzling problems posed by the superfluid form of liquid helium called Helium II, the state assumed by liquid helium below 2.6 degrees.

A recent NBS development in the field of atomic instrumentation is the omegatron (within the glass tube), which makes use of cyclotron resonance frequency of ions to discrimate between particles of different masses. Although the omegatron operates on the same fundamental principle as a large cyclotron, the heart of the instrument is little larger than a package of cigarettes.

These few examples of the present program of the National Bureau of Standards can only suggest the nature and scope of the work. They are typical of the work of the fifteen divisions of the bureau: Electricity, Optics and Metrology, Heat and Power, Atomic and Radiation Physics, Chemistry, Mechanics, Organic and Fibrous Materials, Metallurgy, Mineral Products, Building Technology, Applied Mathematics, Electronics, Ordnance Development, Radio Propagation, and Missile Development. The growth that I have implied has come about because, in the more familiar fields, science has demanded greater precision in measurement, better standards of measurement, more sensitive instruments, and more precise values of the properties of various materials and of physical constants. In newer fields, such as atomic and nuclear physics, low- and high-temperature physics, radio propagation at the higher frequencies, and high polymer research, science and industry have made new demands on the bureau. Meanwhile, the present emergency has confronted us with many pressing problems of the military services.

Evolution of the Quantum Theory

E.U. Condon

Dr. Condon (Ph.D. California, 1926), who has been director of the National Bureau of Standards since 1945, has just been elected president of the Philosophical Society of Washington, the capital's oldest scientific society. The University of Delhi and the New Mexico School of Mines have recently awarded him honorary of D.Sc. degrees, and the Royal Swedish Academy of Engineering Sciences has elected him a corresponding members in recognition of his contributions to science. His article is based on a paper presented in the Section B symposium on "Fifty Years of Quantum Theory" held during the December 1950 meetings of the AAAS in Cleveland, Ohio.

The great new reformulation of quantum theory, which we know today as quantum mechanics, occurred mostly in 1925 and 1926. I shall not attempt to deal with the events of the past few years, on the ground that it is too soon to try to see them in historical perspective. But so great was the range of subject matter opened up for physics in the quantum-theoretical problems of the period 1925–40, that I shall have difficulty in even taking the barest notice of them in a short article.

It is important at the outset, I think, to stress that, contrary to impressions in some quarters, theoretical physicists are a conservative lot, and do not accept radical revisions of their thinking without a very thorough exploration of all possible alternatives. This conservatism is not the product of intellectual inertia or laziness. I think it is because in classical physics we have inherited an immensely powerful and fruitful tool, as evidenced by its past successes. It is natural to wish to exploit its power to the fullest, not merely because it is something with which we are already familiar, but also because in that way we assure ourselves of building the most unified structure for theoretical physics as a whole.

Therefore, throughout the early period of classical physics, especially in the period before the Bohr atom, leading physicists were most reluctant to go along with the discrete energies and the discontinuous processes that were inherent in the ideas of Planck and Einstein. We have a tendency sometimes to remember the physicists of a previous generation only for their lasting contributions to the physics of our own time. In this way we are likely to ignore the fact that they were quite conscious of basic problems of principle that they themselves were unable to resolve. Since we tend only to publish on those subjects where we think we have a positive result, the literature of this phase of physics tends to be meager, except where personal memoirs and biographical notes are available.

For example, a perusal of Lord Rayleigh's *Scientific Papers*, collected in six large volumes, reveals almost nothing dealing with quantum theory, yet he is known to us for the Rayleight-Jeans law, which so emphatically requires the quantization of energy in the radiation field. That he was concerned by such questions, however, we see from a brief letter of Rayleigh to Nernst (*Scientific Papers*, Vol. 6, 45), written in 1911, in which he is troubled by failures of the classical theory of equipartition of energy. He writes:

Perhaps this failure might be invoked in support of the views of Planck and his school that the laws of dynamics (as hitherto understood) cannot be applied to the smallest parts of bodies. But I must confess that I do not like this solution of the puzzle. Of course I have nothing to say against following out the consequences of the [quantum] theory of energy – a procedure which has already in the hands of able men led to some interesting conclusions. But I have difficulty in accepting it as a picture of what actually takes place.

How many of us can echo the feeling expressed in that last sentence! Let me quote him again:

We do well, I think, to concentrate attention upon the diatomic gaseous molecule. Under the influence of collisions the molecule freely and rapidly acquires rotation. Why does it not also acquire vibration along the line joining the two atoms? If I rightly understand, the answer of Planck is that in consideration of the stiffness of the union the amount of energy that should be acquired at each collision falls below the minimum possible and that therefore none at all is acquired — an argument which certainly sounds paradoxical.

Nowadays, when every undergraduate physics student learns about critical potential experiments and is taught the experimental facts at once as well as their interpretation in terms of quantized energy levels, it is hard for us to understand what a great step it was to make such a deep-seated revision of fundamental ideas as the quantum theory called for.

The hesitancy did not arise because the new ideals are so difficult to master, once you decide it is necessary to make such a radical break with classical theory; it was rather due to a strong desire to preserve an underlying unity in the structure of physics — just think how foolish all of us would feel if even now someone, by a clever flash of insight, would see a way of incorporating all the phenomena for which we use quantum mechanics into some model based on classical mechanics!

But let us not continue discussing the flavor of the situation without first reviewing a few of the facts about the growth the development of quantum mechanics.

In the few years preceding 1925, the central problem of theoretical physics was that of arriving at a more exact formulation of quantum mechanics than that afforded by the Bohr-Sommerfeld quantization rules and the general guide lines of Bohr's correspondence principle. Bohr himself, who had made the most decisive contributions to atomic theory thus far, was the most insistent critic of the inadequacies of his work — not simply in that it failed to give correct numerical results for special problems such as the ionization potential of helium, but also in that it failed to give a definite basis for calculation of intensity of spectral lines, or a method of dealing with collision problems, or of dealing with chemical bonds and with many other things.

The particular form of classical mechanics known as Hamiltonian was proving most suggestive and powerful in dealing with general questions, and various papers were being written in which tentative trials were being made of ways to modify the Hamiltonian dynamics so as to introduce the discrete energy levels and abrupt quantum jumps so characteristic of quantum physics. A good deal of this was in the nature of probings for a better general treatment of problems that were special enough that there could be, on the one hand, hope that the results could be controlled by specific experimental confirmation but at the same time also general enough to afford guidance for an all-embracing theory.

In this direction the most fruitful step proved to be the development of a quantum theory of dispersion, by adroit modification of the classical theory of perturbation of a conditionally periodic system by the periodic perturbations of the fields of a classical electromagnetic light wave. This was done by Kramers and Heisenberg and led to the association of a quantum-theoretic dipole moment amplitude with each quantum jump, this amplitude being a measure of the strength of the spectral line in emission, absorption, and dispersion. A slight extension led to the theoretical prediction of the Raman effect, but, oddly enough, so far as I know, no one took this seriously enough to look for it. When it was discovered a few years later in India by Sir C.V. Raman it was as a by-product of his intensive studies of light scattering rather than as a result of the theoretical prediction.

The next step was that taken by Heisenberg in proposing to deal with the ensemble of these dipole amplitudes as quantities subject to special mathematical rules of quantum algebra. Immediately it was recognized by his associates, Born and Jordan, that the quantum algebra toward which he was groping was simply the algebra of matrices, and that his quantum rules could be expressed as a particular statement of noncommutation of certain of the matrices involved.

All this was very formal and devoid of direct physical-pictorial content. How much more emphatically would Lord Rayleigh have said to all this: "But I have difficult in accepting it as a picture of what actually takes place."

The matter was made all the more difficult by the fact that the mathematical theory of matrices was quite unfamiliar to most physicists — for some reason they had not used matrices as a tool for dealing with several topics in classical physics for which they are particularly appropriate. The difficulties of physical understanding were thus enhanced by the need to assimilate an unfamiliar mathematical method.

Most of this was going on in Göttingen, where much of the theory of infinite matrices had been developed in pure mathematics by David Hilbert. There is a story that the founders of matrix mechanics turned to Hilbert for help, and that the advised them that the problem of diagonalizing matrices was always associated with the more familiar problem of finding the allowed values of some differential equation's boundary value problem. But this tip was not followed up in the excitement of the times, and thus, probably, the Göttingen group missed discovering wave mechanics.

In Paris, a young physicist named Louis de Broglie wrote a doctor's thesis in 1924, which was published in the *Annales de physique*, and to which practically no one in the main stream of theoretical physics paid any attention at the time. De Broglie considered the connection between the wave and particle aspects of light quanta from the point of view of special relativity. The most remarkable suggestion that his studies produced was that perhaps other "particles" than light quanta were somehow associated wave motions — in particular that the motions of electrons were somehow governed by associated wave motions — and he gave an interpretation of the Bohr-Sommerfeld quantization rules as being the condition that an integral number of these waves be contained inside a periodic orbit.

Shortly thereafter Erwin Schrödinger became acquainted with these ideas on a visit to Paris and found in them the inspiration for his basic papers on wave mechanics, which appeared in the *Annalen der Physik* during the spring of 1926. Also in 1926 there came the discovery by Davisson and Germer, at the Bell Telephone Laboratories, of diffraction of an electron beam scattered by crystals in complete confirmation of the suggested association of a wave motion with problems of electron motion.

Then came the recognition that the wave-mechanical differential equations of Schrödinger were simply an alternative formulation expressing the same mathematical content of the matrices of Heisenberg. The two theories were recognized as simply two representations, or aspects, of the same underlying material, which as yet existed more as mathematical formalism than as clearly formulated physical principles.

By the fall of 1926 Born provided a decisive step in postulating the probability interpretation of the Schrödinger wave function, or, more correctly, the square of its magnitude was interpreted as giving the relative probability of finding the electron in various positions in the physical situation described by a particular wave function. This was necessary in order to give meaning to the formal mathematical calculations by which he gave a wave-mechanical description of the quantum mechanics of collision processes.

Within a year, more general formulations of the transformation theory of quantum mechanics were worked out, principally by P.A.M. Dirac, in England, and by John von Neumann, in Germany. These were natural extensions of the mathematical formalism which included both the matrix-mechanical and the wave-mechanical formulation in a general framework. An essential part of their physical content rested on the adoption of Born's probability interpretation of the wave function.

During the years 1926–27 the principles of the theory became pretty clear, and a host of papers were written applying the new methods to every branch of atomic and molecular theory. Born himself laid the foundations for treating collision problems, Sommerfeld revived the electron theory of metals with appropriate adoption of the Fermi-Dirac statistics for the free electron gas, Heitler and London laid the foundations for the theory of the covalent chemical valence bond, and Heisenberg recognized in exchange interaction of electrons the basis of strong coupling between elctron spins, which makes ferromagnetism possible. Many other developments of equal importance could be cited. All these specific results applicable to old phenomena, and some of them giving totally new insights into new phenomena, rapidly gave people confidence in the essential correctness of the new mathematical formalism.

The formalism has been so successful, and has continued to be so fruitful in so many parts of atomic, molecular, and nuclear physics, that today it is an indispensable tool for physics. And I think it is fair to say that nearly all the theoretical physicists not only accepted and used the formalism but also adopted the probability interpretation of the wave function.

The general basis of the statistical interpretation of quantum mechanics was, in the fall of 1927, greatly clarified by Heisenberg's analysis of the inherent limitations involved in carrying out certain measurements in the field of atomic physics by the quantum nature of the interactions between the phenomena observed and the observing equipment. This point of view was much more profoundly analyzed and carried much farther in its philosophical implications by the studies of Bohr.

There was one physicist of importance and, so far as I know, only one, who thought deeply on the subject, who did not accept these views as having the fundamental validity that is generally claimed for them. That was, and still is, Albert Einstein. I remember how in 1926–27, when I was studying these things as a National Research Fellow in Göttingen and Munich, we heard that Einstein had refused to accept the statistical interpretation.

"*Der lieber Gott würfelt nicht,*" he remonstrated. ("The good Lord does not shoot craps," to translate into good American.) In this brief sentence Einstein summarized his conviction that a statistical theory of this type could not be more than a temporary makeshift, reflecting an inadequacy of our present knowledge rather than being a fundamental attribute of the world in which we live.

He still holds this conviction, despite the many successes of quantum mechanics in interpreting phenomena, and despite the lack of success attending efforts that he has undoubtedly made in the past twenty years (even though principally occupied with other problems) to provide what to him would be a more satisfactory kind of interpretation of quantum mechanics, or an alternative development that would include the tangible successes of quantum mechanics. He stands pretty much alone in his view, for the general tide of physics has been running strongly in the statistical direction for these past twenty years, but I would not want to say certainly that he is therefore wrong. It could be that future developments might carry us once again in the direction of more specifically deterministic theory — although it is hard to see now any way in which this is likely to come about.

In Born's recent book, *Natural Philosophy of Cause and Chance*, he quotes a letter from Einstein dated December 3, 1947 in which Einstein recognizes that the "statistical intrepretation . . . had a considerable content of truth," but he goes on to say that he is "absolutely convinced that one will eventually arrive at a theory in which the objects connected by law are not probabilities, but conceived facts as one took for granted only a short time ago."

With characteristic modesty, however, he concludes this strong affirmation of faith by saying:

Zur Begründung dieser Uberzeugung kann ich aber nicht logische Gründe, sondern nur meinen kleinen Finger als Zeuger beibringen, also keine Autorität, die ausserhalb meiner Haut irgendwelchen Respect einflössen kann.

("However, I cannot provide logical arguments for my conviction, but can only call on my little finger as witness, which cannot claim any authority to be respected outside my own skin.")

For my own part, I have never been repelled by the lack of objective determinism that Einstein finds so unsatisfactory. In the first place, I have always felt very keenly aware that the material we deal with the science is much more subjective than most of us like to admit. It is by no means a complete account in terms of orderly causal relationships of all the sense data we have, but only that part of it which we have thus far been successful in codifying in such terms. It is true that such a tremendous amount of sense data has been codified this way in the past three centuries as to give rise to the hope that all can be so treated. On the other hand, think for a moment how much such data have so far resisted codification in this way.

In terms of such a perspective, how can we have the complete faith that Einstein apparently has in "the perfect rule of law in a world of something objectively existing"? I agree with Einstein that many of the successes of physical theory lend support to this view, but, after all, is it not considerable extrapolation from present knowledge to draw the inference that all of nature must be describable in this way?

Whether all the data of experience can be codified in terms of orderly causal relationships. I do not know, of course, but there is no question but that it is useful to have such organization of knowledge carried to its furthest limits, and therefore if an affirmation of faith in the ultimate possibility of doing a complete job stirs us on to greater efforts, I suppose it is a good thing. But this gives to the issue the status of a "pep" talk by the coach in a football game, or any other such rallying cry, which we know from past experience does not always correspond to a critical statement of truth.

If I may stray a little farther in the direction of expressing personal views, I will say that I think the history of science, especially on the metaphysical or philosophical side, has been marred by the tendency human beings have toward facile generalizations, unwarranted extrapolations, and that kind of oversimplification which fails to qualify what would be true if properly qualified. How many pages of disputation have been filled with discussions of free will and determinism! In my view, nothing that physics has to say has any bearing on this one way or the other as an issue about human behavior and human conduct. It was an unwarranted extrapolation in the first place to suppose that we knew so much about classical mechanics, as applied to living tissue, just because we understood the planetary motions fairly well, that we were justified in saying that our knowledge of mechanics strongly supported a view of extreme mechanistic determinism.

This being so, it is, to my mind, equally irrelevant to the basic philosophic issue, when the statistical determinism introduced by quantum mechanics in place of the exact determinism of classical mechanics is used as a basis from which to argue support for the alternative of free will in human behavior. To me it seems simplest to interpret my actual life experiences in terms of a considerable degree of freedom, although the logical possibility of the alternative view is conceded. But on this general question there is one thing on which I have a profound conviction, and that is that the issue of the deeper interpretation of the meaning of quantum mechanics in the world of atomic physics probably has nothing whatever to do with the issue of free will and determinism as affecting human affairs.

If we now turn our attention to the specific accomplishments of quantum mechanics, it is not possible to separate these from the whole stream of achievement in physics in the past quarter century. Every branch of atomic, molecular, and nuclear physics, including the theory of metals and solids, has been profoundly affected. Quantum mechanics has greatly influenced the development of chemistry, not only in clarifying molecular structure through the interpretation of spectra, but also in interpreting the equilibrium thermodynamical properties of materials by use of the partition function, and also through the use of quantum mechanical theories of chemical rate processes in the hands of Eyring and his associates. Astrophysics has been greatly helped by the more exact theory of atomic spectra and through nuclear reaction rate theory.

It is interesting to note that the entire theoretical structure of nuclear physics is cast in quantum mechanical terms. This branch of physics is so new that is has never been handled any other way. The first application was the interpretation of α-particle natural radioactivity in terms of the barrier leakage, or tunnel effect, as done by Gamow in Göttingen and by Gurney and me in Princeton in 1928.

This same theory of barrier leakage, but applied in reverse to barrier penetration from the oustide, made it seem probable that nuclear reactions could be made to occur at lower voltages than would be expected from classical theory. This suggestion became a stimulus in the early 1930s to more rapid progress in this field than might otherwise have occurred. The theory also suggested, as Gurney pointed out, the notion of resonance penetration, making clear why nuclear reactions go more favorably at certain preferred values of energy than at other values either above or below the favored one.

In attempting to build a general theory of nuclear forces between protons and neutrons, quantum mechanics gave the general concept of saturable exchange interaction a meaning that is quite foreign to classical ideas. Relativistic quantum mechanics in the hands of Dirac not only suggested the existence of positrons, but made clear why these particles have only a transient existence, and gave the foundation for calculations concerning the processes by which they are generated and by they are annihilated. The relativistic theory further gave the basis for the calculation of high-energy Compton scattering processes, as well as pair production processes, which play a decisive role in interpretation of the new physics of the billion electron volt region and higher that is realized in cosmic rays and in the more powerful particle accelerators recently built.

From quantum mechanical theories concerning exchange forces came the concept of Yukawa of these forces having their origin in a field associated with a new kind of particle, the meson. In a short time mesons, too, were discovered, not one, but several kinds, the interrelationship of which with each other and with the other fundamental particles is probably the most important problem of fundamental physics today.

In spite of this long record of remarkable successes, however — and the things here briefly mentioned are by no means all that ought to be discussed — the record is not one of complete and unbroken success. From the very beginning it was recognized that it should be possible to treat the electromagnetic field in a quantum mechanical way in order to give a completely adequate description of such phenomena. The method by which this might be done seemed clear: for many years it had been the custom to describe the dynamics of the electromagnetic field for purposes of statistical mechanics in terms of an analysis into orthogonal wave functions, whose amplitudes were treated as generalized coordinates in the Hamilitonian sense. What was more natural, therefore, than the treat these coordinates as quantum mechanical variables in order to get a quantum mechanics of the electromagnetic field?

This step was taken by Dirac very early, and it had at once a number of success; among them it gave a correct account of the relation between the spontaneous and induced radiation transition probabilities and a better calculation of dispersion and light scattering. But, when it was applied to more abstruse problems of radiation theory, it was quickly found that the theory often gave infinite results because of the occurrence of divergent integrals in the calculations.

This rather fatal defect has plagued theoretical physics continuously for almost the entire period under review. Throughout the 1930s and early 1940s it was customary, though no one liked the situation, to work problems with this faulty theory, and to try to get reasonable results from it, trying to invent some sort of procedure whereby the range of integration on divergent integrals would be made finite in order to get finite results. This would have been a distasteful and arbitrary procedure even had it given correct results — which it did not do.

This situation has been to some extent relieved in the past five years by some of the more recondite developments in field theory, especially those associated with the names of Schwinger and Feynman in this country, but this is a topic that cannot properly be reviewed here, other than to say that

MAX PLANCK, 1858–1947

Tita Binz, Heidelberg

there is some basis for hope that real progress in quantum electrodynamics is being made.

If I were to try to sum up, I think the position today might be stated something like this:

1. Most physicists believe that the Bohr-Heisenberg analysis shows the fundamental basis of the theory to lie in real limitations on the separability of the object world and the effects produced by the observer watching it, and that therefore the statistical nature of quantum theory is basic. But Einstein thinks otherwise, and he might be right in the long run.

2. The nonrelativistic quantum theory for low-energy problems has had so many successes in the simpler problems that a strong feeling exists that all problems of atomic and molecular structure, and even of the crystalline solid state, could be successfully treated this way, except for the great computational labor in dealing with all but the simpler ones.

3. The relativistic theory and the quantum electrodynamics, although beset with difficulties, have enabled approximate treatments of many problems to be given, which are in general accord with the facts and which serve as guide lines around which to organize the experience of high-energy particle physics.

It has been an exciting period of enormous fruitfulness for the development of physics. If the next fifty years bring as much new knowledge as has the half century just passed, then the physics of the year 2000 will be as strange and unforeseeable for us as the physics of today would seem to the physicists of 1900.

Some Thoughts on Science in the Federal Government . . .

E.U. CONDON

The following is an address given by Dr. Condon on September 25, 1951, less than one week before his resignation as director of the National Bureau of Standards became effective. His talk was prepared for delivery at the National Academy of Sciences in Washington.

As my nearly six years of service as Director of the National Bureau of Standards draw to a close, it seems that an important final part of that service would be to set down some over-all views concerning the scientific work of the Federal Government growing out of that experience. Our governmental institutions are so close to us that I had some experience with them before entering Federal service full-time, especially during World War II, and likewise I expect to have association with such matters in the future while in private employment.

It seems to me that the scientific research activities of the Government are on the whole good but nevertheless, like all things human, capable of improvement, and it is to some suggestions for improvement that I will principally turn my attention.

The first general point I wish to make is the very obvious fact that the whole complex of modern material civilization arises from application of scientific knowledge. All modern engineering and industry, agriculture and medicine is based on the results obtained by consciously planned laboratory experimentation within the last three centuries, and largely within the last century. It is this new type of activity which has in the last century made greater changes in our material ways of life than have occurred in thousands of years before. The improvement of the conditions of life through the lightening of burdens by the development of mechanical power from flowing water and from fuels, the improvement of our homes and clothing by modern products of applied science, the more effective production of foods and the use of refrigeration for their large-scale preservation and wide distribution, the increased knowledge of nutritional principles, the improvement in all kinds of techniques of medicine and

---

E.U. Condon is director of research and development for the Corning Glass Works, Corning, N.Y. He was director of the National Bureau of Standards from 1945 to 1951. He has been scientific advisor to the Special Committee on Atomic Energy of the U. S. Senate since 1945.

surgery — all these may be counted as great blessings to manking resulting from the cultivation of science and its application to our material
needs.

Even greater perhaps than all those material benefits, however, is the
benefit that comes from the freeing of men's minds and spirits from the
oppressiveness of superstitious belief and the growing realization that we
live in a world of law and order that is intelligible to us if we will but
discipline ourselves to the effort necessary to understand its structure and
workings. Certainly this spiritual blessing, in common with the material
blessings already mentioned, should combine to produce in all of us the
recognition that scientific study is one of the most rewarding fields of
human endeavor possible in the world today.

Science is a method by which we learn to know in ever wider ways and
with ever greater precision about the world in which we live. The study of
science can make genuine and wholesome contributions to character
development not the least of which is an uncompromising demand for truth
and honesty in all the affairs of life and a proper humility before all the
many wonders which surround us. But great as I think are the values which
science has brought and will bring to humanity, I would not wish to leave
you with the impression that man can live by science alone, for science
does not provide him with the ethical guidance nor the spiritual insights
which are needed to realize our ideal of the good life.

Not all of the consequences of this enormous increase in man's
knowledge of the world have been beneficial nor can it be said that we are
effective in making the fullest use of the knowledge we already have. We
have been slow to bring about a widespread distribution of these benefits to
all of the population of even a wealthy and favored nation like the United
States. While steady progress is being made — at a lamentably slow pace
— the fact is that we have done very little toward slum clearance in our
major cities or toward providing adequate schools and hospital service for
all of the population. We are doing very little to assist the underdeveloped
countries to bring the benefits of modern applied science to improve the
welfare of the hundreds of millions of their population.

We talk of bold new programs in this direction, and we feel uneasily that
much more needs doing than we have undertaken so far, and still we do
essentially nothing about it. Our carelessness here is storing up great
trouble for us in the future. We in America and in Western Europe are a
small minority among the world's peoples. Other hundreds of millions of
persons, chiefly in Asia, have caught a glimpse of what modern science can
do for them and they are determined to have it. If we help them we can
have their friendship as equals. If we do not, they will get these benefits for
themselves anyway in the course of time, and on terms which will involve a
great deal of strife and difficulty for us. It is true we have done much to
assist in the reconstruction of Western Europe, but we have done
practically nothing to assist the development of Asia and Africa. We have

not even made effective plans for relief and reconstruction in the devastated areas of Korea.

The effort in this direction that I feel is necessary will be very great but it is my sincere conviction that effort of this kind is the most important thing we can do to preserve and extend the kind of Christian democratic civilization which we believe in. I believe that this kind of constructive effort to assist in bringing the benefits of modern science to the whole world is the only kind of effort which will accomplish the construction of the kind of world in which peace and goodwill can reign. I do not regard this required effort as a burden but as a great opportunity which has been presented to us which we should be grasping with eagerness and enthusiasm.

While it may be necessary, under present conditions, to use our scientific knowledge and our industrial productive capacity largely for building up our military strength, I am convinced that we are, perhaps unconsciously, placing too great an emphasis on this, as if it would give us the means of solving the difficult social problems with which we are confronted. All that we can hope for from military strength is that it will enable us to preserve a situation in which Western civilization will have an opportunity to share its wealth-producing techniques with the other peoples of the world, instead of having them snatched from us by angry hordes of men who outnumber us ten to one and who will have come to resent bitterly the seeming hypocrisy of our attitudes toward them. I will not therefore go so far as to say that under present conditions the building up of military power on which we are again engaged is now avoidable. But this course of action by itself may prove fruitless unless it is accompanied by a very great program — one whose scale of effort is at least as great as that we are putting into building up our armaments — that is designed to help all peoples of the world who are willing to work with us, to achieve the benefits of modern science which we enjoy. If we do this we shall derive great spiritual benefit from the increased happiness of these millions of God's people and material benefits from our participation in the contributions which their intelligence can bring to our unsolved problems.

There is another aspect of recent tendencies in development of military armament which we need to consider very carefully. War at best is an evil thing in which peoples resort to force to impose their will on each other instead of using love and compassionate efforts at mutual understanding to arrive at a solution of their difficulties. The opening years of this century were marked by all kinds of efforts in the way of agreements for the humanitarian treatment of prisoners, in agreements to confine the fighting to organized military forces, and even in agreements to avoid the use of certain particularly horrible weapons such as dum-dum bullets. In the two world wars of recent years, and in the military rearmament which is now going on, such ideas have become quaintly old-fashioned.

No longer do we give the slightest consideration to the distinction between military and civilian populations. In World War II both sides gave very little regard to avoiding destruction of the civilian population of their enemies, and enormous damage was done to other than strictly military objectives. A minute part of this terrible destruction was made by the use of the bombs based on the fast neutron fission of uranium and plutonium. The loss of life in Japan alone due to fire raids using napalm was much greater than that due to atomic bombs.

A large part of our organized effort in modern science goes today into putting enormous teams of men to work on developing even more deadly and destructive weapons than the world has ever seen before. We talk openly of germ warfare and nerve gases and we almost never hear of these being criticized as inhumane and revolting to the consciences of Christian men and women. No, we hear them criticized because it is difficult to produce germ cultures or gases in sufficient quantity or concentration to wipe out the whole population of a city as their proponents would say is possible and therefore we should devote our attention to the creation of some other fiendish thing like the hydrogen bomb. This, in turn, we hear criticized, not in terms of revulsion that men would use such things against each other, but that maybe its destructiveness is too conventrated and that the same effort put on more conventional types of atomic bombs would enable destruction to be carried out over an even greater area.

At San Franciso a few weeks ago the President spoke unspecifically of fantastic new weapons too horrible even to describe. The press was thereby filled with all kinds of science fiction speculations about what these horrible new wonders might be. Within a few days Congress increased the already enormous appropriations to the Air Force by five billions. In a matter of hours the Congress gave five billion for fantastic new weapons of which it knows next to nothing — the same Congress which after long debates finally cut one billion dollars out of the foreign aid program, the same Congress which by its long delays did much to nullify the effects in promoting goodwill of our finally supplying a credit (not a gift) for $190 million for grain to alleviate severe famine in India; the same Congress which refuses to provide $300 million in Federal aid to our overcrowded and inadequate school system, the same Congress which has lopped off the paltry appropriation of $14 million for the National Science Foundation which was intended to give some slight nourishment and sustenance to the fundamental scientific research on which rests the whole structure of modern industry, agriculture, and medicine.

Some may think that in referring to $14 million for the National Science Foundation as a paltry sum I speak like one of those terrible bureaucrats who has no regard for the burdens which the taxpayer must bear. I am concerned about taxes but I also want us to show some sense of proportion. Congress is this year spending $60 billion of new money or a total of about $100 billion of available funds on the Department of Defense. It has just increased this by another $5 billion for "fantastic" new weapons which the newspapers say can "conquer the atmosphere," whatever that means. It is spending $6 billion on foreign aid much of which is for rearmament rather than economic development.

Included in the military appropriations is about $1.5 billion for military research and development, a staggering sum of money which, if invested at 6% interest, would produce annually as much money as Congress has appropriated to the National Bureau of Standards in the entire fifty years of its existence. But it cannot spare $14 million next year for strengthening basic scientific research.

Today every activity of Government is being adjudged solely on the basis of its contribution to defense. I doubt whether such vast sums can be spent wisely for the purpose intended, and whether it is wise to put so much of our reliance on military strength while thinking so little about peaceful and constructive solutions of the difficult domestic and international problems before us.

If so much of our scientific effort is directed toward military weapon development, it must necessarily mean neglect of the basic science on which future progress must be built and neglect of the application of modern science to improving human well-being in our own and other parts of the world. There is another reason why we might be disturbed at the extent to which science is devoted to military purposes today. Although it seems to be very little in evidence at the moment I believe that deep in the consciences of men and women there is a horror and revulsion at the terrible means and methods of modern warfare which will some day find expression in a new and powerful and constructive determination to live together peacefully, and effectively to renounce war as an instrument of national policy. If in the years to come science and the scientists are closely identified in the public minds as the wizardry and the wizards who have made all the fantastic new weapons of mass destruction that Governments are now so eagerly urging them to produce, this horror and revulsion of war may, in that illogical and irrational way that so many things go in politics, be extended to science and the scientists. If this were to happen it would be bad not only for the scientists, but it would be bad for society, for a rejection by society of the method and power of scientific inquiry will stop progress in understanding and tend to retard the extension to all mankind of its beneficial applications. If men's consciences reawaken to the absolute necessity of abolishing warface, then there may be serious danger that science may be the baby which is thrown out with the bloody bath which is War.

This situation poses very difficult problems for scientists in general and especially for those in official positions in our Government. Speaking personally, all of my friends know with what strong conviction I hold the general view which I have tried here, rather inadequately I am afraid, to outline. When I came to Government service at the close of World War II,

I hoped and believed that there was to be an era of peace in which fundamental research in science would flourish and be supported by society as a whole as a worthy intellectual activity and for the constructive benefits to man's well-being which it can bring. At that time, only six years ago, the United Nations had just been born and many of us believed that the experience of wartime alliance had taught the lessons which would gradually enable the growth of a mutual confidence and trust between Russia and the United States and other principal nations of the world which would remove any basis for future war of major proportions. In such a setting one could hope for a steady reduction of national armaments, with the enormous economic waste which they imply, and their replacement by an international police force. In such a setting we hoped that all efforts in the field of atomic energy would go to peaceful purposes in chemical and medical research and in making available new sources of power.

At this time it seemed that Congress and the people of the United States, impressed by the contributions which applied science had made during the war, were prepared to support a National Science Foundation in a really adequate way — by this I mean to the extent of several hundred million dollars a year — and that science in other countries would be aided by a major program of the United Nations Educational Scientific and Cultural Organization as well as by local efforst in those countries.

During my first year in Washington, 1945–46, my attention was largely taken up with assisting the Special Senate Committee on Atomic Energy of the 79th Congress, as scientific adviser, when it was developing the Atomic Energy Act of 1946 by which the present Atomic Energy Commission was established.

During that first year the Senate held extensive hearings on proposed legislation for the National Science Foundation and passed a bill, but this was allowed to die in the House when the situation became confused by behind-the-scenes lobbying of those who insisted on a large part-time board for the Foundation. Otherwise the National Science Foundation probably would have started operating five years ago with an annual appropriation of about two hundred million dollars. If this had been allowed to happen we would have been incomparably better off than we are today from every point of view. Fortunately, the vacuum thus left was quite well filled by the enlightened scientific research program of the Office of Naval Research. This was conducted as liberally and as intelligently as any purely civilian program could possibly have been conducted and has made a wonderful contribution to the development of basic science in America during the post-war period.

Soon after that first post-war year it became clear that expenditures for scientific research for military purposes would be maintained at a high level and expanded above the minimum reached in the demobilization period.

Work in this field has always been an important part of the program of the laboratories of the National Bureau of Standards. The Bureau has a long history in meeting such military needs, having first developed the opitcal glass industry in World War I, having initiated the atomic bomb project in World War II, and also having played a large part in the development of proximity fuzes, having developed the only fully automatic guided missile to be used in warfare thus far, and having done much to improve knowledge of long-distance radio transmission on which the continuity of military communications depends. This latter service was initiated during World War II and organized as a permanent service in the Bureau during the first post-war year. Congress has been willing to support this work reasonably well and has made provision for splendid new laboratories for the radio work of the Bureau to be built in Boulder, Colorado. This radio work is, however, essentially the only new activity of the Bureau for which it has been possible to get direct financial support from the Congress during the post-war years.

In this period, to be sure, and particularly during the last year, there has been a great expansion in the level of activity of the Bureau. But this has not been by direct Congressional support, but rather by doing project work in Bureau laboratories for the armed services and with funds provided by them from their own appropriations. For example, this fiscal year the Bureau will operate on a total budget of some 60 million dollars, less than $10 million of which is directly appropriated by the Congress, nearly all of the rest being paid by the military for work done for them. To get some idea of the disparity of figures involved it is interesting to note that this year the Bureau will spend on electronic ordnance developments alone about 50% more money than the $14 million which the House has refused to give the National Science Foundation for Federal support of basic science.

The amount of military work done by the National Bureau of Standards has increased almost by a factor of seven during the time that I have been Director. Provision has been made for expanded facilities for such work in Washington, in Boulder, Colorado, where large new Bureau-operated laboratories are being built for work of the Atomic Energy Commission, and also in Corona, California where some unused former Naval hospital facilities have been converted into a splendidly equipped development laboratory for guided missile work of the Navy. Another Bureau development of military importance has been the establishment of a department of applied mathematics with facilities both in Washington and Los Angeles, and the development of an important electronic digital computer, the SEAC, which has been in service for more than a year on military problems. These are just highlights of a program which involves dozens of research projects of specifically military interest some of which relate directly to fantastic new weapons which cannot even by mentioned. I think therefore that the National Bureau of Standards is in a stronger position than ever before to make important contributions to military needs.

Turning to the fundamental support of the civilian program of the Bureau the situation is far from satisfactory. The National Bureau of Standards is a Cinderella whose Prince Charming has yet to come along. In spite of its long record of splendid accomplishments, its scientific program was crippled by severe budget cuts in 1933 as one of the economy acts of the Roosevelt administration. Valiant efforts were made by my predecessor, Dr. Briggs, to hold an effective staff together in spite of this short-sighted action but the Bureau is still suffering from the effects of that decision.

Except for the expanded radio work the direct support available for the Bureau in the post-war years has remained nearly constant, as expressed in dollars, and therefore has declined steadily in real purchasing power for goods and matericals. This is a most serious situation, for it has occurred at a time when there has been a steady growth in the amount and complexity of the needs for standards of precise physical measurement.

Every kind of physical quantity is being measured, in connection both with scientific research and with more accurate control of industrial processes, with greater precision than before, and over a wider range of extreme conditions, and the need for exact calibrations of measuring instruments arises from a much greater number of research laboratories and industries than ever before. This has put a burden of work on the National Bureau of Standards with which it is quite unable to cope within the framework of its present appropriations. Try as we will we have not been able to keep up with the demands for such services. The result is, of course, that much money is wasted by others in duplications of calibrating set-ups which the Bureau should have and that many scientific jobs are done with a lower grade of accuracy than desirable and than would be possible if the National Bureau of Standards were allowed to render an adequate service.

I confess that I do not know how to do a better job of bringing this need to the attention of the Government. It has received a great deal of my attention in the last five years but with essentially no results. I hope that my successor in office will be able to do better on this than I have. Here it is important for him to realize that not all of the difficulty is with Congress. The budget of the National Bureau of Standards has to pass four hurdles before it is approved. It must first be approved by the budget officers of the Department of Commerce. It comes before them as a peculiarly difficult-to-understand technical item which amounts to less than two per cent of thetotal budget requirements of that Department. Since it is such a small part of the Departmental budget itis only natural that these budget officers have no scientific and technical background. At this stage efforts to get even what increase is necessary to keep abreast of the declining purchasing power of the dollar are pretty well nullified because these men are working under a general over-all limitation as to what the Department of Commerce itself may have.

After the Department of Commerce has finished its consideration, the Director and his staff must write up the whole thing again in great and specific fiscal detail for the Bureau of the Budget. This supposed to show that the whole program of proposed work has been very thoughtfully considered. Having filed all this data with the Bureau of the Budget, several hours are spent explaining the needs to staff officers of the Bureau of the Budget. Here again because scientific research is diffused over the whole structure of the government one is dealing with individuals who have very little background either in the over-all needs of the Government for scientific research, or in the accomplishments of the National Bureau of Standards in particular, or for the methods and aims of physical science in general.

This process goes on intermittently during the first half of the fiscal year preceding the one for which the budget is being prepared. Out of it comes an official determination by the Bureau of the Budget of what each governmental agency will be allowed to ask for in going before the Appropriations Committees of the House and Senate. The end result of this process when carried out for all the agencies of the Government appears in a large document which is printed and transmitted to Congress as the President's Budget. This is now official, and sometime in the spring the Director and his staff are summoned down to present the Bureau's part in the President's Budget to his subcommittee of the House Appropriations Committee and then to the Senate Appropriations Committee. Before doing this, however, his own staff of budget officers have had to rework completely the elaborate document by which the plans for the coming fiscal year were submitted to the Bureau of the Budget.

It is hard to convey any idea to persons outside of the Government of the extent to which the working agencies are called on to supply over and over again statistical reports about their work which over essentially the same ground in slightly different forms.

Each agency sends up a large budget document to the Congress for the use of the Appropriations Committee in advance of the hearings. At the hearings the questioning often indicates that the Congressmen have very little understanding of the particular scientific needs of a technical agency and that perhaps they have not had time to look at the contents of the elaborate budget document which was prepared for them.

Some of my most treasured memories of Government service are connected with incidents which occurred in these appropriations hearings. One feels rather nervous and tense on these occasions for on their outcome hinges the whole fate of the Bureau's work.

One time while waiting our turn outside the committee room, the budget officer of the Patent Office came out of the door looking pale and fell on the floor of the hall in a dead faint. We bustled around administering first-aid and when he came to partially he muttered deliriously, "It's awfully hot in there." Later, when it was my turn to go in, I found that he was right. That was during the Eightieth Congress at a time when the Alsops referred to the House Appropriations committee as a bunch of blind men pruning a jungle.

I remember one time one Congressman had me quite upset because he was scowling through the whole of my presentation. When it came his turn to ask questions he asked me, "Doctor, where *is* the National Bureau of Standards?"I told him it was out on Connecticut Avenue and he said excitedly, "Is *that* what that place is?" and became quite friendly.

On another occasion a Congressman was questioning the chief of the Bureau's radio division, who had been talking about the scarcity of space in the radio frequency spectrum for the many needs of communication services. He said: "Doctor, I understand that among you scientists there are two theories: some say space is finite, others say it is infinite. I want to know, where do you stand?" The witness started to explain the limitations of using very low and very high frequencies but the Congressman interrupted him to say, "No, I mean space, you know, *space*," making a large and globular gesture toward the part of the three-dimensional continuum in front of him.

The witness squirmed and looked at me for guidance, quite willing to make it finite or infinite for the sake of the budget, but I could only indicate with a gesture that I did not know which was the preference of that particular Congressman. So he gulped hard and said, "I think it's infinite." "Thank you very much, Doctor, that's all I wanted to know", replied the Congressman and passed on to another topic.

When these hearings deal with science they are apt to be rich in non sequiturs. For example, only yesterday I was reading the Senate defense appropriation hearings (p. 1177) where an Army colonel is asking for funds for an electronic computer for logistic planning. A Senator asks him: "Now, is there any relationship between the number of equations that have to be developed and the time the machine is in operation?" And the colonel replies: "Electricity travels 186,000 miles per second, sir, so it is an infinitesimal difference."

There would be no point in describing this procedure in such detail unless I had a suggesting to offer. I do have.

I am convinced that the over-all importance of scientific research in Government has become so great that it requires careful attention and study by a new standing committee of the Congress. It is at least as important as atomic energy which has a permanent Joint Committee, affording an organized means by which Congress can study these problems. A similar means is needed for scientific research broadly if we are to get intelligent acton and focus attention on unwise actions or inactions. Such a committee would study and deal with legislative problems affecting scientific research.

In addition it would be very desirable if the Appropriations Committee of the Congress would find a formal way to give some unified over-all attention to the scientific research requirements of the Government. A legislative committee on science in the Congress would not be enough unless the Appropriations committees were also prepared to have a look at the whole program of all the large variety of specialized agencies in the government which are doing scientific work.

The most natural way for the Congress to deal with science in a unified way would be for the scientific agencies of Government to be gathered up into what would be in effect a single Department of Government. I believe that the general importance of scientific research in the Federal Government has become so great that this should be done. If it were not considered desirable to establish a new Department of Scientific Research then I would recommend that the Smithsonian Institution be used for this purpose. I believe that the new Department or enlarged Smithsonian Institution should include all of the scientific agencies of Government including the major military research laboratories, the research laboratories of the Atomic Energy Commission, the National Bureau of Standards, the National Institutes of Health, the laboratories of the Department of Agriculture, the Weather Bureau, and the Bureau of Mines, the Geological Survey, the National Advisory Committee for Aeronautics, and, of course, the National Science Foundation.

Whether a new Department of Scientific Research in the Executive Branch or an enlarged assignment of responsibilities for the Smithsonian Institution represents the better proposal I am not prepared to say. But it seems clear to me that a unified administration of the scientific affairs of the Government, including unified treatment of them by the Bureau of the Budget and by appropriations and legislative committees of Congress, would surely be an improvement over the present situation. I am inclined to favor the adaptation of the Smithsonian for this purpose over the creation of a new Department, for the reason that each cabinet member is on the board of the Smithsonian and thereby the relation of science activities to the other government activities they support would be preserved while giving scientific research as a whole a coordinated administration.

The suggestion that the Bureau of the Budget should have a special staff for study of the needs of scientific research is not a new one, having been made as a recommendation in the Steelman report, entitled "Science and Public Policy." But it has never been acted upon, I suppose, because of the difficulty of finding properly qualified individuals to do the job and the Budget Bureau may feel that it is better to do it not at all than to do it badly.

If there is any merit in the general suggestions I am making I would like to see the Bureau of the Budget call on the National Academy of Sciences for a study and recommendations and also to ask the Academy for its help in reviewing the budget of the existing agencies. The Congress too should recognize the many ways in which it could get help on scientific problems from the Academy and call on it for help more often on large broad issues than it has in the past.

As part of such a plan, the National Academy of Sciences, the National Research Council, the American Association for the Advancement of Science, and the specialized scientific societies would retain the independent status which they have now but would work in close cooperation with the new science administration to make sure that the Government's research program is effectively carried out in a way best suited to serve the national interest in relation to the professional needs of the scientific work in the universities and in industry.

One of the most remarkable omissions in the report of the Hoover Commission on the reorganization of the Government was its almost complete lack of any recommendations for improving scientific research in the Government. This is hard to understand for surely the men who developed that report appreciate the importance of science today in Government, and cannot have felt that the present diffusion of responsibility over many separate agencies is a form of organization which cannot be improved upon.

It seems to me all the more important that a unified central body for science in Government be set up because research is a very fashionable thing these days and every new agency feels it must do research in order to have status in the world of bureaucracy. While it is very difficult to get adequate support for the established research agencies, it is always possible to set up a research program as a small part of a general need to which assent has been given and by indirection to obtain vaster sums of money than the established agencies can get for research. For example, it would be extremely difficult the way things are now to get a modest increase in the funds available to the National Bureau of Standards of research on synthetic rubber in spite of a splendid record of past achievement, whereas a quite substantial amount of support is carried along by the Government as an incidental to the operation of the Government-owned synthetic rubber plants. But I am convinced that when the work is carried on in this way, with uncertainty as to its continuance, and therefore an unusually high personnel turnover, it is not nearly so effective as if it were part of an over-all coordinated scientific program supported on a more stable basis.

Another example of an agency of Government which has recently entered the field of science is the Department of State. It has established a science liaison office and looks forward to having scientific attaches serving in various of our embassies in leading capitals of the world. I believe that there is an important service to be rendered in fostering international relations in the field of science. But I do not believe this can be done effectively under circumstance where it is just one minuscule activity under the supervision of men who are so busy with so many other matters that they are unable to give it their attention. All such activities of the United States Government could probably be better handled by a general science agency, of the kind suggested.

Another recent venture in organization of Government science that many feel could be improved is the Research and Development Board of the Department of Defense. This was established by one section of the National Security Act of 1947, the law which established a single Department of Defense and was intended to be the means for bringing about a close coordination of the scientific research and development work sponsored by the Army, the Navy, and the Air Force. Experience has shown that it has not been a very effective tool for doing this. I think that this outcome might have been foreseen from the outset and for the reason that the Research and Development Board was set up as purely advisory body, without operating responsibilities. Operating responsibilities for research continue to belong to the three services and their individual bureaus. Because the RDB lacks direct responsibility it is not an attractive place for scientists of real ability to work, so it has been unable to attract staff of sufficient competence to cope with the very difficult problems presented by an extremely complicated situation. I am convinced that the RDB cannot perform a useful function as long as it functions in a purely advisory way, and that the situation could be greatly improved by putting all of the military research laboratories completely under civilian management of a Department of Scientific Research or a new Smithsonian Institution.

Next I would like to make a few observations on the Federal Government as an employer. Uncle Sam is a reasonably good employer so far as salaries, retirement plan, vacations, and the like are concerned. But the salaries paid for positions of major responsibility are in no way commensurate with the rewards which can be obtained in private life for similar services. Some curious inequities develop in this way. The tax position of many corporations is such that it costs the Government more in decreased tax revenues paid by the private employers to have a man work in private industry than the salary which the Goverment will pay that man to work full-time for the Government.

The curious thing about the low salary scale which the Government pays to scientists is that one way and another the Government is finding it impossible to compete with itself in securing the services of the men it needs. Many private employers of scientists use them on Government contract work on a cost-plus basis so the Government pays the man's full salary at higher than Civil Service rates as part of the cost of the contract. This possibility has, in the post-war years, led to a new development which is having devastating effects on the ability of Government operated laboratories to recruit qualified staff. More and more there is a tendency to assign Government research programs to *ad hoc* groups organized as private corporations solely for the purpose of taking a Government contract and even in some cases for the purpose of staffing and operating a Government-owned facility. In this way Government money is used to pay salaries in excess of Civil Service rates and all manner of operational red-tape is avoided, but the Government finds itself paying much more for the same services than it would pay if the work were done in its own laboratories This is not good for the morale of loyal Government workers. The proper remedy would be to improve the rules affecting the Civil Service instead of inventing ways to evade them.

Aside from questions of salary alone, some members of Congress so often give expression to attitudes of contempt and distrust toward the thousands of dedicated, conscientious, and intelligent citizens in the Government service that they have to be quite thickskinned indeed not to have their morale in some measure impaired by such treatment. No private employer would think of saying the kinds of things about his employees which are often said about Government employees and expect to retain any of their loyalty or devotion. This situation has been greatly aggravated in recent years by the use of dishonest smear tactics in Congress giving rise to an artificial hysteria which has led to widespread injustices toward Government employees carried out in the name of the Loyalty Program. Everyone in Washington knows dozens of stories of great suffering caused by silly and trivial accusations in this connection. For example only recently I heard of a labor relations expert who was employed by our Government in Japan to work to diminish communist influence in labor unions there, who was officially commended for his work, and then later had to defend himself before a loyalty board against the charge "that when you were in Japan you evinced a great interest in Communism". I know of a case of a woman who was accused of disloyalty on the grounds of sympathetic association with her own husband. I know of another who was charged with acquaintance with a scientist who is in fact the man who is entrusted with a major role in the hydrogen bomb development.

Not the least of the evils associated with the actual functioning of the Loyalty Boards is the slowness with which they operate. Often a person is kept in a state of nervous suspense for months after a hearing is held before he gets word of a decision. In general the processes are carried out in an altogether too formal and unsympathetic manner. No man can become a psychoanalyst until he himself has been analyzed. I think the situation would be improved if no one served on a Loyalty Board until he had a laboratory course in the Golden Rule by having himself been given a protracted experience with a Loyalty Board.

In conclusion let me thank you for your courtesy in listening to this rather too dogmatically expressed recital of opinions of one who sincerely believes in the importance of Government service and of science in the modern world of affairs and who only hopes that some of these thoughts may make a slight contribution toward working out some improvements. No one who has ever been entrusted with major governmental responsibility can fail to be impressed with the importance of the American government and of strengthening the American contribution to the welfare of all the peoples of the world. I am not an old soldier and I hope not to fade away. I leave the Government service happy in the friendships and experience it has given to me and hoping that I may still in private life be responsive to the duties of citizenship in this our beloved America.

A Half-Century of Quantum Physics

E.U. CONDON

Consulting Physicist
Berkeley, California

The presidency of the American Association for the Advancement of Science is one of the greatest honors that American scientists can confer on one of their colleagues, and I am grateful for the opportunity to have served in this post. This is a particularly happy occasion* for me, for the setting is the auditorium in Wheeler Hall, where as an undergraduate I heard many a lecture, and this is the 30th anniversary of the awarding of my A.B. degree in physics by the Unviersity of California, our host institution for this meeting of the AAAS.

Before and during my undergraduate student days, I used to work this campus as a reporter for San Francisco and Oakland newspapers. Thus began a long and pleasurable acquaintance with the local chairman of our meeting, President Robert Gordon Sproul.

Of course I cannot talk to you entirely from first-hand knowledge of a half-century of quantum physics. I was born in Alamogordo, New Mexico, in 1902, which was 2 years after Max Planck introduced the quantum idea into physics in Germany. Aside from the fact that I was pretty young at the time, I do not think Planck's theory attracted much attention at the time in Alamogordo. Now, of course, my home town is quite conscious of atomic physics, and its chamber of commerce has placed signs on the highways leading into town which proclaim that the town is the birthplace of atomic energy. They refer, of course, to the fact that the first atomic bomb was exploded near here in the summer of 1945.

I first heard of physics when I was 12 years old and bought a high-school textbook by Carhart and Chute for 15 cents in the old DeWitt and Snelling bookstore in Oakland. The following year I became pretty deeply involved in what we now call the atomic age. I had been reading *The New Knowledge* by Robert Kennedy Duncan, which was a popular book on atoms and radioactivity by the man who founded the Mellon Institute for Industrial Research. That was in 1915, the year in which San Francisco

---

This paper is based on the address of the retiring president of the American Association for the Advancement of Science, at the annual meeting, 28 Dec. 1954, in Berkeley, Calif.

celebrated the opening of the Panama Canal by holding the Panama Pacific International Exposition. Another boy and I discovered that the state of Colorado had as part of its exhibit a large pile of 10 or 20 tons of raw carnotite ore. This is a brilliant yellow sandstone, which today is being sought all over the mountain states by prospectors who rent their Geiger counters from local drugstores.

My friend and I managed, as boys will, to acquire enough carnotite from that pile to fill a shallow cigar box. With it we could take, using overnight exposures, shadow pictures of keys and other metal objects which were made by the gamma rays emitted by the radium and uranium content of the ore.

Once a boy of 13 has become this deeply involved with modern atomic physics, there is likely to be no hope that he is good for anything else. From then on it is impossibly hard for his teachers to get him interested in reading Gayley's *Classic Myths*.

Practically all the important progress in physics in this century is bound up with quantum ideas. Moreover, it has been a half-century in which physics has developed at a revolutionary pace that is totally unprecedented in the world's history. Therefore all that I can do here is to pass the main ideas in rapid review, perhaps lightening the story with an anecdote here and there, and hope to stimulate a wider interest in this exciting subject. Everything I say is well known to the physicists However, the ideas are complicated, and they may experience the academic delight of catching me in a mistake or two.

Quantum Ideas

By quantum physics we mean all parts of the science that involve a peculiar universal constant, known as Planck's constant, h, where

$$h = 6.55 \times 10^{-27} \text{ erg sec.}$$

So defined, quantum physics involves nearly all of physics and chemistry. It also involves a good share of astrophysics. Moreover, quantum ideas have required a good deal of searching into the philosophic foundations of physics.

The quantum idea was first introduced into physics in 1900 and 1901 by Max Planck in connection with the study of the radiations emitted by hot solid bodies. Throughout most of the 19th century, such radiation, including visible light, had been regarded as a wave motion. But, in developing the theory of radiation from hot bodies, Planck found it necessary to assume that light energy is not emitted and absorbed continuously by atoms. Rather he supposed that it was emitted and absorbed in definite little bundles of energy, or quanta.

Many experimental properties of light pointed to its being propagated as a wave motion. There is nothing remote or esoteric about these experiments. Take a silk umbrella and look through the fabric at a distant street light. In addition to a central white image, you will see a series of colored images extending out from the central image in two mutually perpendicular series in directions related to the warp and woof of the fabric. These are caused by interference of light waves which go through different interstices between the evenly spaced threads of the fabric.

A diffraction grating is an accurately made device for observing these spectra more accurately. By measuring the angle of spread between them and the central image, one can find the wavelength of the waves, and, by knowing the velocity of the waves, one can find the frequency or number of oscillations per second that occur as the wave passes a fixed point.

In this way, one finds that the wavelength for violet light is about 3×10^{-5} cm and that the wavelength for red light is about twice as great, or 6×10^{-5} cm. Thousands upon thousands of these wavelengths have been measured to at least 6 decimal places. These form the largest and most precise body of experimental data in all physics. Since the velocity of light is 3×10^{10} cm/sec, it turns out that the frequency of violet ligth is about 3×10^{14} cy/sec.

On Planck's view, light of frequency n cy/sec is emitted and absorbed in quanta of energy equal to hn, which is therefore about 6.5×10^{-12} erg for violet light. For X-rays the frequencies are some 10,000 times greater, and the quanta are therefore some 10,000 times greater.

The reasoning that led to this result was so complicated that Planck himself was not fully convinced of its validity. Physicists are all an extremely conservative group of people, at least in matters having to do with their own science, and they were reluctant to accept the radical quantum idea on such slender evidence.

In 1905 Einstein showed how clearly and neatly the main facts regarding the photoelectric effect could be understood if the quantum view of light were favored over the wave view. In the photoelectric effect, electrons are emitted from a metal when light shines on it.

Early experiments showed that increasing the brightness of the light caused more electrons to be emitted but did not increase the energy of motion with which the emitted electrons came out. On the wave view, one would think that a bigger wave would shake the electrons harder and make them come out with more energy.

Experiment also showed that the energy with which the electrons were emitted increased linearly as the frequency of the light was increased. This result was not at all understood in terms of the wave theory of light.

Einstein pointed out that on the quantum view, if 1 light quantum goes to 1 electron, then greater brightness means more quanta and therefore more emitted electrons. Planck had already found it necessary to suppose the energy content of a quantum to be proportional to the frequency of the light wave, and thus a natural explanation is provided of why the energy of the emitted electrons increase linearly with the frequency.

Wave-Particle Duality

Thus was born the wave-particle duality or dilemma of modern physics. Light, on going through a series of closely spaced slits, behaves in ways that have found only qunatitative explanation on the wave theory. Light, on falling on a metal, liberates electrons in ways that have found satisfactory explanation only in terms of the quantum or corpuscular theory. From here on, the subject began to develop at an ever-increasing rate.

When atoms are excited in a gaseous discharge tube, such as is used for advertising signs, the kinds of light emitted consist of sharply defined frequencies characteristic of the gas atoms inthe discharge tube. If light is emitted in quanta of definite amounts, this must mean that the atoms are capable of existing only in states of definite energy values. The differences in these allowed, or quantized, energy values are the energies of the light quanta emitted by an atom in passing from a state of higher total energy to one of lower total energy. In 1913 Niels Bohr built his successful theory of the hydrogen atom on a combination of this quantum idea with the general picture of the nuclear atom that had been developed experimentally by Ernest Rutherford. Soon afterward, James Franck and Gustav Hertz performed experiments in which they showed the reality of these quantized energy levels in atoms by finding that electrons can give up quantized amounts of energy to atoms only on colliding with them, and that these quantized amounts are closely correlated with the sizes of the emitted light quanta.

In 1912 another discovery of major importance was made. Since the discovery of X-rays in 1896 by Wilhelm Roentgen, there had been speculation on whether these were a wave motion or a stream of corpuscles. Attempts at diffraction experiments gave negative results with a sensitivity indicating that, if they are a wave motion, the wavelength cannot be more than about 10^{-8} cm. This is just about the distance apart of layers of atoms in a crystal, which gave Max von Laue the idea that perhaps the regular arrangement of atoms in a crystal would diffract X-rays in the way that the rulings of a diffraction grating diffract light. The experiment was successful. Thus two new branches of physics were born. By use of a crystal of known structure, it was now possible to measure the wavelengths of the characteristic X-rays emitted by various atoms, so spectroscopy was extended to the X-ray region. By use of X-rays of known wavelength, it was possible to infer from the nature of the diffraction pattern how the atoms are arranged in crystals of unknown structure. Thus a powerful tool was provided for the study of the structure of solid matter.

All this served to point up the disturbing puzzle of the dilemma on whether X-rays and light were really a wave motion or really a stream of corpuscles, for it seemed to be something like both and yet on one could see how it was possible for it to be both in any sense. Only W.H. Bragg, writing in *Nature* in late 1912, hinted at a combined outlook. He wrote:

The problem then becomes, it seems to me, not to decide between the two theories of X-rays, but to find . . . one theory which possesses the capacities of both.

On Bohr's model of the atom, the electrons revolve around the nucleus like planets going around the sun in the solar system. Although the theory was immensely successful in correlating spectroscopic facts, it threw no light on the fundamental nature of the valence forces that hold atoms together in molecules. In Berkeley, G.N. Lewis developed a rival theory based on a static model of the atom in which electrons had favored locations at the corners of a series of cubes surrounding the nucleus, the eight corners corresponding to the length of the short periods in the periodic system of the elements.

When I entered the University of California as a freshman in 1921, the Bohr atom was being taught in the physics department and the Lewis atom was orthodox doctrine in the chemistry department. Now both departments are preaching the same kind of atom, which resembles neither of its forerunners and combines the best features of both. The things I am talking about are so old that if they are mentioned anywhere it is probably in the history department.

The early 1920's were an exciting time to be studying physics. We had these rival atomic theories, each with its inadequancies and uncertainties. Some things were lacking. In Livermore, California, there was only the rodeo and on Charter Hill nothing but the Big C and a few grazing cows.

In 1923 the wave-particle dilemma became even more acute. Arthur Compton, in St. Louis, discovered that X-ray quanta have momentum as well as energy. When X-rays are scattered by matter of light atoms, it is found that some of them are scattered, but that the scattered X-rays consist of smaller quanta than those which struck, and the shift toward smaller quanta is greater, the larger the angle of deflection through which the X-rays are scattered. All this was exactly in accord with the idea that the X-ray quanta were scattered by colliding with electrons by exactly the same rules of conservation of energy and momentum that are applicable to the collision of two material particles, such as billiard balls.

In that same year, 1923, Louis de Broglie in Paris published his now-famous doctor's thesis, in which he suggested that the wave-particle duality might extend to the behavior of electrons as well as to light and X-ray quanta. Up until this time physicists felt sure that a beam of cathode rays was simply a corpuscular stream of electrons moving in accordance with Newton's laws of motion, as corrected in the high-energy region for relativistic effects.

De Broglie suggested that the relationship between the wavelength of the wave aspect of an electron and the momentum of the particle aspect of the electron ought to be the same as that already found to hold for X-ray quanta, namely, that wavelength equals Planck's constant divided by momentum. This suggestion made possible a simple interpretation of the existence of discrete energy levels in atoms, which in Bohr's theory was simply postulated in order to get agreement with spectroscopic facts.

We are all familiar with the fact that a stretched string in a musical instrument vibrates freely at a particular frequency such that the length of the string is just equal to half a wavelength of the wave of that frequency which might travel on the string. Then it can vibrate also at double this frequency, so the length equals two half-wavelengths, or at triple the fundamental frequency so the string's length equals three half-wavelengths, and so on. Similar rules apply to the modes of vibration of other continuous bodies such as the stretched membrane of a drum. De Broglie argued by analogy that, if the motion of electrons was somehow governed by an associated wave motion, then the allowed orbits in an atom must be governed by mathematical restrictions similar to those which determine that vibrating bodies can vibrate only in a certain discrete set of modes of vibration.

It turns out, on these views, that the de Broglie wavelengths of electrons which have been accelerated by a potential drop of a few hundred volts will be of the same order as that of X-rays. This suggests that electrons, too, ought to show diffractive scattering by the regularly spaced layers of atoms in a crystal. In 1927 electron diffraction was discovered in New York by C.J. Davisson and L.H. Germer, working with the scattering of low-energy electrons by a single crystal of nickel, and independently that same year by G.P. Thomson in England, who worked with the scattering of higher energy electron beams by polycrystalline materials. These experiments fully confirmed the idea that electrons are scattered from crystals like a wave motion having the wavelength that was predicted by de Broglie. At the same time a new tool for crystallographic studies, supplementing that of X-ray diffraction, was made available.

A few years later it was shown experimentally that beams of hydrogen molecules and of helium atoms were also governed by de Broglie wave principles when scattered by crystals. This was done by Otto Stern, now a distinguished resident of Berkeley, who was then professor of physics in Hamburg, Germany.

In consequence of these experimental discoveries and many associated theoretical developments, physicists now believe that the wave-particle duality applies to all things in nature, be they light quanta, electrons, protons, or entire atoms and molecules. With larger things, the wavelength becomes so small that the wave aspect escapes observation, which is why all ordinary motions appear to be governed entirely by the particle formulation originating in Newton's laws of motion.

Matrix Mechanics

In 1925, Werner Heisenberg in Göttingen discovered a new mathematical way of treating problems in atomic physics. It was called matrix mechanics because quantities which in Newtonian mechanics are represented by ordinary numbers are represented in this theory by an abstract kind of mathematical entity known as a Hermitian matrix.

This theory caused physicists a lot of trouble. Up to then practically none of them had ever studied matrix algebra. It is true that the mathematicians knew about matrices but, under pressure from the physicists to teach them only what they needed to know, the mathematicians had not talked about matrices when physicists were around. Max Born, the 1954 Nobel prize winner in physics, was in Berkeley from Göttingen in 1925 as a visiting professor, and he lectured on matrix mechanics. What a rough time he gave us as we tried to grasp the strange new ideas of matrix mechanics.

Then, in the spring of 1926, what a relief it was when Erwin Schrödinger's rival wave mechanics came on the scene, and we could avoid the difficulties of matrix algebra. And what a surprise it was in the summer of 1926 when Carl Eckart, in Pasadena, and also Schrödinger himself discovered that the two theories were identically the same. They were simply dressed up in such totally different mathematical costumes that it took some time before their identity was recognized.

In the early fall of 1926 I left Berkeley to study the new quantum mechanics with Born in Göttingen. There the great mathematician, David Hilbert, used to delight to tell us how he had told the Göttingen theoretical physicists of the close relationship between matrix algebra and certain boundary value problems of differential equations. If they had followed up this lead they might have discovered wave mechanics before Schrödinger.

In those days Hilbert used to say, "Die Physik wird zu schwer für die Physikern" — physics is becoming too difficult for the physicists.

In 1927 the pace of discovery in theoretical physics was probably greater than in any other year in the history of the science. Every issue of the leading journals had at least one paper of great importance. There was the more general formulation of the laws of quantum mechanics that was made principally by P.A.M. Dirac in England and John von Neumann in Germany. There was the development of the quantum theory of the radiation field by Dirac and the relativistic form of the quantum theory of the electron, which led to the prediction of the existence of the positively charged electron or positron, that was discovered a few years later by Carl Anderson in Pasadena.

Arnold Sommerfeld laid the foundations for the whole modern theory of metals and semiconductors by applying the quantum mechanical methods to the treatment of the free electrons in a conductor. W. Heitler and F. London applied quantum mechanics to the theory of the covalent chemical bond between two hydrogen atoms and showed that this atomic theory could at last meet the needs of the chemists. This gave rise to a wide program of developments, which resulted in the award of the 1954 Nobel prize in chemistry to Linus Pauling of the California Institute of Technology.

Heisenberg showed how the new quantum theory could account for the extremely strong interactions between the electrons in iron, cobalt, and nickel, which give rise to the strong magnetic effects shown by these elements. Many other discoveries of great importance were made among which may be mentioned the final clarification of the low-temperature heat capacity of gaseous hydrogen. It had long been known that this had something to do with quantum restrictions on the rotation of hydrogen molecules, but David M. Dennison showed the solution of this problem leading to the discovery of two stable forms of hydrogen gas known as orthohydrogen and parahydrogen.

Things were happening at such a pace that all the physicists, young and old, were suffering from acute mental indigestion. In the spring of 1928 when I taught a course in quantum mechanics for the first time at Columbia University, I remember that the late Bergen Davis summed it all up by saying,

I don't believe you young fellows understand it any better than I do — but you all stick together and say the same thing.

Statistical Theories

Going back a bit, it was in the fall of 1926 that Max Born took a decisive step in supplying the hypothesis that provided a general basis for interpretation of the mathematical formalism of quantum mechanics. We had a mathematics of wave motion that was somehow associated with the motion of the electrons or other atomic particles. The big question was What is the basic relationship between the associated wave motion and the behavior of the atomic particles?

Born's answer, which was largely the basis of the award of the 1954 Nobel prize in physics to him, was that the theory does not and cannot make precise predictions about the motion of the particles, but that it can make only predictions about the relative probability of appearances or motions of different kinds. In particular he postulated that the square of the amplitude of the de Broglie waves at a particular place gives the relative probability of finding a particle in that place. This is a radical and revolutionary idea in its implications, and fundamental disputes among physicists still rage regarding its basic meaning. Nevertheless, it must be realized that this idea of a statistical interpretation of the wave as describing probabilities of behavior of the particles has now stood the test of time for more than a quarter-century and lies at the foundation of all modern atomic physics.

Physical science got its start with the precise astronomical predictions resulting from the dynamical theory of the solar system. These many quantitatively verified results exercised a dominating influence on physical thinking. All physics was assumed to be reducible to a fully deterministic description of motions, such that, given a full description of the situation as of now and sufficient calculating skill, one could calculate precisely what will happen at all times in the future.

Prior to 1926 statistical theories had been used in physics. Statistical methods were used to give an over-all average description of the heat motions that give rise to the thermal properties of matter. But in all such theories it was supposed that there really exists an underlying fully deterministic reality, and that statistical methods are used by choice for simpler descriptions rather than by fundamental necessity.

The questions now arise: Is there really an underlying fully deterministic description of the phenomena of atomic physics that has so far eluded our observations and theory-making because of some basic incompleteness that may be remedied in the future? Or, on the other hand, is that some inherent limitation in the world and our possible ways of observing it such that our knowledge of events is fundamentally restricted to observations and conclusions of a statistical character?

In the fall of 1927, Heisenberg provided an analysis of the processes of measurement that strongly favors an affirmative answer to the second question. Later analysis by Bohr in 1928 extended these ideas. The essence is that on an atomic scale the processes of observation necessarily introduce uncontrolled disturbances, and it is these which give rise to the over-all uncertainties that make fully deterministic knowledge impossible. If one refrains from observing, he makes no disturbance but remains ignorant of the data needed for deterministic calculations. Observations can be arranged in ways that increase the precision of knowledge of one variable but only at the price of introducing more uncertainly into the knowledge of a complementary variable.

The analysis of Heisenberg and Bohr provides a deep insight into the nature of limitations on knowledge of deterministic behavior, which seem to be truly fundamental. Most physicists today accept these views and regard the statistical element of the theory as an intrinsic feature of the world in which we live. Classical determinism on this view is an ideal limit toward which our knowledge can approach in large-scale phenomena where the quantum limitations become unimportant corrections.

But one physicist of outstanding importance steadfastly thinks otherwise. He is Albert Einstein. At the very outset he expressed himself by saying "Der lieber Gott würfelt nicht." In American vernacular we would say "the good Lord doesn't shoot craps."

Born's book, *Natural Philosophy of Choice and Chance* quotes a letter Einstein wrote in 1947 in which he says, "the statistical interpretation . . . has a considerable content of truth." However, he goes on to say

I am absolutely convinced that one will eventually arrive at a theory in which the objects connected by law are not probabilities, but conceived facts as one took for granted only a short time ago.

With characteristic modesty he concludes then by saying

Zur Begründung deiser Überzeugung kann ich aber nicht logische Gründe, sondern nur meiner kleinen Finger als Zeuger beibringen, also keine Autorität, die ausserhalb mainer Haut irgendwlechen Respect einflössen kann. [I cannot provide logical arguments for my conviction but can only call on my little finger as witness, which cannot claim any authority to be respected outside my own skin.]

Whether all the data of experience can be codified in terms of fully deterministic relations I do not know, of course, but unquestionably it is useful to have such organization of knowledge carried to its furthest limits. The history of science is filled with facile generalizations and the kind of oversimplification that fails to qualify what would be true if properly qualified. Think of the many pages of disputatious writings on free will and determinism!

In my view physics has nothing to say on this one way or the other as an issue related to human conduct. It was an unwarranted extrapolation in the first place to pass from the planetary successes of classical mechanics to extreme mechanistic determinism for human actions. It is equally incorrect to argue from the statistical determinism of quantum mechanics any support for the idea of free will in human behavior.

Nuclear Physics

By 1927 the principles of quantum theory as we know them today were pretty well developed. In the 27 years since then the ideas of quantum physics have been so closely identified with all the progress that has been made in physics and chemistry that it is not possible to discuss quantum physics separately from progress as a whole in these sciences.

The entire theoretical structure of nuclear physics is cast in quantum mechanical terms. This new branch of physics has never been handled in any other way. The application of quantum mechanics to problems of the internal structure of the nucleus was initiated in 1928 with the discovery of the theory of alphaparticle radioactivity by George Gamow in Göttingen and independently by the late Ronald Gurney and myself in Princeton.

This theory provides one of the most extreme examples of the use of probability ideas. According to classical mechanics, it is not possible for a particle to be in places where its total energy is less than its potential energy. In quantum mechanics this impossibility is changed into an improbability. An alpha particle in a uranium nucleus collides with the wall surrounding the nucleus some 10^{20} times a second. According to quantum mechanics it has a very slight chance of getting through the wall, even though it does not have energy enough to get over it. This chance is extremely small, amounting to only about once chance in 10^{36}. In consequence, the alpha particle remains in the nucleus on an average about 10^9 years before the spontaneous disintegration occurs. Nevertheless, the statistical feature of the theory shows up in that some uranium atoms disintegrate in a very short time, whereas others have lasted for many thousands of years without disintegrating.

This same theory of barrier leakage applied in reverse indicated that light elements could be made to undergo artificial transmutations using particles accelerated with voltages much lower than had been estimated to be necessary. This gave a strong stimulus to the experimental investigation of nuclear reactions which began in the early 1930's.

Quantum mechanics has also provided the concept of saturable exchange forces between fundamental particles, an idea that is foreign to classical ideas but appears to be essential in the further development of the theory of nuclear structure. Relativistic quantum mechanics, as I have already mentioned, provided the prediction of the existence of the positron and provides the theoretical basis for calculations of many of the basic processes that occur in the region of high-energy physics — that is, the physics of particles having energies of several hundred million to billions of volts, a branch of physics that is extensively studied here in Berkeley.

From quantum mechanical theories concerning exchange forces between protons and neutrons, H. Yukawa in 1936 was led to postulate the existence of a hitherto undiscovered kind of particle, called the meson, intermediate in mass between the electron and the proton. Experiments in recent years have shown that there are in fact many kinds of mesons, with complicated interrelationships, whose study is today one of the most important topics in fundamental research in physics.

In spite of all these successes and many others too numerous to mention here, the record is not one of complete success. Very early in the modern period, namely in 1927, Dirac took the decisive steps toward the development of a quantum theory of the electromagnetic field and had a number of significant successes with the theory as he developed it. Heisenberg and W. Pauli extended the theory, and many others have worked on it.

This theory, or rather this family of theories, in various forms, however, suffers from a fatal defect that many of the important problems of physics have no solution. When the solution is carried out, they lead to divergent integrals that give infinity for a formal answer to a problem that ought to have a finite solution. A large amount of study has gone into efforts to remove this difficulty but with little success. Therefore the quantum theory of the electromagnetic field remains today in an unsatisfactory state. Probably the difficulties can be overcome only by some radical revision of the fundamental ideas that is as revolutionary in its nature as the ideas of the present theory seemed when they were first developed in 1927.

The past half-century has been an exciting period of enormous fruitfulness in the development of physics and chemistry. Today a greater effort, measured both by adequacy of the equipment and numbers of well-trained men, is going into the investigation of the fundamental nature of matter than ever before in the world's history. We may expect therefore that the next 50 years will bring a development of our knowledge and our ideas that is even greater than has occurred in the first half of the present century. If this happens, the physics of the year 2000 will be as strange and unforeseeable by us today as the physics of today would have seemed to the physicists of 1900.

What is this New World?

EDWARD U. CONDON

May I begin by complimenting your program committee on the vision which led to "Values We Live By — Choices We Make" as the general theme for this meeting. In this same spirit I thank them for assigning me such a big topic as "What is this New World?". I could never have had the temerity to suggest that I could deal with such an important subject, but having been asked to do so, I have found it challenging to try.

Everything that can possibly be said on this subject is constantly being said, over and over again, in newspaper editorials, in magazines, in books, from the pulpit, in college and high school classrooms, in discussions everywhere. Therefore, mine is a task of selecting, of sorting out, and by such selection giving emphasis to the things that seem to be of primary importance for us.

It is well known that the methods of empirical science as applied during the past three centuries to the study of the physical world in which we live, have given us a power to harness natural forces to human purposes far beyond the wildest dreams of people in ancient times or in the Middle Ages. This is the first big general fact of the new world.

The scientific knowledge thus far accumulated has produced a technological revolution which has given us electric power, rapid transportation and communication, improved food production and preservation, improved understanding of medicine, health and sanitation and many other things usually regarded as good. These vast changes in our way of life have necessitated new forms of social organization, especially in large-scale corporate enterprises, for getting human beings to work together. And the products of such new technical enterprises have made possible changes in social organization, especially the large-scale urbanization of life.

Moreover this scientific and technological advance has given a vast new content to our job as teachers, for today no one can claim to be cultured or educated who remains ignorant of the principles which have given rise to these changes.

Whether we think that this movement in human history is good or bad there is no denying the reality of it or the vastness of the change in the pattern of human life that it has produced. I will venture the opinion that man's increased power over the forces of nature is neither good nor bad in itself. Knowledge has given us the power: What is good or bad is the wisdom or lack of wisdom we show in how we use that power. Knowledge can be condemned as bad only on the premise that men are so depraved that they will make bad uses of the power it gives them. Discouraging as things may seem at times, I do not think we are justified in taking such a pessimistic view of the situation.

We have only to look back a little more than one hundred years to reach the time when electricity was just a plaything and a curiosity, not yet applied to any human purpose, when there was no anaesthesia in the operating rooms, when diseases like typhoid fever, now nearly extinct in our country, were able to wipe out whole towns, when infant mortality was enormous, and there was essentially no rational basis for the treatment of mental illness. Thus a vast amount of progress has been made in the past century and a half in learning how to improve the conditions of human life.

The second major fact about this new world is the appalling non-uniformity of distribution of the benefits of such knowledge among the peoples of the world. This non-uniformity manifests itself in two ways: non-uniformity as between individuals or families or social classes inside a particular political unit such as the United States of America, and an even greater non-uniformity as between peoples belonging to different geographical regions of the earth. If we believe that there is a power for good in the applications of scientific knowledge to improvement of human life, and if we believe, as I think most of us do, that human life is sacred, then we must feel terribly dissatisfied that such non-uniformities continue to exist. And we ought to be giving all possible attention to efforts for the removal of these inequalities.

For terrible inequalities do exist and many of our social institutions have been largely engaged in efforts not to eliminate, but to maintain them. Their existence in turn is a principal cause of the frustrations and perversions of human aspirations which lead in children to bad behavior, in adolescents to juvenile delinquency, in adults to crime, and in nations to war.

Another important aspect of this second major fact is that everywhere people who are deprived by these inequalities are beginning to struggle for their correction. I am ashamed of the way we have treated the Indians and Negroes throughout our national history, and I am grateful for the forbearance under suffering with which they have sought to overcome these wrongs without resort to violence. But these peoples are becoming aware of their political power and they will not much longer remain quiet under injustice. We have already waited too long and we ought not to delay any longer the adoption of changes that give full opportunity to all of the depressed elements in human society.

Similarly, in the international scene, the history of this crumbling of empires and the increasing self-determination of colonial subject peoples.

This great movement began with our own establishment, less than two centuries ago, of the free and independent United States of America. Soon after that the countries of Latin America threw off their subjection to the thrones of Spain and Portugal. But the great colonial empires by which Europeans controlled the fortunes of vast populations in Asia and Africa have remained. With the freeing of India and Pakistan and Indonesia and the Philippines, the past decade has seen the end of most of the colonialism of Asia. The process of freeing the native populations of Africa is already beginning and will surely be complete long before the end of this century.

These tendencies represent progress toward the wider fulfillment of a better life for all of humanity. They are therefore good. What I find bad in the present situation is the large amount of effort that is being made by those who have hitherto benefited most from our technological riches to keep those riches from the depressed peoples of the world.

This is bad not only for the present evil in retarding the progress of our fellow men rather than offering a helping hand, but it is worse in that such conduct is storing up all kinds of difficulties for the future. If the depressed peoples have to make a violent struggle to gain the freedoms and opportunities that we ought gladly and freely to give them, this will generate another long-lasting period of hatreds and resentments among men. We of the white race are bound to be severe losers in this, for equality of freedom and opportunity is surely coming to all mankind, and if things continue as they have in the past, we may find ourselves a weak and fiercely hated minority among a vast majority of peoples of other races. I do not like to base an appeal for us to do right on fear of the consequences of doing evil, but it is nevertheless an element of this situation that is not to be ignored.

What is there about our present situation that makes men so ungenerous? I think it is that we do not understand the richness of productive power which the new scientific technology has given us. If I have only one apple, and not a very big one, and my mother makes me give half of it to the neighbor boy, so that after I have eaten my half my apple-appetite is quite unsatisfied, it is not commendable, but at least understandable why I resist being generous. But that is not the correct analogy. We have apple orchards that can produce more than enough apples for all of our people, and we can supply others with seeds and stock and cultivation techniques so that they too can have all the apples they can eat. We are in a position whereby we can do things which are morally and ethically right in improving the lot of others, and without depriving ourselves of anything.

On the contrary we already have almost too much productive capacity, and we can help others by sharing our knowledge of the way to use scientific technology to enable them to improve the conditions of their lives.

The second great fact of the present is then that the non-uniformities among the peoples of the world of the benefits of knowledge of scientific technology are bound to be reduced or largely eliminated in the next few decades, *and* that it is our great opportunity to take the initiative in bringing about this better state of affairs, rather than to strive for the preservation of the explosively unjust conditions of inequality.

The method by which science advances technology is new. We have by no means exhausted its possibilities for improving the human condition. Every year new discoveries in science are made which make possible undreamed of advances in technology and in health. Put this fact with the second fact and we can count ourselves as fortunate that we live in a time when the potentialities for human progress are so great. Never before has our capacity and power for good been as large as it is today. But there are also dark clouds, not merely on the horizon, but now hovering threateningly near us. The third great fact of the new world is that never before in human history have men had so great a power to destroy as they have today.

Throughout all of human history, men have resorted to violence to settle their disputes. This method has caused a vast amount of suffering and unhappiness and destruction and these evil consequences are not confined to the times and places of direct violent conflict. Before and afterward, the peoples of the world have lived in a spirit of hatred and hostility, devoting a large part of their productive energies to repairing the damage of the last war and preparing the means to be used in the next. They have raged and hated and have even dishonestly twisted and contorted the ethical doctrines of their religions to give holy approval to their evil conduct.

Over the centuries there has been a steady development — I dare not call it progress — of the technical means for carrying on warfare. Military units increase in size and complexity. Weapons increase in power and range. The two World Wars of this century were vastly more destructive than any previously known. In 1945 the second World War ended abruptly after two atomic bombs had been used in Japan by America to kill more than 100,000 people and to injure another 100,000. In Hiroshima over 90 percent of the doctors were casualties and a month after the bomb only 30 were able to perform their duties. Out of 1780 nurses, 1654 were killed or injured. Only 3 out of 45 civilian hospitals survived and two large army hospitals were made unusable.

Many people once knew these things, and the lesson they teach of the complete futility of so-called civil defense preparation, but also they seem to have a remarkable talent for forgetting them. Those two bombs were of a type now described as "nominal," although they each had explosive power equivalent to that of 20,000 tons of TNT. The Nagasaki bomb used plutonium. The Hiroshima bomb used U235. That the casualties of the second were smaller is due to unevenness of the terrain in Nagasaki.

Both these bombs are now in sense obsolete in that the hydrogen bombs now available have an explosive power equivalent to 15 million tons. The total weight of all bombs dropped on all European countries in World War II was 2.7 million tons. Therefore *one* such hydrogen bomb is equivalent to more than five times all the bombs used in Europe in the entire war. No wonder it is making the British nervous that our Air Force overseas is fly around over them with live hydrogen bombs.

It is not enough to compare these bombs with ordinary TNT, for TNT produces destruction only through the blast of the pressure wave. These bombs also kill people by immediate radiation damage and by delayed radio-active poisonings. Bombs of this kind now exist as part of the regular military armaments of the United States, Russia and Great Britain, to name the countries in the order of their having developed them. These weapons are so large and produce such vast quantities of radio-active poison that such poisons have now been found in the bones of people, especially growing children, all over the world. It is no exaggeration to say that many thousands of persons in the world will suffer agonizing death from bone cancer and leukemia as a result of the poison put into the earth's atmosphere by the hydrogen bomb tests that we have already conducted.

This fact is a terrible indictment of our civilization and of the governments that have accomplished it. Perhaps this situation is tolerated because these deaths will occur in the future over twenty or thirty years and there is no way of identifying which individuals die because of fall-out poisoning.

Within the last year of two the armaments race has gone further in that Russia has given evidence of being able to fire long-range missiles of a size large enough to carry one of these bombs. We have spent a billion or so on a radar network across Canada which is now obsolete because it is unable to detect the passage of the Russian satellites. Our response to these developments is to rush more billions of dollars to the aircraft manufacturers. We have even begun to talk about improving science teaching in the schools for this wrong reason, although money for this may come slowly if at all.

I have a copy of the St. Louis *Post-Dispatch* for Sunday, April 7, 1946, almost twelve years ago, in which the Acheson-Lilienthal report on international control of atomic energy had just been made public. It contains a map of St. Louis showing what vast destruction would be done to this city if seven atom bombs were dropped on various parts of it. All that and much more too would be done today if just one hydrogen bomb — the equivalent of 1000 such atom bombs — were to drop on the city. Take a map of any city and draw a circle with a radius of 10 miles if you want to see what will happen when one such bomb strikes.

Without going into further details, let me say that all of these large-scale atomic weapon developments simple add up to this: The third great fact of the new world is that a major war in the future will mean the total destruction of civilization. It may not mean the complete destruction of every living thing on earth, for that depends on how many bombs of the present style are exploded before the war ends. It might very well mean the destruction of all life. It will certainly mean the destruction of a large part of all life and making the earth almost if not quite an unfit habitation for life for many years to come.

It follows that the most urgent problem before us is to make sure that everybody in the world learns the dangerous situation in which the human race finds itself. People must be made to understand that the next war would be so different from all previous wars that it is dishonest and misleading to use the same word for it. We should all learn not to refer to it as the next war. We must learn to refer to it as the end, the termination of all hopes and aspirations, of all life and love and beauty; oblivion, finis, the great and final blasphemy, the utter negation of everything good.

Ever since the existence of atomic bombs was made public in August 1945 by their use against Japan, a large number of scientists have been trying to educate the public about the true situation. In this work they have had to battle against great obstacles, for there are strong political forces in this country who do not want the public to know how terrible the situation really is. Although I think that secrecy of the kind they seek to maintain is a kind of subversion against our form of government, I at least give these misguided people credit for worthy motives. Nevertheless I think their analysis of the situation is wrong, and that it can lead us to destruction.

These people believe that our political differences with Russia are so irreconcilable, and that the Russian government is so completely untrustworthy, that there is no possibility of making any kind of agreements that lead safely to disarmanent and world peace. They think that there is no security, no future for humanity, other than in the indefinite continuance of an ever-intensifying, all-embracing and all-consuming armanents race in which we must and we shall always "keep ahead." This is called deterrence.

Against this view I would urge the following considerations:

1. That its proponents do not face the fact that failure of the policy will lead to final destruction off all humanity.

2. That there are very grave risks of a catastrophe by accident or error when vast destructive forces are kept on a continuous hair-trigger alert.

3. That it has already been demonstrated that we are unable to "keep ahead," and that right now and for some time in the future we are living at the mercy of the Russians, just as they have been living at ours for the past twelve years.

4. That the policy makes no provision for any future other than continuous intensification of its own terrible imperatives.

Fortunately I think there are evidences that the extreme advocates of this policy are beginning to discredit themselves before the world. They have tried to keep the facts from public knowledge but they have not been successful. As we learn more and more of the truth, we learn to have less and less confidence in the men who try to hide it from us.

On March 1, 1954, he United States tested a 15-million-ton-equivalent fission-fusion-fission bomb. Eighty-five miles away the crew of the Japanese fishing vessel, the *Lucky Dragon*, received radiation burns which resulted in the death of one of them and more than a year of hospitalization for all 22 of them. Marshall Islanders and Americans 330 miles away also were affected by radiation. From the beginning our government officials in the Atomic Energy Commission followed an uncandid policy, minimizing and belittling the seriousness of the fall-out dangers.

During the next two years a similar policy was followed with regard to minimization of the health hazards due to world-wide strontium 90 fall-out from bomb tests. Nevertheless the story could not be hidden. Thanks to scientists of high reliability, with no government connections, more and more people learned of the fall-out dangers. Last spring a petition urging world-wide agreement to discontinue H-bomb tests was signed by over 200 American scientists in an impromptu movement with no organization behind it. Later in 1957 over 9,000 scientists the world over signed this same petition, which was filed with the United Nations secretariat by Professor Linus Pauling about two months ago.

Concurrently we have seen repressive and diversionary tactics used by government officials. In 1955 Professor H.J. Muller, a geneticist of Indiana University, who was awarded the Nobel Prize for his discovery of radiation-induced mutations, was not allowed to present a paper on genetic radiation hazards at the Geneva Atoms-for-Peace conference. In 1957 Professor Pauling was subpoenaed by a Senate Committee, with resultant newapaper headlines of insinuation that the scientists' petition was subversive in intent. In the summer of 1957 a major diversion was provided by a White House announcement following information from two AEC scientists: They could make a "clean" bomb, that is, a bomb that could kill a million people by direct blast but only a smaller number by the greatly reduced radio-active poison generated. This was contingent on the continued testing of H-weapons for the next four or five years.

The clean bomb diversion was soon to be followed by another: the tests could be carried out in a way that could not be detected. In the mean-time the movement to end further H-bomb testing had gained strong support in the London disarmanent meetings of the summer of 1957 in which we were represented by Harold Stassen. The headway towards a negotiated agreement to end bomb tests, with mutual inspection by the nations involved as a safeguard against infringement, has been confused by speeches and public statements raising doubts about inspection procedures, and again by an action of the AEC in early March. In September 1957 the AEC made an underground test in Nevada for a small atomic bomb, less than one-tenth the explosive equivalent of the one used on

Hiroshima, hence about one ten thousandth that of a modern hydrogen bomb. One of the objects of the test was to find out how far away earthwaves from this test would appear on seismographs. On March 6 the AEC issued a statement to the press that 250 miles was a maximum distance found. The correctness of this statement was disputed by scientists and it soon became clear that the statement was inaccurate. Many seismographs at much greater distances had recorded the test, including one 2,320 miles away in Alaska. The AEC issued a correction to its March 6th statement.

We now know that the March 6th statement had been sent to Dr. Edward Teller, often the unofficial spokesman of the AEC, and others at Livermore. There the mistake had been recognized and its deletion recommended, although by inadvertance, this was not done. However, its deletion, without the added acknowledgement that an atom bomb explosion of one ten thousandth the power of a modern hydrogen bomb had been detected 2,320 miles away, would have left the public grossly uninformed. Furthermore to have acknowledged this would have done a great deal to upset Dr. Teller's assertion that H-bombs can be tested without detection. Senator Humphrey's statement that scientific facts may be used "to prove a political point" finds endorsement by many thoughtful persons. Public statements must be scrutinized very critically before being taken at their face value.

The fourth great fact of the new world is that the world can not survive indefinitely with the deterrence policy, and that whatever temporary security it may give us is only a short reprieve and that therefore the most important task before us is the negotiation of agreements that will get the world out of the position it is now in. The nations of the world have solemnly renounced war as an instrument of national policy any number of times, after which it seems to me the military expenditures always go up.

But now we are having our last chance. This time we have to mean it. We have first to make disarmament agreements containing suitable technical safeguards against infringement, which I believe to be entirely practicable, and next we must initiate an orderly process of disarmanent and a discontinuance of the development of ever more horrible weapons.

In the first few years after World War II we used to hear a great deal about proposals for world governments, but as the cold war fears intensified it become more and more difficult to arouse interest in or support for such proposals. Some kind of strong international political agency is required for successful administration of the disarmanent agreements which must be made. This ought to take the form of a United Nations police force and inspection agency which as soon as possible would take the place of all national military forces other than a few military bands and color guards for use on ceremonial occasions.

The fourth great fact is then that these things will be done because they must be done. These is no other way to survival and I simply will not believe that all that is good and beautiful on this earth is going to end in a hydrogen bomb holocaust. Surely the peoples of the world will require of their governments that military force be put completely aside and surely all of out super ballistic hyper electronic fission-fusion-fission missiles will become as obsolete and as ridiculous as medieval suits of armor seem to us now.

But before will come to pass we must make sure that everyone is aware that survival is impossible if we continue long on the road we are traveling now. This is a major task for our educational institutions, our churches, and every means we have for deepening our political and ethical insights.

Finally, the fifth great fact of the new world which is reserved for the future. When suspicion and fear and hatred shall have been replaced by trust and friendliness and brotherhood, then all the vast resources of spiritual and material energy which we now waste can be applied to a cultivation of human welfare and happiness among all the peoples of the world. Let us recall the words of the Prophet Micah:

"But in the last days it shall come to pass
That the mountain of the house of the Lord
Shall be established in the top of the mountains
And it shall be exalted above the hills;
And people shall flow unto it.
And many nations shall come, and say,
'Come let us go up to the mountain of the Lord,
And to the house of the God of Jacob;
And he will teach us of his ways,
And we will walk in his paths':
For the law shall go forth of Zion,
And the word of the Lord of Jerusalem.
And he shall judge among many people,
And rebuke strong nations afar off:
And they shall beat their swords
into ploughshares
And their spears into pruning hooks:
Nation shall not lift up a sword against nation,
Neither shall they learn war any more.
But they shall sit every man under his vine
And under his fig tree;
And none shall make them afraid:
For the mouth of the Lord of hosts hath spoken it.
For all people will walk every one in the name of his god,
And we will walk in the name of the Lord our God for ever and ever."

Graphical Representation for Unit Systems

E.U. CONDON

Department of Physics
Washington University,
St. Louis, Missouri

This paper presents some ideas concerning physical systems of units, and a graphical representation of them, which has been found useful in teaching the subject.

We are all quite familiar with the usual presentation whereby various physical quantities are assigned dimensional formulas, usually in terms of three basic quantities, length (L), mass (M), and time (T). Thus, the definition of a physical quantity leads to a statement of its dimensional formula expressed in the form

$$(F) = (\mathcal{L}^\lambda \mathcal{M}^\mu \mathcal{T}^\tau).$$

Often it is not made clear what such a dimensional formula means and sometimes a certain mystical aura surrounds the concept of "dimensions" of a physical quantity. Statements are often made which imply that there is something absolute about a physical quantity which defines its dimensional nature as an intrinsic attribute of the quantity.

The point of view taken here is opposed to this. Instead, the statement of dimensions is taken to refer to the field of possible changes of units which are being admitted to the discourse, and this choice is arbitrary and a matter of convenience. Thus, when an author says "the" dimensions of F are $(\mathcal{L}^\lambda \mathcal{M}^\mu \mathcal{T}^\tau)$, he is not making an intrinsic statement about the quantity, but about the field of systems of units that is encompassed by the equations he writes in the form in which he writes them.

The dimensional formula for any physical quantity is simply a short-hand way of expressing the rule to be followed in changing the numerical value of a quantity when a change is made in the basic units. Thus we say that length has the dimension $(\mathcal{L})$. This means this, and nothing more: that any length whose numerical value is x in a certain unit takes on the changed numerical value x' if we use a new unit of length such that there are L of the old units of length in one of the new, where

$$x = Lx'.$$

For many purposes it is convenient to think of the letters x and x' as standing for the symbolic product of a numerical value and a unit name. Thus x would stand for (x barleycorns) where x is a numerical value, if the barleycorn were being used as the unit of length. In that case, it is desirable to assign a symbolic unit to the factor L as well. Thus if (x' furlongs) is the same length physically as (x barleycorns), the conversion factor L needs to be regarded as including the numerical value and an appropriate symbolic unit: it is (L barleycorns/furlong).

This relationship for lengths of all kinds is conveniently shown on a logarithmic scale. The origin (O) of the scale is marked at the point corresponding to the length which is arbitrarily being called unit length (1 barleycorn). Then a point on the scale whose coordinate is ($\log x$) relative to this origin represents a physical length of (x barleycorns). In particular, the point whose location is ($\log L$) relative to the barleycorn origin is the point which represents 1 furlong. Thus, an origin O represents a length of (1 barleycorn), and an origin O', whose coordinate is ($\log L$) relative to O represents a length of (1 furlong). Then a point P whose coordinate is ($\log x$) relative to O has a coordinate ($\log x'$) relative to O', and represents the *same* physical length:

$$(x \text{ barleycorns}) = (x' \text{ furlongs}).$$

Thus the physical quantity is represented invariantly by a point on this scale; the coordinate assigned to it depends on the arbitrary choice of origin which means the arbitrary choice of the length we call unit length. These trivial relations are shown in Fig. 1.

Any length can be chosen as an arbitrary unit. In this sense we may say that the points on the line in Fig. 1 represent all possible choices of unit length, as well as all possible lengths. Change of unit corresponds merely to a shift of origin for describing the coordinate of the same point, representing the same length.

In the same way, the meaning of the statement that ($\mathscr{F}$) has dimensions ($\mathscr{L}^\lambda \mathscr{M}^\mu \mathscr{T}^\tau$) is that if its value in one system is f and we go to a new system such that there are L of the old length units in one of the new, M of the old mass units in one of the new, one T of the old time units in one of the new, then the numerical value f' of F in the new system is

$$f' = f \cdot (\mathscr{L}^\lambda \mathscr{M}^\mu \mathscr{T}^\tau)^{-1}.$$

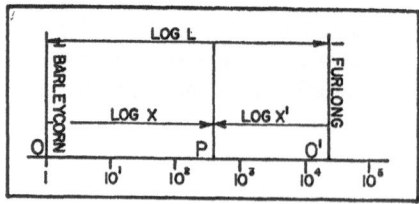

FIGURE 1.

It is evident that these transformations have a simple law of combination. It $L'M'T'$ are the parameters of the transformation from f' to f'', then the parameters of the transformation directly from f' to f'' are LL', MM', TT'. Thus, the transformations are easily seen to have the group property.

Most accounts of the subject act as if it is a self-evident truth that area has the dimensional formula $(\mathscr{L}^2)$. There is, however, a tacit arbitrary assumption here that will now be examined.

There is no doubt that if we perform a similarity transformation on any geometrical figure whereby its linear dimensions are multiplied by a factor L, that the area is multiplied by the factor L^2. But what do we mean when we write

$$A = xy, \text{ instead of } A = kxy,$$

for the area of a rectangle whose sides are x and y? The general form is the latter; the form with $k = 1$ is only valid after we have made the special *arbitrary* choice of choosing as the unit of area the *square* whose side is equal to unity. If, instead, we were to make an equally valid choice of defining unit area as the area of a *circle* whose diameter has unit length, then this unit of area is $(\pi/4)$ times the square unit and the proper formula for the area of the rectangle becomes $A = kxy$ with $k = 4/\pi$. Although the square of unit side is nearly always taken as the unit of area, the choice of the circle of unit diameter to define unit area is actually sometimes found convenient in practice, as witness the use of the circular mil as unit of cross-sectional area of wires.

Thus the expression of area in numerical measure depends through L^2 on the linear parameter expressing the size of the standard figure and also on a factor k which depends on the shape of the standard figure in terms of which we define unit area. From this viewpoint one says that the dimensional formula for area is

$$(\mathscr{A}) = (K\mathscr{L}^2),$$

where the factor K is needed to allow for a possible change in choice of shape of standard figure defining unit area on going from one system of units to another. The only basis on which the K can be omitted is by explicit agreement to restrict the field of changes of units to that in which the choice of standard figure remains the same in all systems of units to be admitted into the discourse.

This point is well known in connection with the distinction between so-called rationalized and unrationalized electromagnetic units, in which different arbitrary choices are made by different people about where they want the inevitable 4π to appear in their equations. About the formula $A = (4/\pi)xy$ for the area of a rectangle some will say "What has π got to do with a rectangle?" The answer is that π *must* appear whenever a circle is intercompared with a rectangle, as here the area of the rectangle is expressed in *circular* units, or in the more usual case, $A = \pi r^2$, when the area of a circle is expressed in *square* units. One can push the π around to different places in the equations by different arbitrary choices, but one cannot get rid of it any more than can state legislatures assign its value.

To cover this contingency, one needs a two-dimensional analog of the graph in Fig. 1 to express all possible choices of measure to represent the area A. For this it is convenient to use two orthogonal logarithmic scales, one for $\log K$ and one for $\log L$. On such a graph an area is represented by an invariant point (P). An origin (O) is located by arbitrary choice of length unit, and a choice of factor k determined by the choice of figure shape that is used to define unit area. The same point P always represents the same physical area, but it has different coordinates relative to different origins corresponding to different choices of length unit and to different choices of shape factor defining the area unit.

Evidently the same kind of remarks also apply to volume. It is a convenient but arbitrary custom that leads us usually to define unit volume as the volume of a *cube* of unit length for its side. We could equally well define unit volume as the volume of a sphere of unit diameter. In that case the correct formula for the volume of a cube of edge x becomes

$$V = (6/\pi)x^3.$$

The same can be said about the definition of unit of each new quantity that is introduced. Each such quantity brings with it an arbitrary choice to be made about its unit. For example, we usually say that the dimensions of velocity are $(\mathscr{L}\mathscr{T}^{-1})$. But this is not a statement about some mysterious intrinsic attribute of the concept "velocity," but a statement about our intention to restrict the filed of systems of units we intend to use to a field in which unit velocity is always defined as that velocity in which unit distance is traversed in unit time. This is a convenient, but arbitrary, choice in which we agree to restrict ourselves to units for which $k = 1$ in the general relation

$$v = kdx/dt.$$

Let us stay for the moment with the conventional choice which makes $k = 1$ by defining unit velocity as that in which unit distance is traversed in unit time.

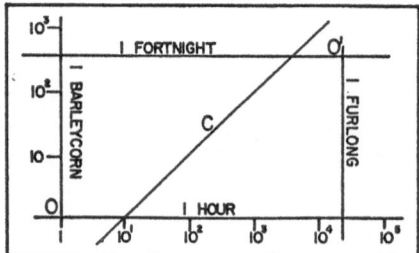

FIGURE 2.

To represent this situation graphically, we need a two-dimensional graph (Fig. 2), with orthogonal axes to represent $\log x$ and $\log t$. The orthogonality is a convenience, not a necessity.

The location of the origin is determined by the arbitrarily chosen units of length and of time, so the points on this plane correspond to all possible choices of units for length and time. On this graph consider the straight line C whose equation is

$$(\log x) - (\log t) = (\log c),$$

which has the intercept $\log c$ on the $\log x$ axis and $- \log c$ on the $\log t$ axis. This line represents the quantity c in the same invariant sense that P represents a particular length on the one-dimensional graph of Fig. 1; namely, this same velocity is represented by this same line in any system of units. A change of the basic units a length and time merely corresponds to a displacement of the origin O to a new origin O'. Relative to the new origin, the same line has intercepts $\log c'$ and $- \log c'$, respectively, where c' is the value of the velocity expressed in the new units. Thus O might correspond to the units being barleycorns and hours, so c is the numerical value of the speed in barleycorns/hour, and O' might correspond to the units, furlongs/fortnights, in which case c' is the numerical value of the same speed expressed in furlongs/fortnight.

The particular line parallel to C and which passes through O represents unit velocity in the O system of units because it has zero intercepts on the axes through O. If O is displaced to another origin along that line, this corresponds to a transformation which has the same unit velocity. Otherwise expressed, the line representing unit velocity is made up of the points representing unit systems all of which have the same unit velocity as does the system O. Similarly, all line C consists of the points representing all unit systems in which the velocity c has the numerical value unity.

In physics, $c \simeq 3 \cdot 10^{10}$ cm/sec plays a fundamental role of great importance in that no matter or energy has ever been observed to be displaced at a higher speed, a fact which plays a decisive role in the theory of relativity. On Fig. 2 this speed represents a limiting line and all actual speeds of energy and matter lie above and to the left of this line. In this sense, only this half of the plane of Fig. 2 corresponds to physical reality. Any actual velocity v is represented on a line parallel to that representing c. The *separation* between these two lines, when measured along the $\log x$ or $\log t$ directions, gives $\log \beta$ where $\beta = v/c$ in the usual notation. The separation between the lines is independent of the location of the origin, i.e., it is the same in all systems of units, a fact commonly expressed by saying that β is dimensionless.

When mass is introduced in order to complete the introduction of the usual mechanical quantities, the graph of Fig. 2 becomes three dimensional if we arbitrarily follow the usual practice of setting each new "k" equal to unity so as to hold to a field of unit systems built on three basic units, L, M, T.

In such a three-dimensional graph, there will be three mutually ortho-gonal axes for the $\log L$, $\log M$ and $\log T$ scales, and the origin O with correspond to a particular choice for these three basic units. Each physical quantity (F) with numerical value f and dimensions $\mathscr{L}^{\lambda}\mathscr{M}^{\mu}\mathscr{T}^{\tau}$ is a particular system which is then represented by the plane whose equation is

$$\lambda(\log x) + \mu(\log m) + \tau(\log t) = (\log f).$$

We see, therefore, that the exponents in the dimensional formula serve (like Miller indices in crystallography) to determine the orientation of the plane, so that quantities of like kind are represented by parallel planes. The dimensionless ratios of quantities of like kind are represented invariantly by the appropriate measure of separation between the planes.

All of the unit quantities, of whatever kind, in a system characterized by the origin O, are represented by the appropriately oriented planes through that origin. Similarly, the plane representing a quantity F is made up of points which represent choices of basic units such that F takes on the numerical value $f = 1$ in such systems.

Two quantities which are of unlike kind, say F and G, will be represented by nonparallel planes which therefore intersect in a line. The points on this line correspond to choices of basic units such that both F and G have the numerical values unity in these systems. Finally, a third quantity H, that is not like either F or G, is represented by a plane which cuts this line, uniquely defining a unit system in which all three of these quantities have the numerical value unity.

It is evident that all of these properties carry over to more complicated systems in which there are N independently chosen basic units. In such cases, the possible choices are represented by points in an N-dimensional space with logarithmic scales, and each physical quantity is represented by a $(N-1)$-dimensional hyperplane, made up of the points representing choices of the N basic units which would reduce that quantity to a numerical value unity. Each time a new kind of physical quantity is defined, a new "k" is introduced which boosts the dimensionality of the graph by unity, and each time such a k is arbitrarily restricted to the value 1, the dimensionality of the graph is reduced by unity.

For example, in the usual system, various special choices are made so as to hold to three basic units for length, mass, and time, so the graph is three dimensional. However, it is convenient in a great deal of theoretical work to choose units so $c = 1$ where c is the velocity of light, so this corresponds to working on the plane which represents c. It is also convenient to choose units such that $\hbar = 1$, where $\hbar$ is the Dirac quantum constant $\hbar/2\pi$, and this corresponds to working on the line of intersection of the planes representing $\hbar$ and c. This leaves, say, the mass unit free to be chosen arbitrarily, but when this is done, the length unit is determined through the Compton wavelength relation, $xm = (\hbar/c)$.

As an example of the introduction of a "k," we may consider the definition of unit electric charge. Having defined the absolute unit of force in terms of LMT^{-2} as usual, and taking cognizance of the Coulomb law between charges whose linear dimensions are small compared with the distance between them, we may define unit charge in terms of the measured force f at distance r according to the equation

$$e_1 e_2 = 4\pi\epsilon f r^2,$$

it being understood that the medium in between is a vacuum. This equation, with various values *assigned* to $4\pi\epsilon$, allows us to measure e in a unit of any size we like. In the cgs electrostatic system, f is measured in dynes and r in centimeters, and the charge unit is *defined* so that $4\pi\epsilon = 1$. In the mks system, f is measured in newtons and r in centimeters, but the unit of charge is taken to be the coulomb, independently defined by a relation derived from magnetic effects, so that $4\pi\epsilon$ becomes a quantity whose value has to be *experimentally* determined. It is $4\pi\epsilon = 10^7/c^2$ where c is the experimentally measured value of the vacuum velocity of light in meters/second.

Let us consider next the question of representation in the graph of uncertain knowledge of an experimentally determined quantity. We know the nature of the quantity so there is no uncertainty about the orientation of its plane, but imprecise knowledge of its numerical value corresponds to imprecise knowledge of the location of the plane. If we know that f lies between f_1 and f_2, then F needs to be represented by a slab to finite thickness between the planes defined by these values. Similarly, if the value is known as a probability distribution, $p(f)df$ being the probability that f is in df at f, this is represented as an appropriate plane slab of density $p(f)$ at the plane f.

The graph has an interesting property which may at first sight seem to be a defect, but which is, in fact, a good quality. The equation for the plane representing f^n is simply n times the equation representing f, since the dimensional exponents for f^n are all n times those for f. Thus, all of the quantities f^n , where n has any value, are represented in the graph by the same plane.

Although no very definite meaning attaches to the word "different" in the phrase "different kind of physical quantity," I think most physicists would be inclined to think of $f, f^2, f^{1/2}$, etc., as being quantities of different kind. Starting with this view, one then must consider it as a defect that the graph represents these all by the some plane. However, we can assign a reasonable meaning to the word "different" in terms of laboratory operations: two observed quantities, f and g, are different if additional observations are needed to pass from knowledge of f to knowledge of g. In this sense f^n is not different from f because when we know f we can find f^n by purely arithmetical operations without additional observations of actual physical phenomena. With this meaning of the adjective, the f^n for various n are not "different" quantities so it is appropriate that they are all represented by the same plane.

When there is experimental uncertainty in the knowledge of f, this is represented by replacing the plane by a slab of finite thickness. When we regard this same slab as representing f^2, we have to caliper its thickness on a double-log scale (like the A scale's relation to the D scale on a standard slide rule) so that a thickness which represents 1% uncertainty in f on these calipers appears as a 2% uncertainty in f^2, as we know it should, so this feature is satisfactory.

By way of conclusion, it may be said that ideas closely related to the ones presented here are current in connection with the problem of solving for adjusted values of the universal constants from discordant experimental results. So far as I know, the logarithmic graph was first proposed in this context by Beth[1] in a short note. This has been further discussed and modified in the work of DuMond[2] and Cohen,[3] who prefer to linearize the equations, which is possible where high accuracy of the data already exists. The viewpoint here is that the graph affords insights into the ideas underlying construction of systems of units that can clarify this general problem and thus reach beyond the more specialized question of adjustment of discordant results on the universal constants.

[1] R.A. Beth, Phys. Rev. **54**, 865 (1938).
[2] J.W.M. DuMond, Phys. Rev. **56**, 153 (1939); **58**, 457 (1940).
[3] J.W.M. DuMond and E. R. Cohen. Revs. Modern Phys. **20**, 82 (1948); **25**, 691 (1953); and *Handbook of Physics*, edited by E. U. Condon and H. Odishaw (McGraw-Hill Book Company, Inc., New York, 1958), Part 7, Chap. 10.

Intermediate Courses in Physics*

E.U. CONDON

Washington University
St. Louis, Missouri

If we are to improve the preparation of physics majors, in keeping with the needs of the times and with the improved preparation that students in increasing numbers will be bringing from their high schools, we must decide on what parts of physics are most important. It is suggested that our courses must pay greater attention to phenomena and less to mathematical formalism. Specific suggestions about the ways in which courses in mechanics, electricity and magnetism, optics, atomic and nuclear physics, and thermodynamics can be modified to accomplish this are given.

This is a period of great change in the situation with regard to science and mathematics teaching in America. A large amount of effort is going into programs for improving the content of high school courses in physics, chemistry, and mathematics. The new curricular content for physics and mathematics especially is being used in a large number of schools this year and will be in many more next year, and the change-over is expected to go forward rapidly in the next few years.

Even though the changes are being adopted rapidly, they are still far from affecting any large fraction of all of the high school students who come to college. Therefore those who have to plan college course content are likely to have the added difficulty for some years to come that there is a greater heterogeneity of preparation of the students than we have ever had in the past.

I think there is no question that the preparation of the high school students who have had the new material is going to be vastly superior to that of the students we have been getting for some decades past. It is therefore essential that we give corresponding attention to how best to deal with these students in college by modernizing and upgrading the content of the courses we offer. But it is going to be difficult to know just how to do this in a situation where only one type of course can be given for economic reasons, since this course has then to serve college students who have had both the old and the new type of high school preparation.

---

* Lecture delivered at the Denver Conference on Curricula for Undergraduate Physics Majors.

I am very much aware of this problem, but I am not going to discuss it further here because it has principally got to be faced by those who plan and organize the introductory courses. Our problem is to consider what ought to be done in physics for physics majors in the junior and senior years.

There is also a great deal of heterogeneity in what is done to the students in the first two years, when we compare different institutions, but fortunately there is not much transferring between institutions at the end of the sophomore year, so a particular institution can pretty well plan its intermediate courses to build on its own freshman-sophomore courses. Even here there can be a good deal of difficulty in locations where the four-year institution gets a large proportion of its student supply from a wide variety of junior colleges.

In order to get started, I am going to make the assumption that the intermediate course structure begins with students who have had at least twelve semester hours of general physics with laboratory, including a semester of elementary modern physics, preferably with some laboratory work. Also I assume a one-year modern course in general chemistry and about twelve semester hours of college mathematics, which will have covered the usual differential and integral calculus and analytic geometry. I will be hoping that the students will have covered much of what used to be called college algebra and also trigonometry while still in high school or will have taken these subjects in a summer session at the college before the start of the freshman year.

Next I am going to assume that about thirty to thirty-two semester hours are available for physics courses in the junior and senior years for the physics major. This permits the taking of two three-semester-hour lecture courses and one unit of laboratory work in each of the four semesters. With all the pressures toward diversified studies in the liberal arts colleges it will seldom if ever be possible to have more of the student's time than that for a major even if we wanted it.

There is another problem which the liberal arts physics major is up against. He certainly ought to take an additional twelve hours of advanced mathematics and at least a course in physical chemistry. If then he must fulfill some humanistic and social science requirements, where will he find the time to take even a basic course in biology? Personally I think that nobody is liberally educated who has not had the equivalent of a college course in modern biology.

Having made these few remarks about the input status of the students, let us briefly consider the kind of output we wish to produce. Last week I took part in a conference at Boulder, somewhat like this one, but devoted to considering the problems of improving engineering education. As happens so often, I came away from that conference strengthened in a conviction that I had before I went there. It is this: that modern engineering for research and development has become so rich in content that it is impossible to give men the kind of training they ought to have without several years of graduate work, preferably organized to lead to the Ph.D. There has been a trend in this direction for some years past, but it needs to be greatly accelerated if engineering is to keep pace with today's needs.

Fortunately the need for going on in graduate work for a good career in physics is go generally recognized that it is not necessary to urge it.

In fact, I think that if there is any mistake we as physicists make along this line, it is an underemphasis on the career opportunities that exist in industrial and government laboratories for men and women who do not continue their formal education beyond the bachelor's degree. I have the feeling that often the young Ph.D.'s on the faculty advise undergraduates that any real activity in the profession is barred to them unless they too get the Ph.D. I think this is wrong.

We should, I think, plan the curriculum for physics majors in a way that prepares them for entrance into graduate school because it always should be our aim to persuade every student that is capable of it to go on for graduate study if he possibly can. This does not make any difficulty because I think that the nature of the subject is such that a suitable curriculum can be planned, and that it will be good training even if the student's formal education terminates at the bachelor's degree. The graduate school objective does add one element to the already closepacked structure of the undergraduate curriculum, namely that the student ought to get off his foreign language requirement in two languages while still an undergraduate.

There is another factor which adds to the closepacking: besides being in the atomic or nuclear age, we are now also in the space age. In the next decade there is going to be a vast range of scientific opportunities to plan and execute experiments using equipment in rocket-launched vehicles, and there will be a vast demand for physicists to do such work. Therefore we ought to make way for more in the way of solar and planetary astronomy, and also for upper atmosphere physics, than is currently being offered in most places.

All things considered, it is clear that the most outstanding characteristic of the situation is that we would like to accomplish a great deal more than we have been accomplishing in the past as part of the physics major for the A.B. or B.S. degree. We are going to be helped in our task if better preparation from high school enables us to go farther with students in the introductory courses than we have gone hitherto. But I think that no matter how we try we are not going to give formal classroom coverage to all of the subjects that we would like to cover.

Therefore we need to scrutinize very carefully what we put in and what we leave out. One dangerous consequence of this feeling of crowdedness is that we are apt to try to cover too much, in the sense of giving some classroom attention to too many different things, and thus become superficial.

I think we need to consider a minimum list of basic principles and techniques which must be mastered. Hopefully we can save enough time by so doing that we can then proceed to give the student more time for project work and the art of getting up a subject by himself. By project work I mean not only a laboratory research project, perhaps in conjunction with the research program of a faculty member or graduate student, but also the thorough looking up and summarizing of the journal literature on some topic. Too many of our students go through to the bachelor's degree without acquiring any skill whatever in the use of general reference books and the original journal literature. I think that time should be allotted during each of the four semesters for each student to prepare a major term paper that reviews the journal literature on some topic of interest. These should involve the preparation of thorough and complete bibliographies and the preparation of a unified critical account of the content of at least the major papers involved. The goal should be for the student to learn how to use the library himself, and not to expect the librarian to find the material for him. There should be practice in public presentation of these papers by conducting a student seminar in which the students give talks of at least half an hour, and preferably an hour, reporting on the material that the paper has covered. These should be scheduled in such a way that there is time for critical discussion of these presentations by faculty and students at the end of each of them. The aim should be to prepare these papers to such a standard that extra copies of them could properly become permanent acquisitions of the department library.

The importance of putting strong emphasis on this kind of independent library work stems from the fact that we can not possibly teach in the classroom all that a student needs to know. This is partly for lack of time, and partly because, believe it or not, we ourselves do not know that much.

There is room for difference of opinion about details. Years ago at Princeton we used to have the departmental students write a junior thesis and a senior thesis, that is, just two papers of the type I have been describing, instead of one each term as I have suggested. Possibly that is better, but I seem to recall that the students had a tendency to delay until late in the school year and then cram too much in late spring. Perhaps it would be better to have two short papers, one in each term of the junior year, due by the end of the school year. Various schemes could be tried. I will leave the point with the expressed conviction that some training of this kind ought to form a serious and important part of the undergraduate program.

Let us turn now to the framework of the courses themselves. Here there is not a great deal of room for variation if one feels that each of the major branches of the subject should be covered at the intermediate level. I will comment on the set we have at Washington University, not because I think it is uniquely superior to others, but simply because it is fairly typical, and I am familiar with it.

In the junior year the students take a three-unit course in mechanics in the first term and fill out the year with a three-unit course in thermo-dynamics in the second term. In addition, they take a full-year (six-unit) lecture course in electricity and magnetism and a two-unit year course called intermediate laboratory. In this they get some mechanics, some electrical measurements, and some modern physics and optical experiments.

Then in the senior year they take a three-unit, one-term course in optics, followed by a one-term, three-unit course in mechanics of continuous media, fluid mechanics, and elasticity. In addition, they take a full-year (six-unit) course called Atomic, Nuclear, and Solid-State Physics, and again a year of advanced laboratory work. This laboratory is mostly study of electronic circuitry for research equipment.

This general plan has been in effect for well over a decade, though naturally the details of handling change from year to year, and the details are what are important. My direct experience with it is confined to having taught the junior mechanics and the senior mechanics of continuous media. I think it offers the students a good grounding in physics, perhaps about as well as can be done in the time available, but we do not gain anything from merely saying that, so I will try to express some specific critical comments.

My strongest criticism is that we are not doing anything of the type of teaching the students to use the library and really get up a subject by themselves, the very thing that I put first in importance. I hope before long we can make some changes that will remedy this serious defect.

My second criticism about the way the courses are taught, in so far as I know about them in detail — and I am just as guilty as anyone for this — is that I think the courses are too mathematical. Junior mechanics tends to become a series of exercises in setting up differential quations of motion for various solvable cases and solving them as isolated exercises. The same is true of the senior course in mechanics of continuous media. Here there is a strong temptation, usually too strong to resist, to turn the whole course into one on boundary value problems of partial differential equations. This is to some extent justifiable as preparation for the mathematical techniques to be used in quantum mechanics and was planned the way it is partly for this reason.

Both of these courses could be improved by including more phenomeno-logical material involving direct contact with observation. We do the differential equations for the derivation of Kepler's laws for planetary orbits, but we do not do the elementary mechanics of the solar system, including the actual observation of planets and correlation with their positions as given in the *American Ephemeris and Nautical Almanac*. We ought also to give detailed discussion of an artificial satellite orbit. We ought to discuss the elementary treatment of relativistic mechanics and take up in detail the dynamics of motion in accelerators and mass spectrometers, including focusing properties, relating this to actual working research equipment such as our own cyclotron or the big accelerator at Brookhaven. The same is true for the relativistic treatment of collision phenomena as they occur in the interpretation of cloud chamber, bubble chamber, and emulsion tracks. We could not go very far with all of these topics, but we could go part way, and the students could then learn a great deal more by writing their term papers on a full treatment of some particular topic of current interest.

I think the same kind of criticism can be levelled at many junior courses in electricity and magnetism. Too often they are watered-down treatments of the field equations of Maxwell. Material of this kind should be held to a minimum in order that a great deal more knowledge can be conveyed of the phenomenology of electromagnetic properties of materials. The student needs to know about the resistivity of metals, the actual properties of dielectrics as to dielectric loss and dielectric absorption and dielectric breakdown, the phenomenology of semiconductors and the ways they are used in rectifiers and transistors, the magnetic properties of materials, including the ferrites, and the phenomenology of superconductivity. He needs to know, moreover, the actual facts about electrical discharge in gases and the phenomena of atmospheric electricity. All of these things need to be presented as aspects of the real observable world, with emphasis on the methods of measurement and the actual numerical values of the parameters involved. It must not be forgotten in planning such a course that the students have had in sophomore year an elementary course in modern physics, so it is perfectly all right to assume some knowledge of the fundamental electrical particles and the general ideas of the structure of matter.

Our senior course in optics tends to build on the course in electricity and magnetism. Again I think it is too much concerned with the mathematics of the wave solutions of Maxwell's equations and not enough concerned with the optical properties of matter. Time can be saved by confining attention with regard to interference and diffraction phenomena to those which help one to understand actual diffraction gratings and interferometers as they are used in spectroscopy. Then one should plunge into spectroscopy, including the processes of excitation of atoms and molecules, fluorescence, phosphorescence, photoconductivity, photography, and photometry. Likewise this course should find time for careful study of the optical basis of the special theory of relativity.

Probably the most crowded course in our set is the senior course entitled Atomic, Nuclear, and Solid-State Physics, as one might surmise from the scope of its title. I think we are attempting to cover much too wide a range of material in this single course. Not only do we include here a great deal of phenomenology that I have suggested would be better included in the electricity and optics courses, but we also spend a good deal of time here trying to give some of the mathematical treatment of elementary quantum mechanics. If the changes that I am suggesting were made in these courses, there would be more time in the senior atomic physics course, in that electrical conduction in gases and atomic spectra would already have been covered. I would then suggest that the course be broken into two one-semester courses that have little to do with each other.

A one-term course in nuclear physics would deal particularly with methods of observation of the actual phenomena of nuclear reactions of various types. Actual measurements of the beta rays and gamma rays, and neutrons and fission and chain reactions carried by neutrons would be included. Here again the emphasis would be strongly on the side of the phenomena and the means of observing and measuring them and the quantitative results. Theoretical models of nuclei and mathematical treatment of strong and weak interactions and processes involved in emission of gamma rays would be given a minimum of attention.

The other term would be a one-term course in solid-state physics. This again should be heavily oriented toward phenomena and the means of observing them. It should start with the principles of X-ray diffraction and include the actual measurements for a simple-structure determination. A good deal of the phenomena of the dielectric and magnetic properties of solids would already have been presented so that the way would be open to spend most of the term on the electronic properties of semiconductors, and of the defect properties of ionic crystals, as well as on the electronic properties of metals. I would want to see all of these topics studied with real data given in the illustrations and a great deal of emphasis on the actual quantitative results obtained in the experiments.

Having said this much in this vein, I am sure you will not be surprised if I say that also I would like to see the emphasis in thermodynamics shifted to a course that might be entitled Thermal Properties of Matter. Here the emphasis would be on descriptive, semi-quantitative, molecular interpretations, and the kinetic properties would receive every bit as much attention as the equilibrium properties. Above all the emphasis would be on how the properties are studied, and what are the numerical results of such studies.

Some of you may be surprised that I have put such strong emphasis on the phenomena and the means of studying them at the expense of decreasing the amount of differential equation work. And you may have noticed that so far I have said nothing at all about quantum mechanics as an undergraduate subject.

The reason for placing the emphasis as I have is this. In the first place, it represents merely a shift in emphasis toward concrete study of phenomena and does not imply complete elimination of the differential equation work. Secondly, it is because I think too many bachelor's degree students do not really learn the observational basis of the subject nor do they get a firm grip on the magnitudes of effects or any idea of the ranges of magnitudes that

can occur. Thus they all know that glass is an "insulator," but few know
that the range of room-temperature values of the measurable resistivity of
different kinds of glass is over many powers of ten, nor have many of them
a clear idea of what is involved in measuring the resistivity of a material of
this kind, where charge absorption and electrode polarization effects
complicate the issue.

I think it is important to get such a concrete impression of vivid reality
while an undergraduate for two reasons. For those that go on to graduate
work, there is time enough to get the more formal mathematical theory in
full while in graduate school, while for those who do not they are likely to
be employed in industry where they will have little or no occasion to use
the mathematical theory, but will be greatly handicapped by not knowing
the facts and how they are found and how to look things up quickly in the
library.

It may be, of course, that I am not well informed about the general
situation with regard to the intermediate courses. Maybe many of you are
already putting the main emphasis on phenomena and the methods of
observation and measurement. If so, I applaud; if not, I urge that you
consider the desirability of moving in that direction.

What is to be said about the undergraduate with a strong bent toward
theoretical physics that shows itself rather early? We do not want to stifle
his interest nor hold him back. The answer, of course, is that he can skip
some of the intermediate work and go at once into graduate courses as
a junior or senior because he can always pick up the other material later
by general outside reading.

And finally, what is to be said about elementary quantum mechanics and
its place in the undergraduate curriculum? Before we can discuss that
properly we have to be clear on what we mean by elementary. In the
graduate schools there is growing up a tendency to regard about two-thirds
of Schiff's book as elementary. Those parts that get into fields and the use
of creation and destruction operators and the use of relativistically
invariant wave equations are then called advanced. If "elementary" is
being used in this way, then I think there is really no place for this much
quantum mechanics in the general undergraduate curriculum. Let those
who are ready for it take it as a graduate course while they are still
undergraduates, but let most of the students wait until they get to graduate
school for it. It seems to me that quantum mechanics to the level of the
little book by Fano and Fano is about as far as we ought to plan to go with
most of the physics majors. We need the idea of energy and other
quantities occurring only with quantized values, and the idea that the
allowed energy levels are determined by something like a wave equation,
and a few simple one-dimensional examples, such as the particle in a box
and the harmonic oscillator. We need the idea of the wave function being
interpreted as leading to a probability function, and that is about all. These

ideas will be introduced and reinforced as we go along without being made the subject of a special course in quantum mechanics.

This about covers the topics that I have in mind. Let me recapitulate. I believe in heavy emphasis on training the student to look up his own materials in the library and to make a systematic, thorough attack on the literature of a topic, and in giving thorough training on the phenomena of physics and the methods of observation and measurement, and on the presentation of theoretical interpretations with a minimum demand on fancy differential-equation wangling. Possibly some of you will want to take issue with this position, and as a result of what you say, I may even want to modify my position, not admitting it, of course, but merely claiming that my original meaning was misunderstood.

60 Years of Quantum Physics

EDWARD U. CONDON

I was invited to speak on the occasion of the 1500th Regular Meeting of the Society, and of course am delighted to be able to come and do it. But those who conveyed the invitation could not refrain from reminding me that I owed the Society a retiring presidential address. I was president in 1951, and it was in the fall of that year that I departed hastily to go to Corning Glass Works to be director of research. That was a very interesting experience, and I am still connected with the glass business, though I am also doing professing. I started my career in experimental physics and lasted one day. When I started work on a doctoral thesis at the University of California in 1925 I had to set up a vacuum system. All experimental physicists in those days had to get a Cenco pump on the floor and glass tubing up to something that was on the table. I started out like all the rest but broke so much glass the first day that they suggested I go into theoretical physics. I told this story at Corning after I became their director of research. Mr. Amory Houghton, chairman of the board, who is now our ambassador to France, said, "Isn't it good that at last you are in a place where you can't possibly break enough glass to make any difference."

Looking back over the various possibilities of things that might be suitable to talk about this evening, I thought it would be interesting to review the historical development of what I now would like to call quantum physics, rather than quantum mechanics, because it has grown and expanded in such a way that it permeates all of modern physics. In fact it is extremely difficult to think of any actively cultivated part of physics that is not directly involved with Planck's quantum constant h. The basic discovery by Planck's was made within a week or two of exactly sixty years ago, so I thought it might be interesting to discuss this subject.

The subject of quantum physics started with the statistical theory of the distribution of energy in the black-body spectrum. The spectrum of radiated energy in equilibrium with matter in an enclosure is commonly called black-body radiation because it is the kind of radiation that would be emitted by a perfect absorber. The active problem in 1900 was the explanation of the distribution of energy in the spectrum.

E.U. Condon is Wayman Crow Professor of Physics and head of the Department of Physics at Washington University in St. Louis, Mo. He presented the address upon which this article is based on the occasion of the 1500th regular meeting of the Philosophical Society of Washington, which was held on December 2, 1960, at the Natural History Museum Auditorium of the Smithsonian Institution in Washington, D. C. His address is included in Volume 16 of the archival *Bulletin* of the Society.

It is interesting to realize that the subject has quite an ancient history. The first application of thermodynamics to black-body radiation goes back to 1859, when Kirchhoff first developed the ideas of radiative exchanges, and the connections between emission and absorption, rules according to which a good emitter is a good absorber, and a poor emitter is a poor absorber. In 1884 the discovery had been made of what we now call the Stefan-Boltzmann law, that the total radiation goes up as the fourth power of the absolute temperature. It was discovered by Stefan experimentally and interpreted theoretically by Boltzmann, making it one of the earliest applications of thermodynamics to radiation after those first ideas of Kirchhoff's

In 1894 came the discovery by Wien of the displacement law, which tells how the distribution of energy over various wavelengths changes with the absolute temperature. The big problem at that time was to try to understand the reason for this distribution. Contrary to the general belief, which has become true in the last thirty years or so, that all physics is really done by young men in their twenties, the discovery of Planck was made when he was at the advanced age of 42. In 1900 he had already put a part of his career of research work behind him and was a professor in the University of Berlin, so that his work on quantum physics was done twenty-one years after he had received his doctorate for a thesis on the second law of thermodynamics. His thesis, it is interesting to note, was done under Kirchhoff and Helmholtz at Berlin. In his autobiography he says that he is quite confident neither of them ever read it.

Thermodynamics was Planck's first love, his principal love throughtout physics. In fact there are many indications that he was rather annoyed with his discovery of the Planck constant of action and did his best for about fifteen years or so, on up to about 1915, to find ways of evading his own discovery and reconciling the theory that he had discovered with classical

Planck   Rayleigh

theory. This resembles somewhat the story that I used to hear from
Professor Ladenburg at Princeton, about Roentgen. Everybody knows
about the great consequences of Roentgen's discovery of the Roentgen
rays, or X-rays. Ladenburg was a student of Roentgen. He said that
Roentgen was annoyed with his X-rays because he did not understand what
they were and much preferred classical subjects. So the upshot of it was
that Ladenburg did a doctoral thesis under Roentgen just a few years after
Roentgen had discovered X-rays, on the subject of the correction to
Stokes' law for a body falling through a viscous medium in a cylindrical
tube, allowing for the finite diameter of the tube and the wall effect. They
had a long pipe filled with castor oil, which is the traditional viscous
material. It reached from the top floor of the laboratory to the basement.
He said nothing ever gave Roentgen quite as mcuh pleasure as to see
the steel ball arrive down at the basement just when the calculation said
it ought to. You can tell by a great deal of Planck's writings and readings
that he felt much the same way about classical physics in relation to the
modern developments.

Lord Rayleigh had published a theory, based on the equipartition-of-
energy doctrine that goes back to Maxwell, Waterston, and Boltzmann,
whereby every degree of freedom in the radiation field should have had the
energy kT. He knew it did not, because that would have given an infinite or
divergent result. But nevertheless that was where the theoretical thinking
of his time led, which served to point up the importance of the quantum
modifications that had to be made.

One of the things that I found interesting in looking back in the history
of this theory is that it has always been referred to as the Rayleigh-Jeans

law, and I had supposed that Rayleigh and Jeans had worked together on it. In point of fact, Rayleigh derived it and made a mistake by a factor of 8, which Jeans corrected in a letter to *Nature*, so that dividing the original Rayleigh formula by 8 was Jeans' contribution.

It was an essential contribution because it is a mistake that we all might make very readily. In counting up the degrees of freedom in the radiation field that are associated with frequencies between v nd $v + dv$, one has to calculate how many integers there are whose squares add up to a certain value, and it is natural to take the volume of a sphere of a certain radius. But in fact one takes only an octant out of this sphere because the integers, all three of them, have to be positive, and that is where Rayleigh went wrong.

The radiation measurements that served to inspire Planck were being made at the Physikalisch-Technische Reichsanstalt by some of the great names of early days of radiation-measurement work: Lummer, Pringsheim, and Rubens. The problem of distribution of energy in the spectrum was thus very much in the foreground and very good measurements were being made.

It was on October 19, 1900, that Planck presented his radiation formula to the German Physical Society at a meeting in Berlin, strictly as an empirical interpolation formula between the Rayleigh-Jeans law, which is valid at long wavelengths, and the Wien law, which is valid at short wavelengths. By interpolating in between, he had been able to find a simple formula that extended across the whole region, but at that time he had no theoretical basis for it whatever.

That night Rubens took the data to which he had access and made a very careful comparison with Planck's formula — a more careful one than Planck himself had made at that time — and found that it represented the data with extraordinary accuracy, much better than an empirical formula usually does. He called on Planck the next morning with a strong conviction that there was some real fundamental truth in the formula and not just an accidental agreement. Planck then set to work to find a theoretical basis for this formula and worked very hard for quite a while. In his autobiography he speaks of this as the most difficult period of his whole life.

Then, within less than two months, on December 14, 1900 — so we are just twelve days ahead of the 60th Anniversary — he presented a paper to the Physical Society of Berlin in which he took the decisive step. By applying the Boltzmann principle for the connection of entropy with probability, which up to that time had hardly been used at all, he was able to work out the spectral distribution of energy that would be in equilibrium with a system of electrical oscillators.

In order to get the desired result, he had to suppose that the energy of each oscillator was built up in finite steps of energy, whereas, in all of physics hitherto, energy had been a continuous variable. To agree with the

Wien displacement law he then had to assume that the finite size of these steps was proportional to the frequence, and so the energy quanta were hv. In that way he arrived at the famous formula, $u_v = [8\pi/c^3] [hv^3/e^{hv/kT} - 1)]$, for the density of the energy in the spectral frequency range between v and $v + dv$ in black-body radiaton at absolute temperature T. As is readily seen, in the limit of hv/kT small compared with 1, this formula transforms into the Jeans formula; in the limit of hv large compared with kT, it becomes the Wien formula and represents the data with great accuracy in between. Additional measurements of the same sort, which were later made with great precision at the National Bureau of Standards by W.W. Coblentz, greatly improved our knowledge of the subject.

One of the most extraordinary aspects of this work of Planck's is the accuracy with which he was able to define these fundamental constants. At that time there was no good value available for Avogadro's number, or for the charge on the electron, and the values that Planck was able to derive were much closer than is usually appreciated. When he first represented the data, in order to obtain a fit with the old black-body data, he had to assume that h was 6.885×10^{-27} erg·sec, and that k, which we now call the Boltzmann constant, was 1.429×10^{-16} erg/°K. The present best value for first number is 6.6252×10^{-27}, instead of 6.885×10^{-27}; and for the second number, it is 1.3804×10^{-16}, instead of 1.429×10^{-16}. At that very first time Planck got Planck's constant only about 4.4 percent too high, and Boltzmann's constant about 3.5 percent too high, relative to the best modern values.

This was actually the first time that the Boltzmann constant had been evaluated. Let me just remind you of its relation to the other basic constants that have so much importance.

The gas constant, R, as we ordinarily know it, per gram mole, is equal to the Avogadro number, N, times k; and the Faraday, F, the amount of charge needed to plate out a gram mole of univalent ions, is equal to Avogadro's constant times the charge of the electron. That is, $R = Nk$, and $F = Ne$.

These molar quantities, R and F, were well known, and good values for them were available in those days, but what was not known was the Avogadro number N. However, if you know any one of these quantities you can get the other, so, as it turns out, obtaining the Boltzmann constant, k, enables one to get N by the first equation, and then, by using that N in combination with the knowledge of the Faraday, F, one is able to get the charge on the electron.

The electron had only been recognized about three years earlier by J.J. Thomson, and while the ratio of its charge to mass was known, its charge by itself was not well known. You will find, in the literature of that time, values published for e, the charge on the electron, ranging all the way from 1.29×10^{-10} electrostatic units, on up to 6.5×10^{-10} electrostatic units, which was given by J.J. Thomson, and a little while later revised back down to 3.4×10^{-10}. In other words at that time one only knew the charge on the electron to a factor of about 5 or 6.

On the other hand if you take the value of the Faraday and the value of k and solve for N from the gas constant and then solve for e, you find, surprisingly enough, that e equals 4.69×10^{-10} electrostatic units, which is only 2.3 percent below the currently recognized value.

Thus, in the space of just a month or two, Planck first found an empirical formula which to this day gives the most accurate representation of the spectral distribution of the radiant energy; second, he found a derivation of that formula. In order to get the derivation he had to introduce the extraordinary idea of energy quantization into physics. Third, he obtained an excellent value for the charge on the electron, which everybody was trying to do at that time.

You might expect that this would cause a great deal of excitement among physicists at that time, but it did not. If you search through the journals you find practically nothing is said about Planck in the years 1900 through 1904. I was very much intrigued, therefore, when just before this meeting Mr. Marton recalled that a search of the records of this Society indicated that in 1902 Arthur L. Day gave a report on Planck's work. Thus, The Philosophical Society of Washington was one of the earliest to pay attention to it.

The first real extension of Planck's work came with Einstein's famous paper of 1905, the paper for which he got the Nobel Prize. (It is important to realize that Einstein did not get the Nobel Prize for the theory of relativity. They might give it to him now if he were around, but they did not in those days.) Planck wrote only one other paper on the subject in that period between 1900 and 1905 and this was mainly an expository paper. There is one brief mention by Burbury, another paper by van der Waals, Jr., and that is all. In those days Planck was almost completely ignored.

In Planck's own autobiography he tells of his own attitude toward the Planck constant, and I thought it would be interesting to read his own words on that, of course translated into English. He said:

While the significance of the quantum of action for the interrelation between entropy and probability was thus conclusively established, the great part played by this new constant in the uniform regular occurrence of physical processes still remained an open question. I therefore tried immediately to weld the elementary quantum of action, h, somehow into the framework of classical theory. But in the face of all such attempts the constant showed itself to be obdurate.

So long as it could be regarded as infinitesimally small, i.e., when dealing with higher energies and longer periods of time, everything was in perfect order. But in the general case difficulties would arise at one point or another, difficulties which became more noticeable as higher frequencies were taken into consideration. The failure of every attempt to bridge that obstacle soon made it evident that the elementary quantum of action plays a fundamental part in atomic physics and that its introduction opened up a new era in natural science, for it heralded the advent of something entirely unprecedented and was destined to remodel basically the physical outlook and thinking of man which, ever since Leibniz and Newton laid the ground work for infinitesimal calculus, were founded on the assumption that all causal interactions are continuous.

He goes on in a more personal vein to say:

My futile attempts to fit the elementary quantum of action somehow into the classical theory continued for a number of years [actually until 1915] and they cost me a great deal of effort. Many of my colleagues saw in this something bordering on a tragedy. But I feel differently about it, for the thorough enlightenment I thus received was all the more valuable. I now knew for a fact that the elementary quantum of action played a far more significant part in physics than I had originally been inclined to suspect, and this recognition made me see clearly the need for the introduction of totally new methods of analysis and reasoning in the treatment of atomic problems.

In spite of Jeans' intimate association with this problem, you find no reference whatever to the Planck black-body law in the first edition of his *Dynamical Theory of Gases*, which was published in 1904, four years after Planck's work. In the Landolt-Bornstein Tables, published in 1905, we find an extraordinary thing, namely, that it gives widely different values for what is often called the Loschmidt number, the number of molecules in one cubic centimeter of various gases under standard conditions. Of course this value should be the same for all gases. But they solemnly give you a table with 2.1×10^{19} for air, 4.2×10^{19} for nitrogen, 7.3×10^{19} for hydrogen, and so on. Apparently Landolt and Bornstein did not believe in the Avogadro number. Planck got 2.76×10^{19} for this number, which as we have seen is a good value.

Josiah Willard Gibbs was America's first great theoretical physicist. He was elected, I find, to membership in the Washington Academy of Sciences in 1900. He died in 1903 at the age of 64. There is no indication in any of his publications or notes that he left behind that he paid any attention to Planck's work. He had puzzled over the problem of the specific heat of polyatomic gases, which everybody was puzzled about at that time, because it has too low a value to correspond with the equipartition law. There is some indication that he found these difficulties with the equipartition law revealed in the specific heat of gases somewhat depressing, and I find an indication of that, perhaps, in an interesting paragraph from the preface to his famous work on statistical mechanics, published in 1902, a year before Gibbs' death:

In the present state of science it seems hardly possible to frame a dynamic theory of molecular action which shall embrace the phenomena of thermodynamics, of radiation and of the electrical manifestations which accompany the union of atoms. Yet any theory is obviously inadequate that does not take account of all these phenomena. [Then comes a wonderful sentence at the end of this paragraph which I think we all ought to realize was written by Gibbs in 1902:] Certainly one is building on an insecure foundation who rests his work on hypotheses concerning the constitution of matter.

Lord Kelvin's Baltimore lectures, which were delivered at The Johns Hopkins University in 1884 but were not published until 1904, had undergone a great deal of revision up to the latter date. The preface to these lectures is very interesting to those who have anything to do with editing or getting things through the press. He admits that he had been working on the revision for all of the nineteen years. I can well imagine that he was a popular fellow around the print shop.

That work includes as its appendix B, the famous lecture to which I am sure you have all heard allusions, "Nineteenth Century Clouds over the Dynamical Theory of Heat and Light." That lecture was delivered in April 1900, some months before Planck's work, and then it was published originally in the *Philosophical Magazine* of July 1901. It makes no reference to the black-body radiation or to Planck's work, although cloud 2 — he had his clouds numbered — was this same concern about the failure of equipartition of energy as evidenced by the specific heats of gases, the same problem that was troubling Gibbs.

Lord Rayleigh's publication of what we now call the Rayleigh-Jeans law was in the *Philosophical Magazine* in 1900. He did not return to the subject again until 1905 when he wrote several notes in *Nature* in which he concedes or agrees with the comment that Jeans had made about being wrong by a factor of 8. It is interesting in that he says that he "has not succeeded in following Planck's reasoning". That is how Planck's work was received by Lord Rayleigh, one of the greatest British physicists. Rayleigh actually published papers actively through 1919, but he seems to have had no more to say on black-body radiation than what he said in that one 1905 paper.

Search through his published papers reveals two more items relating to modern quantum physics. In the 1906 *Philosophical Magazine* he comments on the classical radiative properties of the atom models that resemble J.J. Thomson's. However, he goes beyond them in that he regards the negative charge as distributed more like a continuous fluid and studies it as a normal mode-of-vibration problem. In an editorial note added to his collected papers, written in 1911, he refers back to some old work of an 1897 paper. An interesting thing to me is his comment that all kinds of models of normal modes of vibrations of continuous systems always lead to formulas in which the square of the frequency is written additively as the sum of contributions coming from the different degrees of freedom — from what we would now call the different quantum numbers. Rayleigh was wedded to a classical vibration-theory model where the squares of the frequencies get in because of the second derivative with regard to the time, based on Newton's law of mechanics. Nowadays, when we lecture on quantum mechanics, we just quietly *make* the Schrödinger equation contain the first time derivative so we will not have this trouble, which is the advantage of making up your equations as you go along as compared with getting them from someone like Newton.

Steeped in acoustics as he was, Rayleigh does say: "A partial escape from these difficulties might be found in regarding the actual spectrum lines as due to difference tones from primaries of much higher pitch." That is a well-known device giving physicists license to pass from a square term to a linear term; that is, a small change in the square is linear.

There is still something else in the 1906 paper which intrigues me. Rayleigh devotes a paragraph to the problem of the sharpness of spectual lines despite the random character of the conditions of excitation, and concludes with a paragraph that sounds very modern. I quote:

It is possible, however, that the conditions of stability or of exemption from radiation may, after all, demand this definiteness, notwithstanding that in the comparatively simple cases treated by Thomson, the angular velocity is open to variation. According to this view, the frequencies observed in the spectrum may not be frequencies of disturbance or of oscillation in the ordinary sense at all, but rather form an essential part of the original constitution of the atom as determined by conditions of stability.

Maybe one reads into the statement one's present knowledge of the later developments of quantum theory, but I found it very interesting as a foreshadowing of the way we look at it now.

Even as late as 1911, we find Lord Rayleigh worrying about Kelvin's cloud 2, the specific-heat difficulty, although Einstein had really put that difficulty to rest in 1907. In 1911, Rayleigh wrote to Walter Nernst to express his concern:

If we begin by supposing an elastic body to be rather stiff, the vibrations have their full share of kinetic energy [that is the equipartition law] and this share cannot be diminished by increasing the stiffness . . .

We all know that increasing the stiffness makes the interval between the vibration quantum levels greater, so that they do not take part practically in the equipartition law simply because they cannot get enough energy even to be excited to the first state.

However, Rayleigh goes on:

Perhaps this failure might be invoked in support of the views of Planck and his school that the laws of dynamics as hitherto understood cannot be applied to the smallest part of the bodies. But I must confess that I do not like this solution of the puzzle . . . I have a difficulty in accepting it as a picture of what actually takes place.

We do well I think to concentrate attention on the diatomic gaseous molecule. Under the influence of colisions the molecule freely and rapidly acquires rotation. [He knows this from the specific heat.] Why does it not also acquire vibration along the line joining the two atoms?

If I rightly understand the answer of Planck is that in consideration of the stiffness of the union, the amount of energy that should be acquired at each collision falls below the minimum possible and that therefore none at all is acquired [this is of course exactly what we know] an argument which certainly sounds paradoxical.

This is the end of it for Rayleigh.

So we can see that the acceptance of these ideas was something that came very, very slowly. The examples I have chosen illustrate that very little was stated about the subject at all from 1900 to 1905, and even after that you find the great men of the period hesitant and unwilling to build it into their thinking.

Let us now turn to Einstein's famous 1905 paper, which I must confess I had not read until I got to thinking over the preparation for this lecture. It is one of the papers we all hear about in school and worship, but do not read. One of the odd things about this paper is that "h" is not in it, believe it or not. In the paper Einstein denotes by the letter β what we now would call h/k, and then he writes R/N for what we call k, and thus you find in that paper that the energy of a light quantum is not $h\nu$ at all. It is $R\beta\nu/N$, which certainly takes a bit of getting used to.

The title of his paper is an interesting one: "Heuristic Viewpoint Concerning the Emission and Transformation of Light", indicating, I think, that he meant that there is something in the paper, but he does not quite know what. At least that is what I mean when I say "heuristic". Einstein might, of course, have meant something else. He says:

The energy of a ponderable body cannot be divided into indefinitely many indefinitely small parts, whereas the energy emitted by a point light source is regarded on the Maxwell theory or more generally according to every wave theory as continuously spread over a continuously increasing volume.

Such wave theories of light have given a good representation of purely optical phenomena and will surely not be replaced by any other theory. [He was right in that. They have not been replaced.]

He continues:

It is to be remembered, however, that the optical observations referred to time mean values, not to instantaneous values, and it is quite conceivable that, in spite of complete success in dealing with diffraction, reflection, refraction, dispersion, et cetera, such a theory of continuous fields could lead to contradictions with experience when applied to phenomena of light emission and absorption.

After a little more discussion he makes the key declaration that played such a decisive role in all subsequent developments in which, in one sentence, he says:

According to the supposition here considered, the energy in the light propagated from rays from a point is not smeared out continuously over larger and larger volumes, but rather consists of a finite number of energy quanta localized at space points, which move without breaking up and which can be absorbed or emitted only as wholes.

Oddly enough, though nowadays this paper is quoted purely for the photoelectric effect in the discussion, the photoelectric effect is only one section, paragraph 8, of the whole paper. The first six paragraphs are concerned entirely with another way of looking at the details of the statistical distribution of the black-body radiation law, and the entropy of radiation along the lines of the quantized wave theory.

Finally, paragraph 7 is an interpretation of the Stokes' rules for photoluminescence. In ordinary fluorescence and phosphorescence, Stokes' law, which goes way back to 1860, says that the wavelength of the fluorescent light is always greater than the wavelength of the exciting light, or nearly so. There is some radiation, called anti-Stokes radiation, for which the wavelength is a little shorter.

Why there was so little stress in Einstein's paper on the photoelectric effect, compared with these other things, puzzled me. Then I came to section 8, which deals with the photoelectric effect, and I asked Professor A.L. Hughes, one of the pioneers of photoelectric work, who is at our place — he is emeritus professor in Washington University in St. Louis — how that could be. He told me how very primitive the knowledge of the photoelectric effect was at that time. No vacuum work on photoelectricity had been done, and, even if it had, it would have been done with very poor vacuums, under very poor conditions. In point of fact, no effort had been made to determine the retarding potential required to stop the photo-current in a definite circuit.

What had been observed was that, if you insulate a metal object and shine light on it, it will build up to a certain potential and then stay at that potential. That is, it builds up its own retarding potential and finally prevents the escape of further electrons into the air. Metals differ with regard to the potental built up by certain kinds of light, and it was found that as one went to more and more violet light one got a higher potential. But these were only very crude measurements indeed, so crude that one would hardly think that they might offer any possibility of fundamental understanding.

That, perhaps, is the reason why the photoelectric effect was so little stressed in Einstein's paper. The Stokes'-law argument was much more directly experimental, and, conversely, it seems rather odd to me, as I think about it, that Stokes' law is not more stressed today in teaching the subject.

It was in 1907 that the specific-heat work of Einstein clarified the problem of low-temperature specific heat.

It is fascinating to look up some of the historical information that is available in the literature. I do not mean that one has to go to ancient history, just the history of the last century. For example, in 1904, when the great St. Louis World's Fair was held, various distinguished visitors presented lectures. Lord Kelvin gave a speech suggestive of a sort of inverse neutrino theory. The thing bothering him at that time was — this is a little off the subject of quantum theory, but I think it is interesting — the measurement that had just been made by the Curies of the amount of energy given off by radium per unit time. They had not measured the half life, and the energy given off did not show any signs of weakening, and you know that physicists are great on extrapolation. They said that radium gives off energy *perpetually* — that was the word, perpetually.

So the question was, how could anything radiate perpetually at this tremendous rate? — a rate unheard of when expressed in terms of energies of usual chemical reactions.

Kelvin had an idea which he propounded at this talk; perhaps, he suggested, there was some kind of energy that one could not detect — like the neutrinos — floating around in space, and perhaps radium had the property of absorbing it in that form and then reconverting it, like a fountain, and shooting it out, and that was what was observed. Even in those days, people were perfectly willing to balance the books on conservation of energy in such a manner.

The next major historical event was the development of the Bohr atom model in 1913. At this point, since we are just talking a little bit of anecdotal material about the history of our subject, I will tell a story that I learned from George Gamow. The young Bohr — he was about 26 at that time — came to England from Copenhagen to work in the Cavendish Laboratory. The great J.J. Thomson was at the height of his powers. Bohr came to the great center to study fundamental atomic physics, but within a few months he left the Cavendish Laboratory and went up to Manchester to work under a relatively unknown fellow named Rutherford. The question is, why did he do that ? According to Gamow, Bohr had gotten into trouble with "J.J." because he was a little critical of the Thomson

J.J. Thomson   The young Bohr

atom model, and "J.J." had politely indicated to him that it might be nice if he left Cambridge and went to work with Rugtherford. That is how Bohr went to work for Rutherford, which was advantageous, I think, for all. It was not so good for the Thomson model but it was fine for the future development of physics.

To bring this story up to date, Gamow told me that in 1928, when he worked on the alpha-particle tunneling paper, the basic work which Gurney and I did simultaneously in Princeton, Rutherford sent Gamow to see Bohr and to tell him about this exciting new development. He also wrote him a letter — of which Gamow said he still has a copy — saying, "Please pay attention to this fellow; there is something in it. It isn't cockeyed. You remember how it was with you when you went to 'J.J.' and he wouldn't listen; so now you listen to Gamow." I do not know whether there is any truth in that or not, but at any rate Bohr did listen to Gamow.

Of course the most exciting immediate experimental consequence of Bohr's work, besides the direct interpretation of the spectrum of hydrogen which was well known at that time (I mean the facts of the Balmer series, which went way back into the 19th century), was the interpretation of spectroscopic-term values as being energy levels with the associated implication that controlled electron impact would produce controlled excitation of atoms and molecules. This was the work that was immediately carried further by James Franck and Gustav Hertz, and for which they received the Nobel Prize in 1926.

That work was very quickly taken up here in Washington, in the pioneer work of Paul Foote and F.L. Mohler. At the Bureau of Standards the accountants were rather stuffy, and had rather sharp lines about appropriations and budgets, so all the work on critical potentials for which the Bureau became famous was carried on under a budget number which had something to do with improving pyrometric methods. I am not sure it helped much in advancing pyrometry, but it certainly was a great addition to the development of science.

The period of the second decade of our subject was also characterized by the very first extension of the idea of quantized energy levels to the interpretation of band spectra, rotation and vibration spectra, and infrared.

A curious thing about the atom-model work of Bohr, prior to 1923 or 1924, was that if you look at the then-current papers you get the impression that everybody in the world was terrifically excited about the Bohr model and believed in it hook, line, and sinker, including the electron orbits as they are used in the ads for the atomic age nowadays. Bohr, on the other hand, was constantly making remarks, speeches, and admonitions to the effect that this is temporary and we ought to be looking for a way to do it right.

The great breakthrough, as the modern saying goes, came about in 1924, 1925, and 1926, when the idea of waves accompanying electrons was first

Schrödinger   de Broglie

Born   Heisenberg

published by de Broglie as a doctor's thesis — and was also ignored. I do not know anybody who read that paper until a year or two later. Schrödinger then founded the great discoveries of wave mechanics on de Broglie's work in the series published in the spring of 1926.

Just before Schrödinger's work in late 1925, Born, Jordan, and Heisenberg had developed the matrixmechanics methods. For about a year they were thought of as two rival and distinct theories, until Schrödinger and Carl Eckart, then a young physicist in Chicago, who is now in La Jolla, recognized the mathematical identity of the two theories.

I had the good fortune to get my doctorate in the summer of 1926 when all these things were at their highest peak of excitement, and went to Göttingen to work with Born. These was a young graduate student there named Robert Oppenheimer with whom I got acquainted at that time.

It was an extremely difficult period because the rate of advance was so great, and the whole subject was so obscure to all of us, that it was hard to keep up with the state of affairs. I remember that David Hilbert was lecturing on quantum theory that fall, although he was in very poor health at the time. (He had anemia, and liver extract was then unavailable, so he was eating a vast quantity of liver every day and saying he would rather not live than eat that much liver. His life was saved by the fact that liver extract was discovered just about that time.) But that is not the point of my story. What I was going to say is that Hilbert was having a great laugh on Born and Heisenberg and the Göttingen theoretical physicists because when they first discovered matrix mechanics they were having, of course, the same kind of trouble that everybody else had in trying to solve problems and to manipulate and really do things with matrices. So they went to Hilbert for help, and Hilbert said the only times that he had ever had anything to do with matrices was when they came up as a sort of by-product of the eigenvalues of the boundary-value problem of a differential equation. So if you look for the differential equation which has these matrices you can probably do more with that. They had thought it was a goofy idea and that Hilbert did not know what he was talking about, so he was having a lot of fun pointing out to them that they could have discovered Schrödinger's wave mechanics six months earlier if they had paid a little more attention to him.

I mention some of the occurrences of those years because I do not believe that anybody who did not live through that period can fully appreciate what a tremendous number of things happened then that are still very basic, and have blossomed out into whole areas of physics which now are subjects for courses in themselves.

In 1926 we had the whole wave mechanics, as we know it, and the whole matric mechanics formulated, And just a little before that, we had the discovery of the electron spin and of the Pauli exclusion principle. In 1927 came the whole theory of the chemical valence bond as a perturbation problem in quantum mechanics, with correlations over electron pairs with their spins antiparallel. Almost simultaneously, there occurred the whole development of Fermi-Dirac statistics and its clarification of the problems of metal theory. A few months later, the Dirac papers on the quantization of the electromagnetic field explained, at last, the difference between spontaneous and induced emission and put the two together in a unified theory. Soon after that came the whole Dirac relativistic theory of the electron, which later led to the prediction of the positron.

Shortly thereafter, in 1928, the interpretation of natural alpha radioactivity came as a consequence of the barrier-leakage idea, also an essential element of quantum mechanics and an essential element of its statistical or probability interpretation. I think it is fair to say that the barrier-leakage idea was the opening of the modern period of the application of quantum mechanics to nuclear physics. Nuclear physics, in terms of real specific models, has never had a classical past. Nobody tried in those days to develop specific models of the structure of a nucleus.

Another big year for discoveries was 1932, the year in which Urey discovered heavy hydrogen, which from a nuclear point of view means the deuteron, the year in which the first production of an artificial nuclear reaction was accomplished by Cockcroft and Walton, the year in which the positron was discovered — the antiparticle associated with the electron, as we call it nowadays, and the year in which the neutron was discovered.

In that same decade, a few years later, 1936 saw the development of the Fermi theory of beta decay based on the neutrino hypothesis that had been introduced by Pauli, in almost a joking way, a year or two earlier.

I remember that in the summer of 1937, when we had a conference on beta-decay theory at Cornell University, and a lot of us were having trouble worrying about it, Fermi was in the audience sitting in the back row just smiling and smiling as he usually did. People tried to get him to comment, and he said, "I have always been surprised that people take that theory so seriously." But, as we know, it has turned out to be remarkably correct — that is, the basic formalism which Fermi developed then for accounting for the four-fermion interactions, even in spite of the great crisis it went through in 1957 with the discovery of nonconservation of parity. The basic formalism, as Fermi first introduced it, has beautifully stood the test of time.

The year 1936 is also important to us here because of work done by prominent people in Washington. I refer to the work of Hafstad, Heidenberg, and Tuve in the first real studies of proton-proton scattering, which gave direct evidence of forces between protons other than the Coulomb forces, that is, short-range nuclear forces between protons. The theoretical interpretations of those results were largely done in Princeton by Gregory Breit and myself, in association with Richard Present, who is now at the University of Tennessee. That provided the first evidence of what is now called the charge independence of nuclear forces, because the additional short-range force that was revealed in this way turned out quantitatively to be very close to the force between a proton and a neutron which is revealed in the normal state of the deuteron.

From about 1932 on, the whole field by nuclear physics came into being in a big way, with deuterons available and with machines available, both cyclotrons and Van de Graaff machines. In the latter part of that decade, we began to have the first theories of Bethe and Marshak on the application of specific models of nuclear reactions to the problem of finding satisfactory sources of stellar energy.

I think perhaps I must give up at this point, because the last two decades have seen such an overwhelmingly rapid and vast amount of progress, spreading out into so great many different fields, that one could not possibly, in the short time remaining, do more than just mention it.

We had, in the decade from 1940 to 1950, the whole development of the modern point of view on quantum electrodynamics. It came rather late in the decade, with the discovery of the Lamb shift and the experimental confirmation of the abnormal magnetic moment of the electron, which was somewhat off from the original Dirac theory. We had at last the clarification of the puzzling features of the mesons in cosmic rays, whereby it turned out that there were the two kinds, the pi mesons and the mu mesons, the pi mesons decaying into the mu mesons. The latter part of the decade represented the beginning of public knowledge of fission, and the engineering and political uses of fission.

At the same time, going off in quite another direction, what has turned out to be of equal importance has been the whole wide development of the application of Fermi statistics to electrons in solids, first resulting in the major classification of properties of metals, then of semiconductors, and then finally of really modern tailored effects that led to transistors and other devices.

The decade just passed has corresponded to an enormous further development along these same lines. We have the study of nuclear reactions going on up to higher energies of some hundreds of millions of volts, with predominant interest in the study of polarization effects in nuclear reactions as another way of getting at points of detail; the recognition of the nonconservation of parity; the experimental discovery of the neutrino; the recognition that the Fermi interaction that applies in weak interactions is more general than simply the beta decay, applying also to muon decay and other related processes; and the discovery of the strange particles.

And then finally, as a roundup of mentioning things that we do not have time to talk about, there are the extraordinarily fine extensions that have been made in the last five years of the theory of broad, modern, good perturbation-theory methods for dealing with the many-body problem. They involved not only the better calculation of nuclear models but also, at last, after many years of effort, they are beginning to provide a real understanding of superfluids.

I want to close by remarking that all this started, as I said, almost exactly 60 years ago — barring two weeks — on December 14, 1900, when Planck's constant was first introduced into physics. In the 60 years that have intervened it is now almost impossilbe to find many papers in physics which do not deal directly or indirectly with phenomena that are fully and basically conditioned by the existence of that one universal constant.

Professor A.H. Compton

E.U. CONDON

Arthur Holly Compton died on March 15 in Berkeley, California, as a result of cerebral haemorrhage suffered two weeks earlier. He was sixty-nine.

He had retired last year as distinguished service professor of natural philosophy in Washington University (St. Louis), where he had served as chancellor from 1945 until 1953. He was planning to be active in retirement as professor-at-large between Washington University, the University of California (Berkeley) and the College of Wooster (Ohio), and had gone to Berkeley to deliver a lecture series on "Man, Science and Society".

Compton received his B.Sc. from the College of Wooster, where his father, Elias Compton, was professor of philosophy, and received his Ph.D. in 1916 at Princeton University. The next year he was an instructor at the University of Minnesota, and then was for two years a research engineer for the Westinghouse Lamp Company in Pittsburgh. The year 1919–20 was spent at Cambridge (England) as a National Research Fellow, after which he was appointed Wayman Crow professor and head of the Department of Physics at Washington University, where he remained until 1923. During 1923–45 he served as professor of physics and dean of physical sciences at the University of Chicago, until he returned to Washington University as chancellor in 1945. He was president of the American Physical Society in 1934, of the American Association of Scientific Workers in 1939–40 and of the American Association for the Advancement of Science in 1942. In 1927 he shared the Nobel Prize in physics with C.T.R. Wilson.

His career was marked by an extraordinary range of great accomplishments in physics, in higher education, in war-time scientific research, and in efforts to improve human and international relations.

While a student at Princeton he devised a beautiful demonstration of the Earth's rotation, which ought to be known to all teachers of physics. A toroidal glass tube is filled with water, and mounted so that it can be 'flipped' through 180° about a diameter as axis. Before flipping, the water is at rest relative to the tube, which is turning, relative to the fixed stars, in its

279

own plane with the component of the Earth's rotational angular velocity normal to that plane. After flipping, the water drifts relative to the tube, at a rate proportional to that angular velocity component. By measuring the drifts produced by three successive flips about each of three mutually perpendicular axes, one finds the Earth's angular velocity as a vector; that is, one can infer which local direction is north, what is the observer's latitude, and what is the absolute value of the length of the day.

His first major discovery was the detailed measurement and interpretation of the wave-length change occurring when X-rays are scattered, especially by materials of low atomic number. This is now generally known as the Compton effect. He showed that the loosely bound electrons in the material scatter the X-rays in accordance with the principles of conservation of momentum and energy, as if they consist of a stream of photons, each having momentum of $h\nu/c$, as well as energy, $h\nu$. The energy aspect goes back to Planck and Einstein; but the Compton effect afforded the first clear demonstration that the X-ray photons also carry quantized amounts of momentum.

The next few years were marked by the development of coincidence methods by Compton and A.W. Simon in Chicago, and independently by W. Bothe and H. Geiger in Germany. These experiments showed that individual scattered X-ray photons and recoil electrons appear at the same instant in time, contrary to some views that were then being developed by Bohr, Kramers and Slater in an attempt to reconcile quantum views with the continuous waves of electromagnetic theory.

Compton also discovered the phenomenon of total reflexion of X-rays, and their complete polarization (with C.F. Hagenow), and first obtained (with R.L. Doan) X-ray spectra from ruled gratings. This latter work had an important consequence in leading to a distinct improvement in our knowledge of the electronic charge. By measuring an X-ray wave-length with a ruled grating of known grating space, and then using a crystal to diffract the same rays, one can determine the absolute value of the grating space of the crystal. Combining this with the measured crystal density, it is possible to obtain the Avogadro number, and combining this with the Faraday, the electronic charge is obtained. The outcome was that the Millikan oil drop value had to be revised from 4.774 to 4.803×10^{-10} c.g.s. E.S.U., it being finally recognized that systematic errors had been made in measuring the viscosity of air, a quantity which enters into the oil drop method.

From about 1930, Compton directed his attention mainly to the study of cosmic rays. In the next ten years he was in charge of a major programme involving a world-wide study of the geographic variations of their intensity. This resulted in full confirmation of some observations made in 1927 by J. Clay, indicating a latitude effect on the intensity. The world survey, in the service of which Compton made many long voyages, showed correlation of the intensity of cosmic rays with geomagnetic, rather than geographic, latitude. This opened the way for extensive subsequent studies of the interaction of the Earth's magnetic field with the incoming isotropic stream of primary charged particles.

In 1941 he was appointed chairman of the National Academy of Sciences Committee to Evaluate Use of Atomic Energy in War. In the autumn of 1941 this Committee worked with those responsible for studying this same problem in Great Britain, and also with the S-1 committee headed by Dr. L.J. Briggs, then director of the National Bureau of Standards. This led to recommendations to the United States Government for the setting up of a major effort starting in January 1942. Compton assumed the active direction of the group, which was known by the 'cover-name' of Metallurgical Laboratory of the University of Chicago, which concentrated on development of controlled uranium fission reactors for the production of plutonium. Within the year the first controlled uranium fission reactor was operating at Chicago, this specific project being largely due to the efforts of E. Fermi, L. Szilard, E.P. Wigner, and a host of co-workers. More experimental reactors were designed and built at Oak Ridge, Tennessee, and at the newly established Argonne Laboratory in the suburbs of Chicago. Here was carried out the work which led to the large plutonium-producing reactors built at Hanford, Washington, which produced the plutonium for the atom bomb that destroyed Nagasaki in August 1945.

Throughout the War, Compton also played an important part in the general planning of the atom bomb project, including the setting up of the laboratory at Los Alamos, New Mexico, and in reaching the military–political decisions about the use of the bombs in Japan. He has given a personal account of these matters in his book, *Atomic Quest* (Oxford University Press, 1956).

Compton became chancellor of Washington University in 1945, bringing with him a close associate from the days of his work on cosmic rays, Joyce Stearns, as dean to faculties. He turned with great vigour to the task of re-making a University which had greatly suffered through the depression years and the War years. In eight years filled with hard work he set it along a path toward greatness along which it continues to move. Then in 1953 he asked to be relieved from administrative duties, so that he could return to teaching, and devote himself entirely to problems of the social impact of science and technology, and to ways of promoting world brotherhood and the relieving of international tensions. He was planning to continue active work on such problems into the retirement which ended soon after it began.

He is survived by his wife, formerly Betty Charity McCloskey, and two sons, Arthur Allen, now in the American Foreign Service stationed at Manila in the Philippines, and John Joseph, a professor of philosophy at

Vanderbilt University (Nashville, Tennessee), and by a brother, Wilson Compton, now retired, a former president of Washington State University (Pullman, Washington). Another brother, Karl Taylo Compton, former president of the Massachusetts Institute of Technology, died in 1954.

Lyman James Briggs (1874–1963)

E.U. Condon

Lyman James Briggs was born on May 7, 1874, in Assyria, Michigan. He received a Bachelor of Science degree from Michigan State College at the age of nineteen and two years later a Master of Science degree from the University of Michigan. He became a graduate student under Henry A. Rowland at the Johns Hopkins University in 1895, working as a physicist on a part-time basis for the U.S. Department of Agriculture until he received the doctorate in 1901, after which he spent all his life in the federal service. Thus he served the Government actively for fifty years, with seventeen years more of fruitful work at the National Bureau of Standards after he became Director Emeritus in 1945.

The first period of Briggs's career was devoted to applications of physics to agriculture. In 1906 he organized the Biophysical Laboratory, which was the forerunner of the Bureau of Plant Industry. His principal interest at this time was the study of the water requirements of plants. He established a method of classifying soils, based on determination of the amount of water retained under centrifugation, which is still a standard technique in soil physics. Retaining a strong interest in surface tension phenomena during all of his life, after retirement he devoted much effort to the study of negative pressure in liquids in his laboratory at the National Bureau of Standards. He prepared the article on "Capillarity" for the 1955 edition of the *Encyclopedia Americana*.

Concerning this interest, let me quote from his own remarks on the occasion of a luncheon at the Cosmos Club on May 7, 1954, honoring his eightieth birthday:

. . . . I never felt quite happy considering the soil alone. The growing plant is the marvellous thing. So I was happy to be able later to go into biophysical work in the Bureau of Plant Industry. Dr. Shantz and I studied the water requirement of plants, particularly on the Great Plains, which were then being developed for agriculture. Unfortunately, eastern methods of agriculture were employed by the new settlers. In that semiarid region countless acres of beautiful buffalo grass, which was the greatest stock-food plant in the world, were plowed under ruthlessly. The result is we have had the drought and dust storms of 1930 and the dust storms of 1950, which could have been eliminated if proper methods of cultivation had been insisted upon.

The State of Colorado has recently passed laws which will force farmers to follow conservation methods. If that could have been done thirty to forty years ago, it would have been a great thing for the agriculture of the region.

In 1917 Briggs was transferred by executive order to the staff of the National Bureau of Standards to work on the development of navigational instruments for the Navy. Thus began his long service with the Bureau, where he received a permanent appointment in 1920 as chief of the Mechanics and Sound Division. He became the third director of the Bureau after the sudden death of George K. Burgess in July, 1932.

In the early part of this period, Briggs worked closely with Paul R. Heyl in a collaboration which resulted in the invention of the earth inductor compass. For this they were jointly awarded the Magellanic Medal of the American Philosophical Society in 1922. Briggs was elected a member of the American Philosophical Society in 1935.

Briggs started the program of aerodynamic research at the Bureau, concentrating on development of techniques for air-flow measurements up to the beyond the speed of sound. The National Advisory Committee for Aeronautics (NACA) was established in 1917 with an appropriation from Congress of $5,000, "or such part thereof as may be needed," to conduct this nation's research program in aeronautics. This agency was the predecessor of the present National Aeronautics and Space Administration (NASA), whose budget in the fiscal year 1964 is approximately one million times that initial appropriation to NACA. Briggs was always intensely acitive in the work of NACA and served as its vice-chairman for several years during World War II.

The third phase of his career was that in which he served as third Director of the National Bureau of Standards from 1932 to 1945. This connection began with his appointment as Acting Director by President Hoover in 1932. At that year's end, President Hoover sent to the Senate Briggs's nomination to the directorship, but it was not acted upon because of the impending change of Administration. Later he was nominated and installed as director by President Roosevelt, and thus had the rare distinction of having been selected both by a Republican and by a Democratic president.

It is difficult now to remember and to believe that Franklin D. Roosevelt made campaign promises in 1932 to eliminate the "wasteful extravagance" of the Hoover Administration — he who a few years later was denounced by his opponents as a great spender and whose Administration saw the federal budget pass one billion dollars a year for the first time. Tragically, the National Bureau of Standards was caught in that brief period of severe retrenchment at the beginning of the Roosevelt administration.

The Bureau's budget was slashed by more than 50 per cent in the midst of the severe depression, in what must sadly be recorded as one of the major mistakes of Roosevelt's career. To Briggs fell the task of preserving the Bureau as a great national scientific asset by making that little budget go as far as possible. He was able to keep at work some two-thirds of the career staff, by putting many of them on part-time employment, and he also did a vast amount of good work by personal efforts in aid of those whose employment was terminated at a time when there were almost no jobs open for scientists in universities or in industry in America. A man of lesser character than his might have become embittered or disgusted, but he remained steadfastly at the helm and brought the Bureau through this storm with as little damage as possible.

Thus it was that the Bureau was available as a strong scientific organization when the defense program of 1940 made demands on its services that were soon to absorb over 90 per cent of the attention of an expanded staff.

Briggs was called to the White House in October, 1939, by President Roosevelt, less than a year after the discovery of uranium fission, and assigned the responsibility of organizing a top-secret investigation into the possibility of using fast neutron chain reactions to produce nuclear explosives. Work was started at the Bureau on the thermal diffusion method for isotope separation, by J.W. Beams at the University of Virginia on the use of centrifuges, and by John R. Dunning at Columbia University on membrane diffusion methods. Support was given to the pioneer work on chain reactions by Enrico Fermi and Leo Szilard at Columbia University.

This program became the S-1 committee of the National Defense Research Committee after that was organized in June 1940. Briggs was chairman, G.B. Pegram was vice-chairman, and the other members were H.C. Urey, J.W. Beams, M.A. Tuve, R. Gunn, and G. Breit. A year later the committee was expanded by additon of H.D. Smyth and E.U. Condon. Later the S-1 committee became an adjunct of the Office of Scientific Research and Development. During 1941 the Briggs committee work was greatly expanded. The decision to recommend a major all-out effort to make an "atom bomb" was taken by this committee at a meeting at the National Bureau of Standards on the weekend just preceding the Japanese attack on Pearl Harbor, December 7, 1941.

From that time on the project grew at an enormous pace, and eventually its management was turned over to a specially organized unit of the Army Engineer Corps which provided the scientists with the necessary house-keeping arrangements. In the initial phases of this great growth more than sixty scientists from the Bureau were assigned to Oak Ridge and Los Alamos. The Bureau's principal contribution in its own laboratories was that of perfecting analytical chemical methods for controlling the purity of critical materials in the reactors and in the bomb.

Briggs took an active interest in the many other projects by which the Bureau contributed to the war effort. These included development of the nonrotating proximity fuse, a guided missile project which produced a primitive (by today's standards) device which had service in the Pacific, and the establishment of a major program for forecasting ionospheric conditions affecting radio communications. All of this is told in *NBS War Research — the National Bureau of Standards in World War II*, which was written by Briggs after his retirement. Unfortunately, it was issued only in a small edition in 1949, and so has not been widely read, but it must be regarded as an important source document for the history of the contribution of scientific research to the war.

In addition to his goverment work, Briggs devoted a great deal of personal effort to the work of the committee of Research and Exploration, of the National Geographic Society, especially in connection with the pioneering stratospheric balloon flights in 1935 and 1936.

His service as Director of the Bureau extended until November, 1945, beyond the statutory retirement age, when the writer of this memoir was sworn in as fourth Director. Briggs then took a laboratory in the West Building, where he did research mostly on negative pressure in liquids, and continued to be a valuable consultant to his successors and to many others on the Bureau staff, almost until the end of his life on March 25, 1963.

He was a man of unfailing courtesy, kindness, tact, and consideration. We shall never know the complete story of the many ways in which he helped others scientifically and in their personal relationships. His friendship was a source of great encouragement to me during the difficult days of the persecution of scientists by the late Senator McCarthy and other politicians, which did so much to damage the careers of able scientists, and added so greatly to the difficulties of the federal government in recruiting good men into its service. Later, this same kindly smiling understanding of how to carry on in the face of inexcusably stupid criticism also gave strength and support to my successor, A. V. Astin, during his trial by battery additive when the then Secretary of Commerce was exalting the "free play of the market place" above the results of careful scientific tests.

Briggs should always he remembered as one of the great figures in Washington during the first half of this century, when the federal government was slowly and stumblingly groping toward a realization of the important role which science must play in the full future development of human society.

William Shipley's Barometer (1748)

E.U. CONDON

Department of Physics and Astrophysics
University of Colorado
Boulder, Colorado

In 1748 William Shipley of Northampton wrote to his London friend Henry Baker proposing a type of barometer in which the variations of barometric pressure would be easier to read.[1] His letter to Baker and Baker's reply are printed in the reference cited. The proposal provides an interesting problem in elementary hydrostatics even though the barometer apparently was never constructed. The Shipley proposal for a barometer is shown in Fig. 1 which is reproduced from a sketch in his letter to Baker.

The idea is that the mercury column rises up from the basin B to some place in the enlarged cylindrical bulb A. A lighter fluid floats on top of the mercury, filling A and extending on up to the tube of small bore C, above which there is a Torricellian vacuum as in usual forms of barometer. Changes in air pressure are observed as variations in the height of the column of the lighter fluid in C. Shipley writes: "I suppose that the atmosphere pressing on the extrenal Mercury in the Bason B, and elevating the fluid in the Box A, if it is 16 times lighter than the Mercury, will be elevated in the tube C, 16 times higher from the Box than the Mercury alone would rise by the same Pressure of the Atmosphere."

It is evident that this cannot be entirely correct for it makes no reference to the relative cross-sectional areas at A and C. If the areas do not enter the final result, we may as well set the ratio equal to unity and not bother with the complication of enlarging the cross-section at A. Clearly also if the lighter fluid were of zero density and so did not contribute to the pressure, then a rise of mercury in A would be amplified by a change of fluid level in C by the ratio of the cross-sectional areas.

Baker apologizes for a delay of less than two months in replying, saying that the delay was due "to a Desire of giving you all the Satisfaction possible in Relation to your new-constructed Barometer which I would not do without consulting Workmen as well as Philosophers, and you known that Tradesmen of all Arts are not over hasty in Matters which require consideration and are not of the common Road." He goes on to say that a "most valuable and ingenious friend Mr. Folkes, who after taking some

---

[1] J. Roy. Soc. Arts, London **113**, 829 (1965).

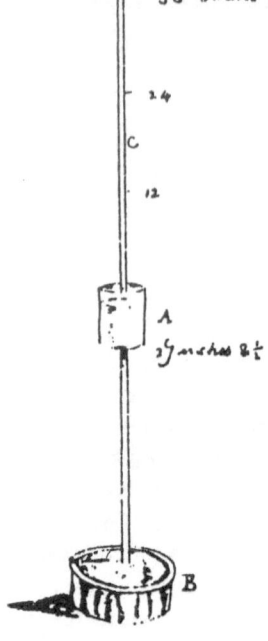

'I suppose the Cylindrical
Box need be no longer than
the difference between the
Mercury when at Highest &
lowest in common
Barometers.'

Shipley's barometer: diagram and note accompanying Letter I.

Days to consider the Matter thoroughly, has set it in a true light by showing
from a way of Reasoning, amounting nearly to a Proof, that the lighter
fluid will not rise as you imagine in the same proportion as the respective
Gravity of one to the other, but that the rising and falling will be in a
compound Ratio of their different Gravities . . . " Baker concludes that
the barometer "being a Thing out of the way, it will I doubt be pretty
expensive, unless you can perform most of the troublesome Work
yourself."

Let H_1 be the column length occupied by the light fluid when it is all in
C, and H_0 be the height of the top of A above the mercury level in B when
the mercury stands at the top of A. The equivalent barometric height in
this condition is

$$h_0 = H_0 + \rho H_1,$$

where ρ is the ratio of the density of the light fluid to that of mercury.

Let α and β stand for the ratios of the cross-sectional area of C to that of
A and B, respectively. When the pressure is reduced, so that some light
fluid flows into A, and the light fluid now stands at $H(<H_1)$, the equivalent
height of the light fluid is $\rho H + \rho\alpha(H_1 - H)$, and the mercury column
height is changed from H_0 to $H_0 - (\alpha + \beta) (H_1 - H)$, so the equivalent
barometric height becomes

$$h = H_0 + [\rho(1 - \alpha) + \alpha + \beta] (H_1 - H_1).$$

Hence variations of h are shown as larger variations of H because the quantity in brackets is less than unity. If α and β are negligible, the gain in sensitivity is that surmised by Shipley, and not in a compound ratio as stated by Folkes.

In his next letter, Shipley thanks Baker for his trouble and mentions that in the meantime he has learned that the idea which he had thought to be original had really been suggested much earlier by Descartes.

Measures for Progress — A History of the National Bureau of Standards

By Rexmond C. Cochrane and
Reviewed by E.U. Condon*

This is a work of great importance to all who are interested in the growth of research in the physical sciences in the United States of America, and its support (and often the lack thereof) by the Federal Government, from the time in which this nation was primarily engaged in agriculture and mining until the post World War II period, in which it leads the world in physical science research, modern technology, and the productivity of its science-based manufacturing industry.

It is highly appropriate that the book is published in this, the sixty-fifth year since NBS was established, when the Bureau has just completed moving into a new $120 million facility at Gaithersburg, Maryland. The new facility was appropriately dedicated in November with a major conference on Technology and World Trade, sponsored by John T. Connor, Secretary of Commerce.

In the author's introduction, Cochrane writes, "Some wonderful controversies have engulfed the Bureau from time to time". The book gives an instructive account of many of them, and affords the reader a view of how it has always been true that there is never a dull moment in the director's office.

It will be realized that the story of NBS is completely intertwined with the story of the development of physical science research in the universities, in industry, and in other government laboratories. Thus, the history is of much wider interest than if it were merely the story of one particular and quite small Government agency.

* Fourth director (1945–51) of the National Bureau of Standards. Editorial consultant, James R. Newman. With a foreword by Vannevar Bush and a preface by Allen V. Astin. 703 pp. Published by the National Bureau of Standards, U.S. Department of Commerce, 1966.

The author is Professor of physics and Fellow of the Joint Institute for Laboratory Astrophysics at the University of Colorado, Boulder, Colorado.
Received 1 November 1966.
This publication is obtainable from the Superintendent of Documents, U.S. Government Printing Office, Washington, D.C. 20402.
The photographs in this article have been kindly provided by the National Bureau of Standards.

Fortunately, Cochrane has been careful to paint the broad picture within which NBS has played its role. Some younger readers will be surprised to learn how many government policy questions which are still being debated today (such as the question whether all science activity of the Government should be assembled into a single Department of Science, under a Secretary of Science in the President's cabinet) were being activity debated decades ago.

Then there is the question of the everyday use of the metric system, which is still being held up in the House Rules Committee by congressmen who proudly announce their inability to comprehend its importance. It is all of a piece with (p. 47) congressional comments made in the House of Representatives on 2 March 1901, when the House was considering the bill to establish NBS. A Georgian representative said, "If I understood it, or if it were possible for me to understand it . . . I might be in favor of it". And "Uncle Joe" Cannon of Illinois, who was then Chairman of the House Appropriations Committee said, "I don't think there ought to be any such bureau organized".

But the very next day the bill was passed, and NBS was born into a scientifically undeveloped nation, authorized to pay up to $5000 a year for a director, and to hire a physicist and a chemist at $3500 each, two assistant physicists or chemists at $2200 each, and various other lesser personnel, including two laborers at $600 a year each, to a total of twenty-one. In fact, however, the first director started at a salary of $4000 a year and slowly worked his way up to $6000 a year when he left the Federal service twenty-one years later. The bill also authorized the expenditure of $250,000 for a laboratory building and $25,000 with which to equip it.

Initially NBS was in the Treasury Department, but in 1903 it was transferred to the newly formed Department of Commerce and Labor.

The first director, Samuel Wesley Stratton, came to Washington in 1899 to be Inspector of Standard Weights and Measures, leaving an assistant professorship at the new University of Chicago, where he was paid $2000 a year, and where he worked under A.A. Michelson. He had gone to Chicago in 1892 from an assistant professorship in the University of Illinois. As Inspector of Standard Weights and Measures he worked indefatigably for the establishment of the Bureau, and was appointed first director by President William McKinley, on the recommendation of Lyman J. Gage, the Secretary of the Treasury. After he took office as ninth president of the Massachusetts Institute of Technology on 1 January 1923, he continued to be associated with NBS as a member of its statutory Visiting Committee.

Stratton's successor as second director was George Kimball Burgess, then chief of the metallurgy division, who served until he died suddenly at his desk in the old south building on 2 July 1932. The third director was Lyman James Briggs, who was nominated for the post by President Herbert Hoover just at the end of Hoover's term, and then nominated

FIGURE 1. Designed and cast by NBS employees, this sundial was dedicated to the Directors of the National Bureau of Standards and presented to NBS three years after the retirement in 1945 of Lyman J. Briggs. It has recently been relocated on the grounds of the Bureau's new Gaithersburg laboratories.

again by President Franklin Roosevelt when he became president in 1933. Briggs served until retirement in 1945 at the age of 71, his term having been extended beyond the statutory retirement age under war powers legislation. I was sworn in as his successor in early November 1945, having been appointed by President Harry Truman, upon recommendation of Secretary of Commerce Henry A. Wallace. I remained until 30 September 1951, and was succeeded by Allen V. Astin, who presently is serving as the fifth director of NBS.

There is something almost frighteningly abrupt about becoming director. I was sworn in one afternoon at 4 p.m. in the office of Secretary Wallance, and from that moment *was* the director. The next morning I took a cab from the Wardman Park Hotel and told the driver to take me to the National Bureau of Standards. As we rode out Connecticut Avenue, he remarked: "Bureau of Standards, eh! They should develop some moral standards and some ethical standards". I agreed that they should, but his remark had only heightened the feelings of inadequacy with which I approached my new job. The situation for me was greatly complicated by the fact that many of the grand old division chiefs who had built the Bureau under Stratton were just at the retirement age, that a great program of war research had to be closed out and the work reoriented to peace-time activities, and that I was soon called on to spend a major portion of my time as Scientific Adviser to the Special Senate Committee on Atomic Energy (79th Congress) which, under the chairmanship of Senator Brien McMahon, drafted the legislation that established the Atomic Energy Commission. In that first year and later, Eugene C. Crittenden and Hugh Dryden, who were soon made associate directors, proved to be sources of great strength in carrying on the work of the Bureau.

Although the spring meetings of the American Physical Society had always been held at the Bureau, and I had not missed attending since I started in 1927, in all of those years I had never been inside an NBS laboratory. At first I was puzzled about why so many of the names of the great NBS old-timers seemed familiar to me. Then I recalled that, in 1916, when I was a high school boy in Oakland, California, and my interest in physics had been aroused by my teacher, William H. Williams, and by the exhibits in the Panama Pacific International Exposition in San Francisco in 1915, I had discovered that one could receive NBS publications gratis simply by writing for them. I had built up quite a collection and read them as thoroughly as I could for someone my age, but had later forgotten about this period.

I pass over the pre-1945 period because I want you to read Cochrane's fascinating account of it. Although I had picked up most of the story during my tenure as director, the book nevertheless gives much detail with which I was not familiar. Optical Society of America members will be delighted (p. 149) with the anecdote told about a North Carolina Senator on the Senate Appropriations Committee who was questioning Dr. Stratton: "You have a request for a physicist qualified in optics. I want to know what you do with that fellow. What is his business?" Stratton patiently explained and, as usual, obtained his appropriation.

Briggs served under extremely difficult circumstances. Roosevelt had campaigned against the "wasteful extravagances" of the Hoover administration, and, when he took office, immediately instituted drastic cuts in the funds available to the Bureau. It fell to Briggs' lot to try to maintain a skeleton Bureau that served the needs of science and industry, while witnessing a whole host of well-funded New Deal agencies come into existence. Despite Roosevelt's lack of understanding or appreciation of the Bureau's work, he did turn to Briggs for the initiation of the atomic bomb project (probably as a simple organization chart reaction) in October 1939 after he had received the famous letter from Einstein, which had been written to Roosevelt at the suggestion of Leo Szilard and Eugene Wigner.

By this time Briggs was sixty-six years of age and had struggled with the terrible financial restrictions on the Bureau for seven years. As Cochrane writes, "A younger man might have seized on the adventure into the unknown promised by nuclear fission, but Dr. Briggs had learned to be cautious. Nor was he at all certain that this was the kind of research, or direction of research, in which the Bureau ought to become involved. He and his committee hesitated". Some good work was supported on a limited scale in 1940 and part of 1941, and then word was received of considerably more progress in England. The committee under Briggs was enlarged by the addition of several new members, including myself, in September 1941.

When we recall that in the next four years approximately $2 billion was spent on the atomic bomb project which brought the war to an end a month short of four years later, it is almost unbelievable to recall that, at the first meeting of that uranium committee which I and the other new members attended, there was no attempt made to explain the present status of the work to us; instead, the first hour was filled by a long discussion of where was the best placed to buy some fused silica dishes that one of the chemists needed, the amount of the cost being under $5000.

Cochrane gives a good account of the stepping up of the project and the Bureau's continuing service to it after it had vastly outgrown the services available at NBS. On page 386 he reproduces the first page of the first report of the Los Alamos Scientific Laboratory which we called the "Los Alamos Primer", and which was declassified in 1963, twenty years after it was written. I must tell a little Freudian story on myself. For years I had the impression that I was the author of that report, and took some considerable personal pride in it. Then when it was declassified and I saw a copy after many years, I saw that it was really based on some lectures by Robert Serber, and that I had merely acted as his amanuensis! The whole memory of it came flooding back, and I remembered vividly sitting in Los Alamos and taking down the notes of Serber's lectures.

As this history shows, there was a time when the Bureau did a large amount of commodity acceptance testing in connection with the purchase of Government supplies, as well as the better-known services in the calibration of physical measuring instruments. Confronted with inadequate funds for the operation of a proper scientific agency and an adequate testing service, I made the decision to curtail the testing service on commodities, after it had become clear that Congress was unwilling to support a proper testing service.

The lack of a proper testing service nowadays, I am convinced, annually wastes hundreds of millions of dollars of taxpayers' money. Vast sums of money are spent on buying paper, and paint, and automobile tires, and all sorts of other things, on contracts which require them to meet elaborately prepared specifications. And then in most cases the government does not carry out tests to see whether the materials supplied conform to specifications. The result is that a great deal of inferior material is supplied: it has always been considered great sport to cheat the Government, usually when a certain type of business man, who is the loudest to raise the cry of bureaucratic interference with free enterprise, has his shoddy cloth, or his

FIGURE 2. The Joint Session on the Unification of Screw Threads, held at the National Bureau of Standards on 18 November, 1948, marked the culmination of the technical development of a common screw thread standard for Canada, the United Kingdom, and the United States. The session was attended by delegations from Government and industry of the three nations involved. Shown signing the Accord, Edward U. Condon, Chairman of the interdepartmental Screw Thread Committee of the United States and director of the National Bureau of Standards, who acted as co-chairman of the Session, and Hume Wrong, Ambassador of Canada.

poor cement, or his no-good paint rejected by a Government inspector. When I was director, we made a careful statistical study of the mileage obtained on post office trucks with various brands of tires, all costing about the same, and we knew that some brands gave three times the mileage given by the others. But such work is highly unrewarding to those who carry it out. It almost seems as if Congress wants the American government to be cheated.

I would not recommend going back to the pre-1945 days when a large part of the NBS energies went into such work; but I would recommend the establishment of adequate testing facilities for checking up on the quality of Government purchases, to be operated perhaps by the General Services Administration. Properly organized, such a government service would pay for itself many times over.

Cochrane gives quite a good account of the AD-X2 battery additive imbroglio of 1953, when, Sinclair Weeks, as President Eisenhower's Secretary of Commerce, abruptly fired Allen Astin as director because, he charged, NBS "has not been sufficiently obejctive because they discount entirely the play of the market place". He does not tell about how I got wind of what was about to happen and came down to Washington from

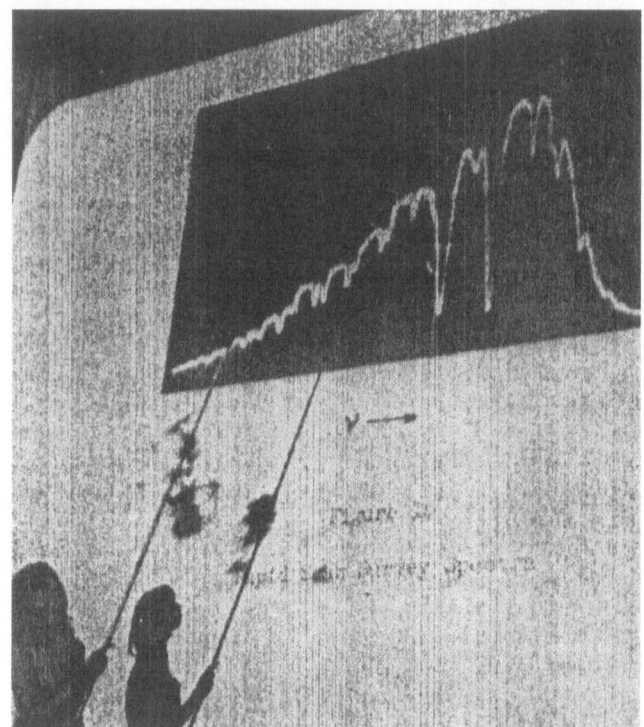

FIGURE 3. Joanne Moore Herr, a graduate student in the Department of Chemistry at Berkeley, photographed by Bryce Crawford, Jr., while making some remarks during the Gordon Conference on Infrared Spectroscopy last August G.C. Pimentel, head of Chemistry at the University of California at Berkeley, will be editor for the feature on Rapid Scan Spectroscopy to appear in the November 1969 issue.

Corning and had lunch with Drew Pearson, as a result of which Drew was able to "predict" Weeks' intention to fire Astin on the morning of the day that he did it, and in this instance, at least, he was 100% correct.

A part of this AD-X2 incident involved long drawn out Senate hearings which run to hundreds of pages in the printed transcript. I have always enjoyed one passage where, in the midst of a discussion of something else, a Senator suddenly remarked that it seemed to him a rather drastic move to fire a man of Dr. Condon's scientific distinction over the matter of a battery additive.

"It wasn't Dr. Condon, it was Dr. Astin", the witness explained to the Senator.

"Oh", replied the Senator, "I thought it was Dr. Condon", lapsing into silence as though to him it did not matter in the least what happened to Dr. Astin!

All such things are wasteful and foolish, but the Bureau goes no doing its great work year in and year out, despite the silliness with which it has to live. Gradually, I like to think, a better atmosphere is beginning to prevail, and Congress is taking a more responsible interest in trying to understand science, as we see in the splendid new plant for NBS that has been dedicated this past November.

Cochrane has produced a fasinating book which should be read thoroughly, and can be read with great enjoyment, by all who are interested in the developing relation of the Federal Government to the physical science establishment in this country. His concluding paragraph:

The present history has tried to show this life force that is the Bureau, acting as individuals and as agency, and the part it has played in the scientific, industrial, and business life of the Nation. As crusader and arbiter, creator and counselor of standards, it works for the future, as it has in the past, for the good society, and by its learning and good will makes itself felt throughout the Nation and the world.

The Past and the Future of the *Reviews of Modern Physics*

E.U. Condon

Joint Institute for Laboratory Astrophysics
Boulder, Colorado

This is the last issue of the *Reviews of Modern Physics* for which I am as Editor responsible. Or irresponsible, as some may prefer to say, because the records show that I was appointed Editor for the period 1957–1959, and my term in office was never extended by the Council of the American Physical Society, so for the past nine years I have been a usurper. But I forgot the limit originally placed on my term, and no one else showed up to serve as Editor, so I merely kept on doing what I could.

On this occasion I want to express my heartfelt thanks to all of those who have served the *Reviews of Modern Physics* in various way during the past twelve years, above all to Sam Goudsmit, Managing Editor for the Society's publications, the editorial staff of the American Institute of Physics which handles all production problems, the many individuals who have refereed submitted papers, the authors of the papers which were published, and the authors of the papers which were not published. Sam and I agree on objectives and have had some sharp disagreements but he, at least, was always a gentleman. Gentlemanliness is a quality that is most appreciated when it is absent, as when one is dealing with the House Committee on Un-American Activities (1948–1952) or with the more ardent believes that UFO's come from more advanced civilizations than ours (1968).

Feeling mellow at this time, I nevertheless cannot bring myself to thank those who promised to write important review papers but did not do so. More about that later.

I am delighted with the new arrangements for the future of the *Reviews of Modern Physics*. I am delighted with the selection of Lew Branscomb as my successor. I am proud of the fact that I recruited Branscomb for the staff of the National Bureau of Standards when he received his Ph.D. at Harvard in the spring of 1951 under Otto Oldenberg, although Lew likes to remind me that I felt the Bureau for Corning before he reported for duty, and did not tell anyone there (so he says and it may be ture) that I had hired him. But he has made out all right at NBS and I am sure he will do a great job with the *Reviews of Modern Physics*.

I am delighted with the extra support for the *Reviews of Modern Physics* which is being provided by the Council in providing for a working board of Associate Editors and for various other innovations which will be announced and put into effect during the coming year.

In the past forty years, American physics has risen from a minor position in the world to one of great productivity and leadership, partly as a result of our positive contributions, partly as a result of the negative effects of Nazi rule in Germany and destruction occasioned by World War II. In the same period there has been a vast rise in the creative power of Russian physics from an even smaller beginning than ours of forty years ago to a position that is now quite comparable with our own. In that period there has been a great expansion of the power of physics for intellectual and technological growth in many other countries, including China. We must hope that in the years to come political adjustments can be made which bring peace and an increasing measure of friendly co-operation to all of the scientists of the world so that the benefits of working together can result in greater progress.

Accompanying the expansion in research activity in physics there has been a vast growth in the size of the physics literature. It is now quite impossible for any physicist to keep abreast of the details of more than that which represents his narrower and narrower specialty. Nowadays we can meet physicists who work in high-energy particle physics with spark chambers who do not keep up with what the bubble-chamber men are doing! But if we regress too far in the direction of intense specialization, we are sure to lose a great deal in the effectiveness of our research, and more than that, to lose a great deal of the intellectual fun that accompanies a wide appreciation of the developing total picture of our knowledge of natural philosophy.

That is why the future of the *Reviews of Modern Physics* is of the utmost importance. It is the journal on which we all mostly rely to keep ourselves informed about research progress in various active areas of physics. Unfortunately, due to a shortage of good critical review writing the *Reviews of Modern Physics* has in the past carried too many specialized technical articles on advanced topics in theoretical physics. This may have gone to the point where many of the younger generation do not understand its role as a *review* journal at all. It will take the new editors some time to turn the ship around. This they must do, because there is a Gresham's law in physics journals, whereby specialized articles being easier to write than good critical reviews, are more often submitted to the editors. Unless they are careful and stern, we will not have a genuine review journal but just another repository for great accumulations of specialized calculations, too many of which have appeared here in the past twelve years.

There is a special art to the writing of a good literature review paper. I believe that it is not being cultivated as much as it should be in our graduate schools. Every student when he starts on his thesis work ought to have to search the literature, *read* the papers which he so retrieves (ridiculous word!) and prepare a critical summary of the present state of the field in which his research is to be done. Later when he reaches post-doctoral status or becomes a young assistant professor he ought to undertake a wider and more sophisticated review of his field. From the best of such materials would be generated a stream of papers which would make the *Reviews of Modern Physics* a much more useful journal than in the past. This kind of development can be encouraged and stimulated by the Council and by the Board of Editors, but it must be regarded as a personal responsibility to be worked on by every physicist who would like to be regarded as part of the research team of the world.

Physicists should pay more attention to the economics of research publication. *Reviews of Modern Physics*, even if it were to double ro triple its subscription price, is the greatest bargain in research material that is being offered today. Physicists often give their writings to publishers of private journals with few small pages and enormous subscription prices instead of to their own journals. Each one of you who takes the trouble to make a small calculation will find that he pays some ten times the price for material in privately operated journals and books than what he pays for the material in his own society journals. These publications thrive on what we supply to them while our own review journal does not get the kind of material it needs.

I urge all of my colleagues to take these criticisms to heart and to bend every effort in the future to help the new editors make of the *Reviews of Modern Physics* the kind of review journal which American physics should be offering to the world.

These ideas are not new. I think we have advanced a great deal, but I also think we have a long way to go.

Forty years ago quantum mechanics was quite new. I was just starting as a young assistant professor at Princeton. The fall of 1928 was a great semester for me. That is when the paper on application of quantum mechanics to what we now call the Franck–Condon principle was written, and the paper with Ronald Gurney on interpretation of alpha radioactivity by barrier leakage, and a paper with Harry Smyth on the critical potentials of molecular hydrogen which led next year to Walker Bleakney's discovery of ionization processes accompanied by violent molecular dissociation, and with the writing with Phil Morse of the first English language textbook on quantum mechanics.

That was the time when Jack Tate, professor of physics at the University of Minnesota and then Editor of the *Physical Review*, first had the idea for a review journal of physics in America. He wrote around to a variety of physicists, young and old, for ideas about whether such a journal was needed and how it should be planned. The file of letters he received in reply was part of the material which was passed on to me by my predecessor, J. W. Buchta. I felt that it should be preserved and passed it on to the historical collections at the American Institute of Physics. Among them was my own hand-written reply (we didn't have much secretarial help then either!) to Tate's request for advice, which is dated October 2, 1928, just forty years earlier than the date of this issue.

It has at least a sentimental interest for me and so as my last act as Editor, I intend to take the space to print it here, because I think it still contains some ideas of value for today:

Palmer Physical Laboratory Princeton University
Princeton, New Jersey

October 2, 1928

Dear Tate:

I have just received your letter on the subject of a review journal. You probably have received mine of yesterday in which I offered gratuitously some remarks on the subject about which you now ask an opinion and therefore know that I am heartily in favor of something to meet this need. [Compare P.S.] I will expand a little on that theme here.

I have been thinking ever since I returned from Germany that the greatest handicap to physical research here is the lack of an adequate literature in English. The market is glutted with American textbooks for freshman and sophomore general physics but there we seem to fade out. There is no question but what our laboratories are better now than those abroad but we lack the literature which brings the young men quickly into step with the research work in the various field. We have relied too much on the German literature in spite of the fact that the education of our students in foreign languages, especially German, is very weak.

There are several directions in which the literature should be expanded, I think.

(1) We should have an expanded book literature somewhat on the pattern of the "Struktur der Materie" series edited by Born and Franck, and something like the annual volume, "Ergebnisse der exakten Wissenschaften."

(2) We should have a review journal of the sort which you are proposing.

(3) We should take steps to see that the weekly journal "Science" is revitalized not only in our own science but in others as well as so that it would become a real organ of science progress in America.

(4) We need an extensive series of books like the "Sammlung Göschen" which would give in pocket size adequate and interesting accounts of various special branches of science.

I will cooperate and aid in any way I can any project which aims to fill this need. As you have raised the question of (2) more specifically I will say that I believe that this is the best entering wedge for the program. I seem to have the impression that Americans generally do not have a very "zusammenfassending" attitude which is the weak point in our scholarship. Too, I think that they do not like to write although you as an editor who has to wonder where the space is coming from to print their present outpourings may have good reason to think otherwise. I mean I have the impression that the young men as a rule are diffident about taking a little chunk of the subject but instead prefer to flirt around over the whole field. To contradict my previous statement somewhat there are too many who take so tiny a chunk in which to specialize that the general viewpoint is quite lost. Therefore there may be some difficulty at the start in developing a good group of contributors who would write on the special topics but with a broad viewpoint. [There still is!]

Until American physicists do recognize that their scholarship is incomplete until they are providing themselves with their own review literature, the promoter of such a journal may have a discouraging time of it. But I hope that it can be promoted and carried through in a successful way.

I suppose you have in mind something like (perhaps a quarterly) the "Chemical Reviews" which has had some rather good things lately in our own field, e.g. Dushman's article on line spectra in the June 1928 number and Pauling's recent paper on the hydrogen molecule. Off hand, I can think of nothing better than a journal patterned along these lines. It would be much more flexible than the National Research Council Bulletins, and as you say, less formidable. Publication could be prompter and the individual papers could be one-man or two-man enterprises which I think is a great help. I asked a friend about Dushman's paper and the reply was "It's all right for the ignorant." I think it could receive no better compliment and could not justify itself in any better way. For we are all ignorant as soon as we wander a little out of the lines in which our own largely accidental graduate training fitted us for research. This is especially true of the younger men. The "Physiological Reviews" is another such quarterly which I find intensely interesting at times. "Science Progress" is occasionally interesting but too diffuse and confuses its function with that of the abstract journals. (By the way, "Science Abstracts" is hopeless for theoretical physics and something should be done about that sometime. [This has been greatly changed.])

I know nothing about the financial requirements of such a thing. I suppose it is expensive. Maybe one of the university presses could be persuaded to assist.

What do you know about "Science"? Couldn't Cattell be made a grand old editor emeritus pretty soon and have the thing be really run by a strong board of contributing editors who solicited general review articles in the fields of their respective sciences? [This has happened.] That would be a great help, too, I think.

That's about all I think of off hand except that I personally am very glad you are taking the thing up and assure you of my enthusiastic co-operation when ever you would want it. Looking forward to seeing you at Thanksgiving if things work out so that I make the trip, I am,

Sincerely yours,
Edw. Condon

P.S. I looked up and saw on my desk unmailed the letter I wrote you yesterday so am putting them together in one envelope.

So finally, farewell to the Editor's chair at *Reviews of Modern Physics*. I am glad the new Editor and the new board is "taking the thing up and assure (them) of my enthusiastic co-operation whenever (they) would want it."

UFO's I Have Loved and Lost

E.U. CONDON

Professor of Physics
University of Colorado
Boulder, Colorado

Our program committee suggested that the members might like a brief account of my experiences in conducting a study of Unidentified Flying Objects, from late 1966 to the summer of 1968, at the request of the U.S. Air Force. The full report has been published under the title, *Scientific Study of Unidentified Flying Objects*, in paperback by Bantam Books, and in hardback by E.P. Dutton and Co.

Throughout human history men have been seeing strange and terrifying apparitions in the sky. The literature dealing with such experiences is enormous. The word "spectre" is used generically to describe phenomena of this type. This world's earliest use cited in the *Oxford English Dictionary* is in the title of a book by Z. Jones published in 1605 *A Treatise of Specters or straunge Sights, Visions and Apparitions appearing sensibly unto men*. The word "spectrum" is cited first in 1611 in a passage which said, "Walsingham hath written of a fatal Spectrum or Apparition . . . where sundry monsters of diuers colours . . . were seen." Sixty years later, Isaac Newton used this word to describe his decomposition of sunlight with a glass prism in these words, "the Sunbeams . . . passing through a glass prism to the opposite Wall, exhibited there a Spectrum of divers colours."

From these two uses of the word *spectrum* come naturally the two meanings which the *OED* gives for the word "spectrology" which are: (1) the science or study of spectres, and (2) the scientific study of spectra. The *OED* cites as an example of the first meaning an 1820 quotation from Washington Irving's *Sketchbook*: "The gloom of religious abstraction, and the wildness of their situation . . . had filled their imaginations with the frightful chimeras of witchcraft and spectrology." And of the second, an 1862 quotation from the *American Journal of Science*, "The attention of the French scientific world is wholly fixed on spectrology, for thus do they designate the experiment with the spectroscope of Bunsen and Kirchhoff."

I am the second man in human history to have written a book on spectrology in both of these two distinct meanings — our distinguished colleague Donald Menzel was the first.

Modern interest in UFO's stems mainly from the observations of Kenneth Arnold, a Boise, Idaho, businessman on June 24, 1947. Flying near Mount Rainier in Washington he reported seeing some objects skimming along which he described in a manner that led newspapermen to call them "flying saucers." Although not all objects later reported are saucer-shaped, this term is often used generically, but the term UFO is preferable. The Air Force studies anything seen flying in the sky which might present a defense hazard, and so the Air Force has been concerned with the thousands of reports of sightings of UFO's that have come to them in the nearly twenty-two years since this first modern report.

From such study they concluded long ago that no defense problem was involved in these reports from the public. The amount of attention which the Air Force gave to the problem after the first four or five years has been minimal.

In the early fifties the story of UFO's began to appear in sensational pseudo-science magazine articles and paperback books. These have had a large sale. The book by Franck Edwards, *Flying Saucers — Serious Business*, probably holds the record with more than 1,300,000 copies sold. Several other titles have sold more than 200,000 copies. Our report was given an initial printing of 200,000 copies in the Bantam edition. In the last three years 40,000 schoolchildren have written the Air Force asking for UFO data.

The principal source of the widespread interest is the contention of some writers that at least some of the things seen may represent flying craft from other civilizations, either elsewhere in the solar system, or even from a planetary system associated with some other star.

We must be extremely careful about our language. Some UFO's *may* be such visitors, it may be postulated, and some writers go so far as to say that they *actually are*. To discover clear, unambiguous evidence on this point would be a scientific discovery of the first magnitudem one which I would be quite happy to make. We found no such evidence, and so state in our report. But it is not true to say that we "proved that flying saucers do not come from outer space." All that can be said is that, of the cases we looked into carefully, we found no evidence in support of the hypothesis of their extraterrestrial origin.

We concluded that it is not worth while to carry on a continuing study of UFO's in the manner which has been done thus far: that of going out into the field to interview persons who say they have seen something peculiar. The difficulty about using objective means of study lies in the rarity of the apparitions, their short duration, and the tendency of the observers not to report their experience until long after it has ended. When a known object is the source of many reports, as in the case of the Zond IV reentry of 3 March, 1968 (see *Scientific Study of Unidentified Flying Objects*, pp. 571–581), there is an extraordinary disagreement among the descriptions of what was seen by different observers of the same event. This result shows that no great certainty attaches to the specific details of *any* of the reports.

These difficulties led us to conclude that it is quite unproductive of results of scientific value to study UFO's in the traditional manner. But, contrary to popular belief, we do not rule out all future study, We say (p. 2):

Although we conclude after nearly two years of intensive study, that we do not see any fruitful lines of advance from the study of UFO reports, we believe that any scientist with adequate training and credentials who does come up with a clearly defined, specific proposal for study should be supported.

This conclusion has been bitterly denounced by the flying-saucer buffs who have been making money from sensational writing and lecturing to gullible audiences, and collecting dues from the membership of their pseudo-science organizations. One prominent professor of atmospheric physics has been giving speeches in which he advocates that the federal government spend on UFO study amounts of money which would "dwarf" that spent on the space program.

Even though nearly a year has gone by since my work in this field ended, I continue to be astonished at the fervor with which many people hold views that are totally unsupported by objective evidence of any kind. Many people seem quite incapable of recognizing any distinction between what *might be* so and what *actually is so*. Some of the believers are charlatans, in my opinion, who profess belief in order to collect royalties from writing and fees from lecturing. But others are deeply sincere.

We ran into many more interesting cases than we could include in the report, already criticized by many for being too thick. There was a young airman, second class, at an Air Force base in New Mexico whose nineteen-year-old wife died suddenly of a heart attack. They were members of a flying-saucer cult which gathered round and decided that the woman's spirit had gone to Venus on a flying saucer, and that she would want her body back when she returned. So they wrapped it in a sheet and stored it in a barn rather than having it properly buried. The police learned of this by a mysterious postcard from a woman in Spokane, Washington. At first they thought the card was a hoax, but investigation proved that the young woman's body had been by this time stored in that barn for about three weeks. The young airman had seen lots of flying saucers but had not reported any of them, saying: "I didn't know the Air Force was interested!"

In the spring of 1967 I was visited several times by a well-mannered man who claimed to be acting as agent for the Third Universe (we are the First, and the Second is inhabited by beings that resemble polar bears, he said). He said he was authorized to negotiate a contract with the United States Government by which they would teach us to make interstellar flying saucers for $3 billion. The first billion was to be paid after a demonstration to government officials at Dulles Airport, the second after a major national laboratory had been built and our scientists and technologists had learned how to make flying saucers, and the third after they had trained our flight crews in interestellar navigation. He was specific down to the point of naming the bank in Arlington, Virginia, where the three billion was to be deposited.

He wanted me, in the interim, to pay him $3,000 as "earnest money" to be deposited in a particular bank in Western Colorado to the account of his organization, which was called the "Omnific Intelligence Continuum." Inquiry to that bank revealed that there really was such an account. Asked the size of the account the banker cautiously said, "Small sums go in and out." Asked about membership of the organization the banker told me, "So far as I know Mr. Xxxx is the only member." Since part of Mr. Xxxx's story is thus verified, ought we now to believe everything he tells us?

Just to play bureaucratically safe, a fully account of this proposition was written up and transmitted to the Air Force and to the Office of Science and Technology. So far as I know, it was never followed up by further contact with this man, nor did he come to see me again. Therefore, I do not know whether he could have delivered on his end of such a bargain if one had been made.

The most vivid lesson that I learned from such experiences is what a narrow, wobbly line there is between real science and pseudo-science. So far as the public is concerned most of the science which they know about they do not understand. Very few people can state clearly the grounds for belief that the earth goes around the sun, rather than *vice versa*, or for that matter, for our belief that the earth is a ball rather than flat. Coming to more modern instances, who among the many investors in the profitable semi-conductor industries have the slightest idea how a transistor *really* works?

In the given circumstances most of the scientific ideas that are accepted by the public are accepted entirely on faith. To most people completely lacking any basic understanding of underlying principles, the proposition that the configuration of the planets and stars at the time of our birth determines the course of events in our lives, seems no more unlikely or preposterous than many of the well-established truths of science which they do accept without understanding them. There are some 10,000 astrologers in America who make their living practicing astrology and only about 2,000 astronomers who live by practicing astronomy. If celestial matters were decided democratically by the members of both professions lumped together, then the "real" astronomers would always end up as a depressed minority.

Flying saucers and astrology are not the only pseudo-sciences which have a considerable following among us. There used to be spiritualism; there continues to be extrasensory perception, and psychokinesis, and a host of others. Hanson W. Baldwin in the *New York Times* has told how the

Marines at Camp Pendleton are trained for Veit Nam in the use of dowsers made of bent wire coathangers as a means of locating tunnels and other underground works of the Vietcong.

Recently a visitor from a navy research installation told me that some admirals had purchased from an inventor a wholly worthless invention which it was claimed could detect submerged submarines by a radarlike reflection of electromagnetic waves. This could not possibly work because of the conductivity of seawater, and it did not work when expensively tested. A test had to be made to satisfy the admirals and certain congressman. (During World War II, the National Bureau of Standards at the insistence of several congressmen, tested a crackpot scheme for making rubber from garbage.) A Russian spy in this navy laboratory got hold of the invention and our test results and sent them to Russia. The Russians did more work on the invention. Although they understood at once that it must be foolish, they thought they might be wrong because the Americans had spent so much time and money studying this crackpot invention. We know that they did, because an American spy in their laboratory sent to us the story of their work.

These and many other examples that could be given show that we have failed rather miserably to give even to so-called educated people some feeling for the way in which science investigates a subject, and the way in which scientists subject their observational material to critical evaluation before reaching conclusions. The thing that most people are least able to do is to refrain from drawing conclusions when there is not enough evidence at hand to warrant drawing conclusions.

In ancient times, the future was foretold in many ways that have gone out of favor, such as by examining the entrails of sacrified animals, or basing omens on the study of the flight of flocks of birds. Cicero practiced this latter method. Before you smile, bear in mind that these views have never really had as much scientific study as have the UFO reports. Perhaps we need a National Magic Agency to make a large and expensive study of *all* these matters, including the future scientific study of UFO's, if any.

In conclusion, let me say that where corruption of children's minds is at stake, I do not believe in freedom of the press or freedom of speech. In my view, publishers who publish or teachers who teach any of the pseudo-sciences as established truth should, on being found guilty, be publicly horsewhipped, and forever banned from further activity in these usually honorable professions. Truth and children's minds are too precious for us to allow them to abused by charlatans.

Education for World Understanding*

E.U. CONDON

The most important problem facing humanity in the world today is that of obtaining and securing a state of peace with justice and freedom for all of the peoples of the world. Today we can be greatly heartened by the fact that on Friday evening U Thant was re-elected to a second five-year term as secretary-general of the United Nations by unanimous vote of the member nations.

This is especially heartening because it represented a real test of senti- ment for the peace program of the secretary-general, who said, after having been re-elected:

"The threats to peace in many parts of the world and more particularly in Viet Nam are for me a continuing source of anxiety and even anguish. I was glad to know a few days ago that there is general agreement in regard to a brief pause in the fighting in Viet Nam on the occasion of Christmas and other holidays. Is it too much to hope that what is made possible for just a couple of days by the occurrence of common holidays may soon prove feasible for a longer period by the new commitments that peace requires, so that an atmosphere may be created which is necessary for meaningful talks to be held in the quest for a peaceful solution?

"I shall seize every occasion to recall that this war must be ended and I will continue to regard it as my duty to make every effort on a personal basis to help promote a solution which will bring peace and justice to the people of Viet Nam."

Let us pray that hundreds of millions of men, women, and children everywhere will rally to the support of the secretary-general in his endeavors for the survival of all of us.

The escalation of war in Viet Nam that has occurred in the past two years, and the prospects of a still greater increase in the killing during the next year, and the warnings that we are getting from our leaders that the war may continue for ten or twenty years more, and might involve us in a

---

* Condon has given different versions of this talk on several occasions, and has distributed copies of it widely to public officials. This version was presented at the morning service of the First Unitarian Church, Denver, December 4, 1966.

land war with China — all of these things can easily give rise to feelings of despair among us. We need to keep resolutely faced in the right direction, the quest for peace, and recall the words of the late President John F. Kennedy in a speech that he gave to the United Nations Assembly just five years ago on September 25, 1961, when he said:

"We need to have a peace race to end the arms race, before the arms race ends the human race. Today every inhabitant of this planet must contemplate the day when the planet may no longer be habitable. Every man, woman and child lives under a nuclear sword of Damocles, hanging by the slenderest of threads, capable of being cut at any minute by accident, miscalculation, or madness."

There is no single or simple solution to the great problem of peace that faces us and so I cannot give you all of the answers in one brief talk. But today I do want to enlarge on one group of ideas which I sincerely believe could be easily implemented and which could in the next few years make an enormous contribution to the establishment of a peaceful world.

In the summer of 1963, a former president of Columbia University, Dwight Eisenhower, made a speech in Stockholm to the conference of the World Confederation of Organizations of the Teaching Profession. In this speech he advocated that all nations should join in establishing a School for World Understanding and suggested that the UN General Assembly should take this on as a project. He said in that speech: "World enlightenment will speed the day when the burdens of armament and fear of others will be removed from the backs and hearts of men." He recommended that the school be staffed with "an international faculty of scholars, whose concern would be objective truth purged of national and regional bias, hatred and prejudices."

The only fault I have to find with his proposal is that he described something too small, and too narrow in scope, to be adequate for the real need. He envisioned a school dealing only with world history, diplomacy, politics, international communication and teaching. He proposed a student body of two or three thousand students taking a two-year course, presumably after completion of a normal four-year liberal arts college course.

Thus, for the whole world he was thinking of a school, narrow in the scope of its subject mater, and very small, much smaller, in fact, than the Denver Center of the University of Colorado,though one might hope that the combined efforts of the nations of the world might give it better facilities than that school has. Even so, he seemed to think that his proposal might be regarded as burdensomely expensive, because he said: "However high the price of a school for global understanding might be, it would still be a minute fraction of the moneys now spent by governments against global war."

Comparison with armament costs having been thus invited, let us see how true that it. Let us see how his project can be expanded to something like the proportions it needs to have, and how, even after this is done, it would still cost only a trivial fraction of what the world now spends to prepare us for war.

It is estimated that nations of the world are now spending about $140 billion a year on armaments. The United States alone spends about half of that vast sum. It is hard to give precise figures because they are increasing at such a rapid rate. Three years ago the world total was $120 billion a year.*

Suppose that the governments in the United Nations would agree to pay to the UN a tax of 1% of their arms budgets to support the Schools for World Understanding. This, too, would represent an investment in security, and a much better one both morally and practically, than one that is based on almost total reliance on indiscriminate mass bombings, and napalmings, of whole populations. In this way we would have available about $1.4 billion a year.

We who are in education are not used to thinking of sums like that. The operating cost of the most lavishly operated college in the world, the Air Force Academy near Colorado Springs, is only about $12,000 per student-year. The budget of the University of Colorado is about $70 million a year, but it takes care of some 20,000 students, or an average of about $3,500 per student-year. For rough planning purposes, let us budget $5,000 per student-year, which is lavishly high, and totally out of line with civilian practice, especially with the amount needed for university operation in any of the underdeveloped countries.

But using this figure, and applying it to only half of our available funds, that is $700 million a year, we have funds enough to pay for the operating costs of training every year 140,000 students. Assuming they are given a four-year course, we have the funds available to reach an annual output of 35,000 students a year who are especially trained for working on the front of international relations and strengthening of world peace on every front.

In the case of the well-developed countries, I would not propose the creation of new colleges and universities, but would, instead, recommend expansion of faculties and facilities of the best ones we now have. In the areas of the world that are not now as well off as we are with regard to capital facilities, I would propose to use the other $700 million a year to build a world-wide chain of major UN universities.

The capital investment in plant and facilities of a brand new major university that offers graduate and undergraduate training in medicine, agriculture, science and engineering, as well as liberal arts and fine arts, is about $100 million, when planned to take care of a student body of some 10,000 students. These can not be built in one year anyway. Capital construction would require about four years. So I would propose that we

---

* By 1970 the world total had increased to $200 billion to year. — E.U.C.

start building about two dozen of these. Actually with $700 million a year for four years we could build 28 of them, and thus provide the world with capital facilities for training 280,000 students after four years time. After the capital construction period is over we would then have the remaining $700 million a year to subsidize another 140,000 students a year.

All of this we could do, you remember, with just one per cent of the world's persent armaments budget. All of this we Americans alone could do with an expenditure of less than one fourth of one per cent of the gross national product. In more current terms, we could do it with about what we spend every two weeks on the present war in Viet Nam. So let no one say that we can not afford it.

The particular allocation of the $1.4 billion a year that I have proposed is not necessarily the best one. I merely give it on a for-instance basis to help us visualize the magnitude of the possibilities that are open to us. People are not used to thinking of figures as large as one per cent of the arms budget — hardly as much as the bookkeeping errors involved in keeping track of military expenditures.

I think it is essential that the Schools for World Understanding include all of the arts and sciences and professions, rather than being narrowly schools of international polities.

When we Americans speak of educational co-operation, I am afraid that we usually have in mind some magnanimous arrangement in which we teach the poor unfortunates of this world who are not Americans all about the blessings that would be theirs if they only adopted our ideas, attiutdes and modes of behavior. I have in mind that an important part of the total plan would be to counteract this tendency by bringing to America as visiting professors a large number of the leading scholars of Latin America, Asia and Africa, as well as Europe, to teach in our colleges and universities. This is something that could be done to good advantage right away by the major philanthropic foundations. All of us would benefit immensely by the establishment in our colleges of a hundred or more such posts which could well be called Listen Yankee Lectureships on World Affairs. Other similar exchanges are needed elsewhere, but it would be tactless to mention specific places where there exist centers of "regional bias, hatred and prejudices" (to quote Eisenhower) that need a little purging.

Finally, let me say that it is quite possible that new developments may well lead soon to the abandonment of a number of outlying military bases, possibly at Guantanamo, possibly in Turkey, possibly elsewhere. Instead of just abandoning these properties, they provide us with wonderful opportunities to get started at once with our Schools of World Under-standing.

Our naval base at Guantanamo, for example, is on a property that is 45 square miles in area, on a beautiful location. Why not start the process of beating our swords into ploughshares right now, by offering to transfer our interest in this base to UN auspices, for the purpose of establishing there a great UN university which would be devoted to the cultivation of the best interests of an alliance for progress in friendship between Latin America and the United States? I am confident that U Thant could win the enthusiastic cooperation of Fidel Castro for a plan by which Cuba became the host to a UN activity which would make Guantanamo the intellectual and spiritual center for co-operation in this hemisphere.

History moves in jumps, never smoothly onward. The re-election of U Thant means that we can now take new initiatives toward peace. I have merely tried to sketch one direction and one possibility.

Let me close by quoting for you the message given by the great Czech educator, Jan Komensky, in his book, *The Angel of Peace*, published in 1667, in which he addresses himself to the English and Dutch negotiators working on a treaty to end a war between their countries. He charges them thusly, in a manner which might well serve as a charge to those who are entrusted with the large negotiations today in the more dangerous situation of today:

And you, ambassadors of peace, that you may live up to your name, do not only consider human, but also divine, plans; take account not only of what is asked of you by your kings, but also what is asked of you by the King of Kings; let your aim be, not war, but peace. Do not write your agreements and treaties only on parchment, but also on your hearts; do not confirm them only with silver seals but also with the great name of God; do not take oaths in deceptive human language, but from the depths of the soul, which is witnessed and searched by God in virtue of that truth which is in Christ."

Let us hope that the time is not far off in which all of man's intelligence and creative spirit may be devoted to works of peace and that neither shall they make war any more.

Reminiscences of a Life In and Out of Quantum Mechanics

E.U. CONDON

Department of Physics
University of Colorado
Boulder, Colorado

Introductory Talk at International Symposium on Applications of Quantum Mechanics to Atomic, Molecular and Solid State Theory and Quantum Biology, Sanibel Island, Florida, 22 January 1973, 9 a.m.

Mr. Chairman, Ladies and Gentlemen:

I thank you all from the bottom of my heart for your having planned this symposium in my honor. A delightful custom, which has developed here, is to have the person being honored give some autobiographical reminiscences about his work in science and related topics. I have greatly enjoyed hearing and reading some of these, and now I shall try to make my contribution to this growing body of informal historical material.

As you all know, in addition to specific calculations in our subject, I have devoted a good deal of my time and energy to social and political problems of war and peace and especially to those growing out of the new aspect of militarism presented by the development of uranium and fusion bombs; so I shall also have a few words to say on that.

These remarks are informal in that I believe the things said are essentially correct, but I have not had time to make detailed checks so there may be some inaccuracies of minor detail in my memory of them.

In the *New Yorker* of January 6, 1973 (p. 55) there is a fascinating account of a recent hearing held by the California State Board of Education on the current controversy out there as to how Creationism is to be handled in the high-school textbooks alongside of Darwinism. The reporter said he saw only one creationist carrying a placard, and he said, "It was a chart of the hydrogen atom, and he intended to use it to help him demonstrate that a certainty about the theory of evolution flies in the face of certain laws of quantum mechanics.

"It's hard to explain quantum mechanics in five minutes," he said when he was finally called to the podium, late in the afternoon, "but I'll try."

The situation reminds me of a remark made to me one spring afternoon in 1928 by Bergen Davis, a distinguished older professor of physics at

Columbia, who had been struggling hard to get a real feel for quantum mechanics. Finally he said in despair: "I don't believe that you young fellows undertand it any better than I do, but you all stick together and say the same thing!"

I have called this the conspiracy interpretation of quantum mechanics.

I think personal reminiscences should begin at the beginning. I was born in 1902 at Alamogordo, New Mexico. The Chamber of Commerce of that fair city has placed signs on the highways which say, "Welcome to Alamogordo: Birth-place of Atomic Energy," referring thereby to the existence near there of the test site for the first uranium fission bomb, tested on July 16, 1945.

The signs ought to be changed to read "Birthplace of E.U. Condon (1902) and of Atomic Energy (1945)." In 1948 one particularly stupid congressman thought it was pretty sinister that my parents had picked the bomb test site forty six years earlier as a place for me to born. In 1950 I met the district attorney of Otero County who exclaimed, "Dr. Condon, I know you! The FBI was out here checking up on your story that you were born here!

"What did you tell them?" I replied, "I know your courthouse burned down thirty years ago and you have no records."

"It was all right" he said, "There were some oldtimers around who knew your father and they said a kid was born about that time so we figured it was you."

As well as I can remember, I first learned of the existence of physics as a science in 1913 when as an eighth grader I bought for 15¢ a second-hand copy of *High School Physics*, by Carhart and Chute, two University of Michigan Professors. Prior to that I had read my share of children's books about mechanical and electrical experimentation, but these seldom give reasons for the way things work or the principles on which physical explanations are based. The principles underlying the explanations are the interesting part of physics and I became tremendously interested in the subject.

My high-school years were those of World War I, including the detail that I graduated at mid-year in December 1918 just as the war ended. I had taken all the science the high school offered in the early years, and in the last two years turned to newspaper reporting, first on the student paper, and then in the summer of 1918, in the regular newspapers of Oakland, California.

Thus at the age of 16 I was doing important work that summer, not because of my brilliance as a journalist, but because newspaper reporting is mostly a young man's game, and in wartime the experienced reporters are away at war. I smoked my first cigar that summer when interviewing Charles M. Schwab, president of the Bethlehem Shipbuilding Corporation, whose Oakland shipyards were being tied up by a strike of Boilermakers' Union Local 233. The newspaper work gave me a lasting interest in public affairs and a somewhat cynical appreciation of the many scoundrels who infest our political life. Absorbing as is this subject, the time I have spent on it might perhaps have been better spent on closer pursuit of professional interests in physics.

In the fall of 1918, in the closing months of the war, I held a part-time job on the *Oakland Enquirer* while finishing the two courses needed to graduate from high school.

Men learn the horrors of war in many ways, and the experience of the battle-field is only one of them. In October 1918, a regiment of Oakland boys was put into the front-line trenches in France in the final battles of the war. Some of them were only a few years older than I, and graduates of my own high school. In the final days of the war, these boys were being killed at the rate of about five a day. My steady assignment as a 16-year-old reporter was to go out each day to interview the mothers of the boys on that day's casualty list, and to steal photos or letters whenever possible. Although a telegram from the War Department was supposed to have notified next of kin before the names were given to the newspapers, this did not always happen. In consequence I was often the first one to convey the news to the mother that her son was dead.

Such experiences left a deep-seated scar on me and an urgent need to do what discouragingly little I can toward bringing about peace and disarmament. Such experiences make it essentially impossible to tolerate or to excuse or understand the recent insane behavior of the President of the United States in launching during last December a massive bombing attack on the North Vietnamese people, just as the war there is ending.

In the fall of 1919 I enrolled as a freshman at the University of California in Berkeley, with the intent of becoming an educated newspaperman, and therefore choosing courses in social sciences and literature. But these proved rather dull; so after two months I dropped out and returned to reporting for the *Oakland Enquirer*. After two more years of this I reentered the university, this time majoring first in chemistry, then in astronomy, then in physics, starting on the career which landed me here today some 54 years later.

On the *Enquirer* I specialized in the news of organized labor. The Russian revolution had overththrown the Czarist government in 1917, and communism was even then beginning to make headway in Europe and America as a world revolutionary movement. The dock workers, the timber workers, and the migratory farm laborers, who are still having a difficult fight for justice, were drawn to communism. The California state legislature had passed a strong bill defining criminal syndicalism and making it a felony. The politicians were looking for a place to use it.

On November 9, 1919, I was the only reporter from a conservative newspaper to cover the organization meeting of the Communist Labor Party of California, as it was called then. I wrote lurid and sensational stories about this small group of one or two hundred persons, which resulted in indictments against them, and which required that I had to testify against them, in trial after trial, over the next several years. In this connection I became aware of open boasting by a police detective of his having framed some of the defendants in a matter where I knew the facts to be otherwise.

The effect of this involvement on me was to wipe out any desire to be even an educated newspaperman; so I entered the university and went into physical science largely as a means of escape from the corruption of the world, in addition to the fact that I was genuinely interested in physical science.

The years 1921–1926 were occupied with undergraduate and graduate physics study. In the fall of 1922 I was married to Emilie Honzik.

I work hard and graduated with the B.A. in December 1924, and then went on to receive the doctorate in the late summer of 1926. The period of graduate study exactly coincided with the beginning of the development of what we know as quantum mechanics. In those years Berkeley was strong in physical chemistry with a vigorous research group under the leadership of G.N. Lewis, and strong in celestial mechanics with the research led by Armin O. Leuschner. But the physics department was comparatively weak in a research way, except for the recent addition to the faculty of R.T. Birge, who concentrated on the early development of the quantum theory of interpretation of diatomic molecular band spectra and of Leonard B. Loeb, who has spent a life-time making important contributions to processes of ionization in gases and of conduction of electricity through gases. In addition, William Howell Williams played an important role in my life. He was a West Point graduate, but not a military man in outlook. He had been my high-school physics teacher, had gone away to war, and returned to the university faculty at the war's end. The result was that he taught me theoretical physics at the university as well as having introduced me to the subject in high school.

He worked hard at the study of relativity theory and quantum theory, but I do not think he ever got a result that was published. He was one of those people who set impossibly high standards for himself and he was not helped by the fact that the university administration began to apply pressure on him to get a Ph.D., just any Ph.D., as a condition for his promotion.

In that last graduate year, 1925–1926, Birge gave a seminar on spectra of diatomic molecules. In the course of this seminar he gathered up and presented such observational material as there was on the relative intensity of the bands in a band system in their dependence on the vibrational quantum numbers of the initial and final states. There were no measurements, just eye estimates of plate blackening of uncalibrated plates. This was enough however to show up the main regularities, namely, that for a given v' there tend to be two different ranges of values of v'' at which the bands are strong.

Also in that year Dr. Hertha Sponer was doing experimental work in spectroscopy at Berkeley on a Rockefeller fellowship. She was a student of James Franck and many years later became his second wife. She was in Berkeley as a second choice as she had wanted to go to Cal Tech, but Cal Tech would not admit women at that time. In later years I have enjoyed twitting Lee DuBridge about this, because Cal Tech's predecessor institution was the Throop Institute of Technology in Pasadena, which had a department that instructed women in the arts of home economics, decades before the place became famous for its research in physical science.

One day Dr. Sponer received some proof sheets from Göttingen of a paper which Franck had given a few months earlier to the Faraday Society in London. This is *Trans. Farad. Soc.* **21**, 536 (1926). It dealt with a particularly appealing and simple picture of the way in which light can produce photodissociation of the diatomic iodine molecule. I happened to be loafing around her laboratory and she let me read the proof sheets. I saw at a glance how the ideas in Franck's paper could be extended to cover the case in which the molecule was vibrating in the initial state. This gave the two favored final vibrational quantum number ranges for each value ov v' in their relation to the relative positions and shapes of the potential energy curves governing the vibrational motion in the initial and final electronic states. Moreover I had the extreme good fortune to have good notes on the literature survey of what was known observationally about intensity distributions, from taking Birge's seminar. As a result some simple calculations could be made on each band system for which data were available, about ten in number, and in general they gave excellent agreements between theory and observation.

I wrote this all up over a rather long week-end from Thursday to Tuesday and presented it to Birge as my Ph.D. thesis. Curiously enough the one band system in which there was not good agreement between theory and experiment was the iodine system, as is clearly shown in my published paper (*Phys. Rev.* **28**, 1182 (1926)). However, within a year this discrepancy was cleared up by Wheeler Loomis who discovered that I made a mistake in the molecular constants used for the upper electronic state of I_2, and that when this was corrected, this molecule fell into line.

This was the origin of the pre-quantum mechanical version of what is known now as the Franck-Condon principle. In subsequent years both theory and observation have become a great deal more extensive, but that story is best told in connection with the quantum mechanical treatment, which came in 1927 and 1928.

In those pre-quantum mechanical days it was generally believed that all problems having to do with radiative transition probabilities had to be done as applications of Bohr's correspondence principle, in which one expanded the classically quantized motion in a multiple Fourier series and identified transition probabilities with the amplitudes of the Fourier components of the motion. So strongly entrenched was this mode of attack then, that as I later learned, the referee to which my first paper on this subject was sent by the Physical Review, recommended its rejection because it did not go at the problem in that way. But it was accepted nevertheless by the editor, John T. Tate, who told me he thought there might be something in it because of the good agreement between theory and experiment which it gave.

In September, Emilie and I, with our infant daughter Marie, left Berkeley for the long trip by train and steamer to Göttingen. We took quarters in the Pension run by the parents of Günther Cario, and we first met J.R. Oppenheimer, who was working under Born as a student, having recently come over from Cambridge, England. Also in Göttingen at that time was my old high-school teacher Mr. Williams. We plunged into the task of trying to learn Schrödinger's wave mechanics and the matrix mechanics of Born, Heisenberg, and Jordan. Born had just written his *Zeitschrift für Physik* paper on quantum mechanics of collision processes, and I worked hard on learning that, using the proof sheets which he loaned to me. This was the first paper in which the probability density interpretation of $|\Psi|^2$ was given, in contrast with Schrödinger's early view that the electron charge distribution of a single electron was somehow smeared out in accordance with the $|\Psi|^2$ distribution. Of course, I also met james Franck at this time, but we did not become well acquainted, because this was the autumn in which he received the Nobel Prize (which he shared with Gustay Hertz for initiating the study of atomic and molecular critical potentials). Thus Franck spent more time than usual attending Nobel affairs in Sweden since his first wife was Swedish. By the time he had returned from Sweden I had decided to move on the Munich for the second semester of my year in Germany, so that acquaintance with Frank really started a number of years later when he migrated to America.

At Munich I worked with Professor Arnold Sommerfeld whom I found to be a much more congenial person than Max Born. By this time I had absorbed enough quantum mechanics to undertake an initial formulation of the Franck-Condon principle in wave-mechanical terms (*Proc. Nat. Acad. Sci.* **13**, 462 1927)), and also a crude calculation of the binding energy of the ground state of the H_2 molecule (*Proc. Nat. Acad. Sci.* **13**, 466 (1927) and *Verh. d. Deut. Phys. Ges.* **8**, 19 (1927)). This second paper was quite overshadowed at the time by the calculations of Heitler and London on the same subject, but many years later it was recognized by Pauling as having been the first to approach molecular binding by means of what later became known as molecular orbitals (Pauling, *The Nature of the Chemical Bond*, 3rd ed., Cornell University Press, 1960, p. 23).

In the late spring of 1927 I became acquainted with David Dennison's interpretation of the specific heat of H_2 gas in terms of what we now call ortho- and para-hydrogen and wrote a full account of it to Prof. G.N. Lewis in Berkeley, hoping that he would put the cryogenic group there on the task on getting detailed experimental evidence about the slow transitions between the odd and even rotational states. A little work was done there, but the main work was actually done by the Bonhoeffer group in Berlin.

By spring and summer of 1927, papers in quantum mechanics were appearing at a great rate. In those days a young theoretical physicist was supposed to keep abreast of progress in every area of theoretical physics. I became discouraged and decided that if this were the normal pace of work in my chosen field (which it was not!) then I was not equal to the task. About this time there appeared a help-wanted advertisement in the *Physical Review* for a man to write popular science for an industrial laboratory, the requirement begin stated that the candidate must have newspaper writing experience as well as Ph.D. in physics. I may well have been the only person in America with that combination at that time. At any rate I applied for the position, was interviewed for it in London, and accepted it. It turned out that the position was in the public relations department of the Bell Telephone Laboratories, then in its old quarters at 463 West Street, along the Hudson River in lower Manhattan.

We returned to America and found an apartment near Columbia in October 1927 after having first gone to the British Association meeting which was in Leeds that year. Among other things Heisenberg gave a paper at that meeting which was one of the first presentations of his ideas on the uncertainty principle. I had already been granted a second year renewal of my fellowship, to work with W.F.G. Swann at Yale, but resigned from that on taking the Bell Labs position.

At Bell Laboratories, C.J. Davisson and L.H. Germer had just done the experimental work on scattering of low-energy electrons by single crystals of nickel which led to one mode of discovery of electron diffraction. The other was the work of Sir G.P, Thomson who directed high-speed electrons against various foils in a manner analogous to the Debye-Scherrer method for X-rays. For this work the two groups shared the Nobel prize in physics in 1937. The importance of this work was not at first appreciated in the business management side of the Bell Labs, and I devoted a good deal of attention in the fall of 1927 to explaining to such people that the work was destined to win for the Bell Labs the first Nobel prize to be awarded to an industrial organization. I did some editorial work under the direction of Dr. R.W. King, who had founded the *Bell System Technical Journal* a few years earlier. King was only a few years older than I, and we used to spend long hours together discussing whether we had done the right thing in leaving university work to go into industry.

In that fall I soon found that the American physicists on the Atlantic Coast were by and large having as much trouble understanding and assimilating quantum mechanics as I had had in Germany. The older men had just tossed in the sponge and were not even trying. The profession of theoretical physics was much smaller then than now. As I remember it,

Gregory Breit, John Slater, John Van Vleck, and Edwin Kemble were about the only ones in America who were really active in research in quantum mechanics, although the number was soon increased by the valuable addition of both George Uhlenbeck and Sam Goudsmit to the faculty of the University of Michigan. Carl Eckart at Chicago should also be mentioned, but I did not become acquainted with him until later.

On the other hand, physicists by this time generally recognized the importance of the revolution in thinking about atomic physics that was taking place, so there was a great eagerness to study and learn about it. I soon found myself in demand as a colloquim speaker at various universities of the Atlantic coast and the Middle West, and King encouraged me to accept such invitations, even though they bore little if any relation to the work I was supposed to be doing for the telephone company.

I was asked by George Pegram to be a lecturer in physics at Columbia University in the spring of 1928, salary being available because Michael Pupin was to be on sabbatical leave. I accepted and started on my first regular university appointment by giving two graduate courses, one in quantum mechanics and the other on electromagnetic theory of light. Jerrold Zacharias was the outstanding student in these courses.

Besides giving these courses I traveled around giving colloquium talks on quantum mechanics and also on the Franck-Condon principle. So great was the demand for young faculty who could deal with these subjects that in the spring I was offered six assistant professorships for the fall of 1929: at the University of California, at Columbia at Princeton, at New York University, at the University of Minnesota, and at the University of Wisconsin. With regard to the first of these, some unfriendliness developed which persisted over many years because I had given a tentative acceptance of that offer, but ended up by taking the offer from Karl Compton to go to Princeton, which Birge regarded as an unforgivable affront for decades thereafter.

In the spring of 1928, while I was at Columbia, the University of California was trying to recruit for its faculty as a brilliant young experimental physicist, Ernest O. Lawrence, who was an assistant professor at Yale, and whom I had met the previous year when he was traveling in Germany with Jesse Beams of Virginia. I was asked by my Berkeley friends to go up to New Haven and try to help persuade him to leave Yale and move to Berkeley. Ernest and I had a most enjoyable weekend at New Haven. He was a sporty young bachelor with a new convertible and we rode around in that discussing the pros and cons of New Haven versus Berkeley. He kept saying, "If Berkeley is so wonderful, how come you are going to Princeton instead of going back there?" And I would reply, "I know I can get by in the Far West, but I would like to test myself in the East. You do not have to do that because you have already made good at Yale." East-West rivalry was much stronger then than later. I think what weighed most heavily in Ernest's mind in persuading him finally to go to

Berkeley was a peculiar old-fashioned rule then in effect at Yale: A Ph.D. candidate had to have his thesis work supervised by a full professor. To be sure, this rule was honored in the breach in that assistant and associate professors directed thesis research in reality, with some full professor merely acting as committee chairman. But this rule irked Ernest, and I worked on his annoyance as much as I could to increase his dissatisfaction on this score. The upshot of it was that he accepted Berkeley's offer, taking a rather large cut in pay. My recollection is that his Yale salary of $3500 dropped to $2800 when he went o Berkeley. This was the beginning of Berkeley's great standing in cyclotron physics and experimental nuclear physics.

That first year in Princeton, 1928–1929, was the most productive in my life. For teaching I gave a course in quantum mechanics again, improving the notes of the previous Columbia course, and a junior course in classical mechanics of which the most outstanding student was E. Bright Wilson, then a Priceton senior who the following year went to Cal Tech to work with Pauling and now is Mallincrodt professor of chemistry at Harvard. Philip M. Morse, who had received a doctorate under K.T. Compton, was at Princeton for a first postdoctoral year. He took my course and we worked up the lecture notes into the book *Quantum Mechanics* (Condon and Morse), which was published by McGraw-Hill in the fall of 1929. Another postdoctoral student who was starting to study quantum mechanics was E.C.G. Stueckelberg from Basel, Switzerland. He and Morse worked up a paper which calculated the potential energy curves of the excited states of H_2^+ by standard perturbation theory.

I personally wrote the paper (*Phys. Rev.* **32**, 858 (1928)) which gave a fuller statement of the quantum mechanics of the Franck-Condon principle. In this connection two other papers were written, which appeared together in the *Proc. Nat. Acad. Sci.* of 1928. Winans and Stueckelberg recognized that the continuous emission spectrum of H_2 in the ultraviolet could be interpreted as arising from transitions to the repulsive $1^3\Sigma$ electronic state of H_2 found from the calculations of Heitler and London. H.D. Smyth and I found also that transitions to this state allowed us to untangle some puzzling observations relating to the critical potentials of molecular hydrogen. Taken together the two pieces of work gave the first experimental evidence of the reality of the repulsive $^3\Sigma$ state.

By far the most important piece of work done that year was the development of the barrier leakage picture of alpha-particle radioactivity, done with R.W. Gurney. (*Nature* **122**, 439 (1928) and *Phys. Rev.* **33**, 127 (1929)). The same idea was developed almost simultaneously by George Gamow, then a postdoctoral fellow in Göttingen (*Z. für Phys.* **51**, 204 (1928) and **52**, 510 (1928)). This was the first application of quantum mechanics to details of inner structure of atomic nuclei, and at the same times its success gave a big boost to the probability interpretation of the intensity of the Schrödinger wave which was only being reluctantly accepted in some quarters.

Gurney had come to Princeton on a fellowship from the Cavendish laboratory where he had done experimental radioactivity work with Lord Rutherford. The summer of 1928 he spent a good deal of time in the library reading quantum mechanics. Thus he came acoss a short paper by J.R. Oppenheimer *Phys. Rev.* **31**, 66 (1928)) which I believe is the first publication in which the mathematical property of the Schrödinger wave of being able to penetrate a potential barrier had been recognized. Oppenheimer had told me about this result some months earlier in connection with the particular case that if an unexcited hydrogen atom is sitting in a uniform electric field, after a time the electron will leak through the barrier and move off to regions of high potential, leaving a bare H^+ ion behind. Oppenheimer had used this mathematical property to help gain an understanding of the pulling of electrons out of cold metals by an applied electric field, a topic which was being much studied experimentally at that time at the California Institute of Technology by R.A. Millikan and Charles C. Lauritsen, which is where Oppenheimer had gone after receiving his doctorate from Born in Göttingen.

Gurney was just beginning the study of quantum mechanics and did not have much confidence in what he was learning, but he immediately though that if particles can leak through barriers, maybe that is the mechanism by which an alpha particle can remain in a nucleus for millions or perhaps billions of years and then finally leak out. He first broached the idea to H.P. Robertson who, for reasons that I have never understood, did not recognize its truth and tried to discourage Gurney. A few weeks later when I had moved to Princeton, Gurney tried his idea out on me. I immediately saw that it was consistent with the principles of quantum mechanics, and that very first afternoon we had worked out a crude calculation explaining the Geiger-Nuttall relation in a semiquantitative way. In a few days we had sent off the letter to *Nature* that is cited above, and then proceeded to try to refine our calculations for the longer paper soon sent to the *Physical Review*. That fall I gave a short report on this work to the autumn meeting of the National Academy of Sciences held in Schenectady, and later to the Thanksgiving meeting of the American Physical Society held that year at the University of Minnesota in connection with the dedication of their new physics building.

Gurney was an extremely shy person around the laboratory. He always wore tennis sneakers and would walk along the halls rather clinging to the walls so as to be as unobtrusive as possible. I would work in the old Palmer Physical Laboratory at Princeton with my office door open. If I saw Gurney flit by the open door two or three times, I would know that he wanted to communicate with me. So I would go somewhere and hide for fifteen minutes or so and when I returned to my desk I would find a neat note from Gurney asking a question or telling me a result that he had reached. Most of our joint work carried out at that time in this way. This shyness did not extend outside the laboratory. He was a great enthusiast for English country dancing and before long had organized the young faculty wives into a class for the study of this art where he proved to be a rollicking leader, and it was my turn to be shy.

After that year he spent a year at the University of Tokyo, and then a year or two in India, where he had a cousin who was fiscal officer to one of the maharajahs of great wealth. He then returned to England where he was helped along in his career by Sir Nevil Mott, then at the University of Bristol. Because of his shyness he was not a good lecturer and did not have a proper career in academic physics in spite of his overall brilliance. He did not continue the study of nuclear physics, but switched his interests to physical chemistry, devoting himself mainly to the application of quantum mechanics to phenomena in electrochemistry, the kinetics of electrode processes at metal — electrolyte junctions. On this account he never became very well known among American physicists.

To jump ahead, the summer of 1940 found him in Stockholm, studying biophysics and living with his vivacious wife, Natalie. After the fall of France and the beginning of the Nazi shelling of London, British overseas were ordered home and they made the trip by train across Russia and Siberia, by ship to Japan and San Francisco and came to visit at my home in Pittsburgh. He reported to Sir Charles Darwin, then British scientific liaison officer in Washington, and was assigned to work on war problems at the Aberdeen Proving Ground. Toward the end of the war, the Gurneys were assigned to the Metallurgical Laboratory of the University of Chicago.

At the end of the war, Gurney was recruited for the initial staff of Brookhaven National Laboratory when it was being first organized during the short period that Philip M. Morse served as its first director. He was prevented from accepting this post by the operation of mysterious security charges against him. Later he was hounded out of various positions, aspecially one at the University of Maryland. By this time America was at the depths of its postwar orgy of Red-baiting in which such prominent roles were played by men like Richard Nixon and Joseph McCarthy. Although he had in the meantime been given a security clearance by the Air Force, after a careful study of all of the FBI's "raw data," the then President of the University of Maryland summarily fired him along with some six others on his faculty, with a loud public flourish in which he declared that "I don't want anyone on my faculty about whom there has ever been any doubt!" He stayed on in various stages of partial and complete unemployment until he finally died of a stroke in the spring of 1953.

He was perhaps the least political person I have ever known, so that I have always been at a loss to know why the security people were so hard on him. Perhaps they knew that he was a friend of mine! Or maybe I was given trouble because of knowing him! Or perhaps both factors were at work in an ascending spiral. At any rate, it was disgraceful episode of which Americans should be heartily ashamed.

In the spring of 1929, although only 27 years old, I was offered a full professorship at the University of Minnesota to take the position made vacant by the move of John H. Van .Vleck from Minnesota to the University of Wisconsin. After long delibration and uncertainty, I decided to accept it, and so moved to Minneapolis in the fall of 1929.

In the summer of 1930 I was one of four summer lecturers in theoretical physics at the summer institute of the University of Michigan. The others were P.A.M. Dirac who lectured on quantum mechanics from the proof sheets of the first edition of his great book; Leon Brillouin, from the College de France, who lectured on quantum statistics; and E.A. Milne, the British astrophysicist. I gave an elementary course based on the proof sheets of Condon and Morse.

I was in Minneapolis only a short time before I began to regret having left Princeton. There was no comparing the two places in the degree of stimulation around for a young theoretical physicist, and I greatly missed the Princeton stimulation. The only faculty member in physics of active interest to me was John T. Tate, then editor of the *Physical Review*, who had around him a splendid group of young experimentalists, including Walked Bleakney, Philip T. Smith, and Wallace Lozier. This should have indicated to me that something similar could have been done there in theoretical physics, especially since Van Vleck had had good students while he was there. But I was too impatient for this and, as I now realize, unfairly to Minnesota, proceeded to arrange with K.T. Compton for my return to the Princeton faculty the following year. In addition to Tate there were two faculty members in chemistry who were particularly stimulating, George Glockler and D.S. Villars with whom I did some joint work.

The most exciting things that year for me were collaboration with Tate's student, Walker Bleakney, who came with me to Princeton in the fall of 1930. He had built a mass spectrograph which was ideally suited for the exploration of electron collisions with molecules which produced dissociation with ion fragments. This permitted detailed confirmation of the modes of dissociation based on the Heitler-London replusive $1^3\Sigma$ potential energy curve in H_2 that had been the subject of the previous year's work already mentioned.

The other exciting thing was the appearance of Slater's memorable paper on "The Theory of Complex Spectra" (*Phys. Rev.* **34**, 1293 (1929)). I had studied very little about atomic spectra prior to its appearance, but plunged at once into the study of Slater's paper, being greatly helped by Julian Mack, who had come to the University of Minnesota as a National Research Fellow but left during the spring of 1930 to go to Sweden to work with Bengt Edlen who was just starting his great program of systematic work on ultraviolet atomic spectra. Prior to Slater's paper the subject was mostly handled by the vector model of correspondence principle treatment of coupled angular momentum vectors, a method which always repelled me, so that it was a great relief to learn Slater's methods when they came along.

During the year at Minneapolis, I became acquainted with George Shortley who started to study quantum mechanics with me during his senior year as a student in electrical engineering and who decided to come to Princeton with me in the fall of 1930 to go into graduate work in theoretical physics.

The summer of 1930 I was the visiting professor for theoretical physics at Stanford University, giving a graduate course in quantum mechanics which I mostly bent toward an exposition of Slater's work on complex spectra, and an undergraduate course in modern physics. In both of these I came in contact with excellent students course in modern physics. In both of these I came in contact with excellent students who continued to be close friends for long after. In the graduate course was W.W. Hansen and Russell Varian, the men who played such a decisive role in the development of klystoon tubes for microwave amplification and generation. In the more elementary course my best student was Frederick Seitz, now president of Rockefeller University and then between his freshman and sophomore years. He took several years to finish at Stanford and then came to Princeton to work with me.

I see by the clock that I am giving much too much detail for the time available. The years up to 1934 were spent with Shortley writing *The Theory of Atomic Spectra* which was published by Cambridge University Press in 1935 and is still in demand. I helped Seitz get started as a specialist in physics of the solid state and even supervised his Ph.D. thesis work on matrix representations of the 230 crystallographic groups, although his best known work done as a graduate student was that done jointly with E.P. Wigner on a quantum mechanical calculation of the cohesive energy of metallic sodium. I had just one other Ph.D. student at Princeton. This was Edwin M. McMillan who has recently retired as director of the Lawrence Radiation Laboratory at Berkeley. He did an experimental thesis in which he measured electric dipole moments of molecules by the electric analogue of the Stern-Gerlach experiment, which was never published, despite my many years of nagging him (including this time!) to do so.

During the time I was concentrating on atomic spectra, the interest of physicists in general was turning more and more to nuclear physics. The year 1932 was decisive for this, being the year that Harold Urey discovered deuterium, Cockcroft and Walton produced nuclear reactions with artificially accelerated particles, and Carl Anderson discovered the positron in cosmic rays. It was a period in which cyclotrons and Van de Graaf generators were beginning to sprout in physics laboratories around the world, and theoreticians were turning their attention more and more to nuclear forces and to nuclear models. In fact so great was this trend that they tended to regard atomic spectra as a closed and finished subject, and the book with Shortly sold very poorly at first.

So in 1935 with the publication of the book on atomic spectra I began to turn my attention again to chemical physics and also to nuclear physics. The former involved some early dabbling with infrared spectra of molecules and some more thorough work on understanding the molecular basis of optical rotatory power (*Revs. Mod. Phys.* **9**, 432 (1937)), and with W. Altar and Henry Eyring (*J. Chem. Phys.* **5**, 753 (1937)).

The year 1936–1937 was largely spent working on nuclear physics with Gregory Breit. We made a calculation of the cross section for photodecomposition of the deuteron by gamma rays (*Phys. Rev.* **49**, 409 (1936)), correction, **51**, 56 (1937)). But much more important was the work done jointly with Breit and R.D. Present on the theoretical interpretation of the experimental results obtained by Tuve, Hafstad, and Heydenberg at the Carnegie Institution of Washington on the scattering of protons by protons at energies up to about one million volts *Phys. Rev.* **50**, 846 (1936)). These results showed quite clearly the charge independence of the strong nuclear force between nucleons on which all modern nuclear theory is based.

In the summer of 1937 I was offered the position of associate director of the Westinghouse Research Laboratories in East Pittsburgh. I accepted it and moved the family to Pittsburgh. The Westinghouse Electric Corporation had failed to keep abreast of developments in modern physics during the depression years, although they had strong research groups in applied mechanics which had been developed earlier by Stephen Timoshenko and in industrial applications of high-power gas discharges, such as switchgear, ignitron rectifiers, and the like, which had been developed by J. Slepian. When the management decided in early 1937 to strengthen nuclear physics they started the construction of a large pressure-type Van de Graaf machine designed for them by William H. Wells and approached Lee Du Bridge, then chairman of physics at the University of Rochester, to build up a more general group in modern nuclear physics. After some deliberation and improvement of his support at Rochester, De Bridge refused the offer and when asked by Westinghouse management, recommended me to them. I remained in this position until the fall of 1945 when I was appointed Director of the National Bureau of Standards by President Truman.

But only the first three years of this period involved working peace-time conditions. In that period the nuclear physics program was pushed ahead, John Hipple launched a mass spectrograph program, Sydney Siegel started experimental work on order-disorder transitions in Cu Au alloys, and some work on microwaves with the new klystron tubes from Stanford was launched. With the fall of France and the launching of aerial attack on

Britain by the Nazis, it became clear that physics as usual could not go on. In the late summer of 1940 President Roosevelt established the National Defence Research Committee under Vannevar Bush which began to make plans for mobilizing university and industrial research laboratories for military research work. In my case this involved turning over a large part of the effort of my group at Westinghouse to work on microwave radar in close collaboration with the Radiation Laboratory at the Massachusetts Institute of Technology. Later, in the summer of 1941, I was appointed to the secret S-1 committee which was planning the program on military applications of uranium fission and so took part in planning the program which later evolved into the Manhattan District of the Army Engineer Corps, commanded by General Leslie Groves.

The four years from the summer of 1941 to the summer of 1945 involved a split of activities between that concerned with furthering the microwave radar program of Westinghouse and that of helping with the Manhanttan District project. In this latter context I spent most of my time at the Radiation Laboratory at Berkeley at the request of E.O. Lawrence and with the support of the Westinghouse management. That laboratory was engaged in the development of large-scale mass spectrographs of the Dempster type for separation of U^{235} from U^{238}. These were built at oak Ridge and produced the U^{235} used in the bomb which destroyed Hiroshima in August, 1945.

In September 1945, immediately after the end of the war with Japan, a great many of the physicists who had worked on various phases of the uranium project, became alarmed at the prospect that control over this new field of great economic and military importance would be retained by the Department of Defense. They began to organize various groups of which the Federation of American Scientists was the most effective, and to call on Congressmen and Senators to alert them to the issues involved which were almost completely unknown and not understood because of having been kept secret so long. The groups at Los Alamos and at the University of Chicago were particularly active in this direction.

For myself, I joined forces with Leo Szilard and the two of us spent most of September seeing members of Congrss and officials of the Truman cabinet and subcabinet to discuss the problem of the future of atomic energy with them.

The political atmosphere in Washington was then much different from what it has been during most of the postwar period. We still felt and behaved like allies of Russia. For instances, there was a flourishing organization called the National Council of American-Soviet Friendship which staged a large rally in Madison Square Garden in New York at which Gen. Dwight Eisenhower was the principal speaker! In the summer of 1945 the international conferences which led to the formation of the United Nations Organization had been held in San Francisco, and there was a general atmosphere of optimism that this would prove to be a new and effective instrument for the promotion of international cooperation. The cold war had not yet been invented.

However, strong political forces were at work to spread mistrust and hatred of Russia on our part. The Russians, seeing this behavior on the part of many influential Americans, responded by showing mistrust of us and our motives. The cold war began, it is fair to say, in the late summer of 1946 when President Truman dropped Henry Wallance from his cabinet on the insistence of his Secretary of State, James F. Byrnes.

I took office as Director of the National Bureau of Standards in early November 1945, a task that was made doubly difficult by the fact that I had no prior knowledge of the inner structure of the Bureau or of the civil service in general, and also by the fact that I was asked by Senator Brien McMahon of Connecticut to serve as scientific adviser to the Special Senate Committee on Atomic Energy, of which he was chairman, and which spent all of the year 1945–1946 drafting the McMahon bill, establishing the civilian Atomic Energy Commission. This bill passed the Senate in the early summer of 1946 and was sponsored in the House by Rep. Helen Gahagan Douglas where it was passed later in the summer. I had no direct experience with this part of the procedure because I was in Bikini observing the naval atomic bomb tests known as Operation Crossroads, serving as a member of the President's Evaluation Commission.

The Republicans were given a majority in the 80th Congress in the election of November 1946, and from then on the cold war gathered momentum. They had hit upon the technique of smearing as many of Truman's appointees as possibly being pro-Communists or soft on Communism, and therefore by association such charges were made increasingly often against the President himself. This election brought to Congress an unknown southern California lawyer named Richard Nixon, who proved to be particularly adept at the use of this technique, having used it to win over the incumbent Jerry Voorhees in the California district he represented. He was promptly assigned to membership on the House Committee on Un-American activities where he perfected the technique even more and used it to win over Helen Douglas in the race for a seat in the Senate from California in the election of 1950. In the period 1952–1960 he used it intensively during his two terms as vice president under President Eisenhower, despite the fact that there is much evidence to indicate that Eisenhower disapproved of the worst of these tactics, especially in the exaggerated form in which they were used by Senator Joseph McCarthy finally to attack Eisenhower himself.

In the summer of 1951 I was offered and accepted the position of Director of Research and Development of Corning Glass Works, Corning, N.Y. This occurred toward the end of President Truman's 1948–1952 term, and I knew that I would not last long in Washington if the Republicans won in 1952. More than that, the salary was considerably more than the $15,000 a year then paid to director of NBS, but best of all, the organization was small enough that the director of research could really give thought to the scientific content of the problems before him rather than being continually absorbed in the ins-and-outs of bureaucratic politics.

The Corning organization proved to be the most satisfying, scientifically and humanly, of any with which I have ever had the good fortune to be associated. But these satisfactions were not to be allowed to last. The Republications were still bent on smearing the Truman record by pretending to a concern over the loyalty of his appointees. Eisenhower and Nixon campaigned in 1952 that they would strengthen the personnel security investigation procedures and clean the "reds" out of Washington and elsewhere. They kept their campaign promise to the extent that the procedures were revised and a number of persons were subjected to long and tiresome hearings. In the spring of 1954 one of these was J. Robert Oppenheimer, who was finally deprived of his security clearance for work with the Atomic Energy Commission. Another was myself where the outcome was favorable to me as it had been in three previous loyalty hearings. However, in October 1954 this favorable verdict was arbitraily suspended by a Secretary of the Navy who announced that it required further consideration. This was an obvious political move on the eye of the election which was backed up in public speeches by Vice President Nixon.

I had been under intermittent harassment in this way since 1947 and decided that I would subject myself to it no longer. So I arranged to become a consultant to Corning Glass Works, a position which I still hold. In the spring of 1955 I was offered professorships by the faculties of two major universities, but in both cases the trustees refused to confirm the appointments under pressure from Washington. Finally I was allowed to become chairman of the physics department at Washington University in St. Louis, and later to come to Boulder as a professor and fellow of the Joint Institute for Laboratory Astrophysics. As the cold war slowly died down, the Department of Defense finally granted me the security clearance which had been improperly suspended in 1954 but this, I am proud to say, I have never used.

All of these details are only germane to our subject in that they give some slight indication of the harassments to which many American scientists have been subjected during this strange period of American history which involves not merely personal unpleasantness but a great slowing down in one's ability to continue scientific work.

During the years at Boulder I have resumed the study of atomic spectra in close association with my student H. Odabasi and several papers have been published on self-consistent field calculations. We also have undertaken to write a book to replace the old one on atomic spectra with George Shortley. This is about half completed, but I have been so preoccupied with personal troubles that I sometimes doubt whether this will ever be completed, despite the fact that Oktay Sinanoglu has kindly agreed to help with the remaining part of the work.

I want to thank you again for the honor you have conferred on me in this conference.

Tunneling — How It All Started

E.U. CONDON[a]

Joint Institute for Laboratory Astrophysics
Department of Physics and Astrophysics
University of Colorado
Boulder, Colorado

A personal account of the early history of tunneling and the contribution of Ronald W. Gurney.

In the talks during the day you have heard of the origin of the quantum mechanical idea of tunneling and up-to-date accounts of some of the many and various important applications which it has in atomic and nuclear and solid state physics. I will try to give you a personal anecdotal account of what I know of the early history and in particularly try to give you an appreciation of the contributions of my distinguished colleague, Dr. Ronald W. Gurney.

I first became acquainted with him in Princeton in the summer of 1928 when I first went to Princeton as an assistant professor. Gurney was there from the Cavendish on some kind of post-doctoral arrangement and was just beginning to study quantum mechanics, having been mostly in experimental work on radioactivity under Lord Rutherford (not a Lord yet). I had just moved to Princeton from my first academic post as lecturer in physics at Columbia in the spring of 1928.

Two papers had just appeared that dealt with the leakage of particles through potential barriers. These were by Robert Oppenheimer and R.H. Fowler[1] and Lothar Nordheim.[2] Both had apparently independently discovered this peculiar feature of quantum mechanics of one-dimensional motion, and both had applied it to the question of pulling of electrons out of cold metals with intense electric fields, a subject that was being very much studied experimentally at that time by R.A. Millikan and C.C. Lauritsen.

Gurney had noticed these two papers in the library of the Palmer Physical Laboratory and had had the idea that maybe barrier leakage could account for the emission of alpha particles through the Coulomb barrier

---

[a] Born Mar. 2, 1902; died Mar. 26, 1974.
[1] R. Oppenheimer and R.H. Fowler, Proc. Natl. Acad. Sci. **14**, 363 (1928).
[2] L. Nordheim, Proc. R. Soc. A **119**, 173 (1928).

surrounding the nuclei of naturally radioactive elements. He had first gone to H.P. Robertson with the idea, because Robertson had already been there a year before I came and Robertson had discouraged him after a superficial examination of the matter.

Gurney was an extremely shy individual, but a persistent one, not easily put off with discouraging opinions that were not clear to him. Several weeks went by before he happened to meet me and he came one afternoon to my office and broached the same idea to me. It was at once clear that the idea had possibilities. One could estimate the barrier leakage factor by using a crude Brillouin-Wentzel-Kramers integration of the wave function across an assumed barrier which was a Coulomb potential on the outside. It did not must matter what one used for the shape of the inner part, so we just cut it off at an assumed nuclear radius that seemed reasonable.

We did not have very clear ideas on the leakage of a wave packet. We knew that this was what the problem called for, but we did not know how to do it. We contented ourselves with a steady-state solution based on the properties of the radial function for an energy that gave maximum amplitude inside and minimum amplitude outside. From this we got an early crude formula for the decay constant, and made some rough evaluations from the known Coulomb potential and the known alpha particle emission energies. Within a few days we had written a letter to the Editor of *Nature* dated July 30, which appeared in the September 22 issue.[3] This letter set forth the idea in rather qualitative terms only, and we at once set about to work out more quantitative details. The concluding sentence of that letter was one which appealed to both of us. I read:

"Much has been written of the explosive violence with which the α-particle is hurled from its place in the nucleus. But from the process pictured above, one would rather say that the α-particle slips away almost unnoticed."

I have mentioned Gurney's shyness. He was a young bachelor then who always wore tennis sneakers and he would slink along the wall of the halls in Palmer lab instead of walking down the middle as most persons do. When we were working together I would see him flit swiftly by the open door of my office without seeming to look in. Thus I would know that he wanted to communicate with me, so I would go and hide somewhere for five or ten minutes and then when I came back, I would find a note from him on my desk, in which he had told me what he wanted to tell, or asked what he wanted to ask. Gradually of course we became acquainted and after about a month communication between us took a more normal turn.

He was a great lover of music and an accomplished piano player. Oddly enough in spite of his shyness in the laboratory, he was an enthusiast for English country dancing and soon organized a faculty dancing class in which he was the chief instructor in this art of the young faculty couples in

---

[3] R.W. Gurney and E.U. Condon, Nature **122**, 439 (1928).

Princeton at that time. At this business I was shy as he was in the laboratory, seldom going to such parties.

Although I still have some of the original notes involved in the preparation of our paper, they are incomplete and disorganized. I remember that we worked hard on it through the fall. That year the fall meeting of the National Academy of Sciences was held in Schenectady in October and I gave a paper on the new theory there. Irving Langmuir acted as chairman of the session, and he discussed it quite fully. However he upset me considerably by asking if we could explain the continuous β-ray spectrum of such cases. I was embarrassed because that was the first that I had heard of the continuous β spectrum which later loomed so large in neutrino theories and weak interaction and all that. But the α-particle part was well received by the physicists present. It was my first introduction to the strong group then at the General Electric laboratory in Schenectady, which included Coolidge, Hull, Dushman, in others.

You must realize that even the probability interpretation of the squared wave amplitude of the wave function was a fairly new idea at that time and since there was no knowledge whatever of the structure of nuclei then, the whole thing became a rather sensational application of quantum mechanics to the nucleus and did much to gain acceptance for the general idea that quantum mechanics could be applied to the nucleus. Remember that the neutron was not discovered until four years later so nuclei were made of protons and electrons at that time!

I also went in November to the fall meeting of the American Physical Society and presented a paper there on the same subject giving a little more detail because the calculations had gone further. That meeting was in Mineapolis. The then new physics building of the University of Minnesota was dedicated on this occasion. It was also known that Van Vleck was leaving Minnesota for a professorship at the University of Wisconsin and on this occasion there were preliminary discussions about my being his successor which did lead to my going to Minneapolis in the fall of 1929 for one year, after which I returned to Princeton.

Jack Tate's letter accepting our fuller paper is dated November 28, 1928 and the paper appeared in the February 1929 issue of the *Physical Review*.[4]

In the meantime, of course, Gamow's first appear in the *Zeitschrift für Physik*[5] had appeared almost simultaneously with our letter in *Nature*.[6] George gave numerous talks on the subject in various European and British colloquia, so his contribution to the subject became better known than ours. In those days American physics did not amount to much and the European physicists did not pay much attention to the *Physical Review*.

---

[4] R.W. Gurney and E.U. Condon, Phys. Rev. **33**, 127 (1929).
[5] G. Gamow, Z. Phys. **51**, 204 (1928).
[6] G. Gamow, Nature **122**, 805 (1928).

Gurney's fellowship expired and he left Princeton about early January of 1929. He went to the Institute of Physical and Chemical Research in Tokyo, where he stayed for a time before going to one of the maharajahs in India where he had a cousin who worked as that official's financial manager before later returning to England. I do not have his exact schedule in my records.

I should not have mentioned Robertson's refusal to see the barrier leakage idea, were it not for the fact that I have now to make a similarly derogatory comment about my failure to appreciate an important point in nuclear physics.

Soon after we had mailed off the February 1929 paper in late November, we began to think about the application of barrier penetration ideas to the possibility that nuclear transmutation could occur at lower energies than by classical mechanics because the incoming proton or α particle could leak inward through a barrier instead of needing enough energy to go over the top. In a way we recognized the importance of this idea, but we were both then under the influence of very snobbish ideas that one ought not to make a paper out of anything that was such an obvious corollary to the general idea of barrier leakage. So we did not write a paper on that. Gamow did.[7]

I remember at that time. Gurney also thought of the idea of resonance penetration of the barrier, the idea not merely that the barrier becomes more and more penetrable as the bombarding particle's energy is raised, but that it is penetrated extraordinary easily when the energy is near to one of the quasistable resonance levels at which it could be captured inside the nucleus. This time it was my turn to fail to appreciate the truth as well as the importance of this idea of Gurney's. I failed to do the rough calculations on a rough model of a light nucleus. Thus I failed to realize that down in this part of the system of nuclei the levels are sufficiently broad to be observable. I was hung up on the idea of the extreme narrowness of the levels in uranium and radium, and failed to recognize what a range of variation an exponential factor can give when the exponent is greatly changed.

Because of this failure of mine I talked him out of publishing in December 1928, the idea of resonance penetration in artifical disintegration experiments. When he went to Tokyo and was no longer subject to my bad influence, he did write a letter to *Nature*, dated Tokyo, February 20 (1929) which was published in the issue of April 13, 1929.[8] I had not seen it for years and just yesterday went to our library and made a copy to see whether he had stated the idea as clearly as I believed from memory that he had.

---

[7] G. Gamow, Z. Physik **52**, 510 (1928–29); **53** 601 (1929).

[8] R.W. Gurney, Nature **123**, 565 (1929).

He had. After recalling the general idea of barrier leakage, he writes (and this is three years before the famous Cockroft-Walton experiments on artificial disintegration and before any of the enormous body of later experimental work was done):

"The object of the present note," writes Gurney, "is to direct attention to the possibility of resonance phenomena if we take into account the solutions of the Schrödinger equation which for certain ranges of energy give psi-functions the amplitude of which inside the nucleus is large compared to the outside. For this seems to indicate that variation of the velocity of the incident particle may be accompanied by an enormous fluctuation in the probability of penetration when the energy approaches and enters the range of energy corresponding to one of the possible quasidiscrete levels. A systematic examination of thin films of various elements might disclose such a fluctuation, if the experimental difficulties can be overcome." He continues to discuss the details of the experimental suggestion.

So far as I know this paper has always been overlooked and he has not been credited in the literature with having made this suggestion. Future histories of the subject should correct this error. Although I had remembered that he wrote such a letter, I had not remembered that he had stated the idea of resonance levels so clearly nor at such an early date.

After returning to England, Gurney became interested in electrochemical topics from a quantum mechanical viewpoint and wrote two books as well as many papers on the subject. Then he went to Bristol and began a collaboration with Nevill Mott which clarified many things about ionic crystals when a photographic film is exposed to light. This work led to his publication jointly with Mott of the book entitled "Electronic Processes in Ionic Crystals."[9]

When World War II broke out with the overrunning of Poland by the German armies, Gurney was studying biophysics in Stockholm. He had married several years before. His wife, Natalie, was not at all shy. she is a lovely, active extroverted person who loved to give big parties, and herself a graduate student in political science specializing in Southeast Asia. The older persons present will remember that the winter 1939–40 was called the season of the "phony war" because there was so little action after Poland was conquered, there being a complete lull in military action until June 1940 when the Nazi armies enacted for the British the tragedy of Dunkirk, and followed by a complete conquest of France. Although the North Sea was mined in that phony war winter, Natalie took a ship from Stockholm to London to visit friends and relatives for Christmas 1939 and returned to Stockholm the same way after a holiday.

---

[9] N.F. Mott and R.W. Gurney, *Electronic Processes in Ionic Crystals* (Oxford, New York, 1940).

But by the summer of 1941 the Nazis had overrun Denmark and Norway, invaded Russia and started the blitzkrieg bombardment of England. People then expected that momentarily Britain would be overrun and conquered. Hilter's armies seemed invincible.

The British government ordered its nationals to return home, or at least in this case to get out of Sweden. But by this time, the only way to get out of Sweden was by ship to Leningrad and then by railroad train across Russia and Siberia, by ship to Japan, and on to Ameria by ship from Japan to San Francisco, which is what Ronald and Natalie did. At that time I was associate director of research at Westinghouse in Pittsburgh deeply immersed in getting our microwave radar program going, making almost weekly trips to MIT to coordinate our effort with the Radiation Laboratory which was directed by Lee DuBridge for the NDRC, later the OSRD.

The Gurneys came to Pittsburgh and stayed at our house for a time. Then they went on to Washington and reported for duty to Sir Charles Darwin who was then Britain's scientific liaison officer there. He communicated with London and it was decided that he should go into military physics work at the Aberdeen Proving Ground in Maryland instead of returning to England. It was here that our colleague, Professor Richard N. Thomas, got to know Gurney well for he spent a large part of the war years on ballistics research there, having newly graduated with a BA from Harvard. In my files I have materials attesting to the importance of Gurney's contributions to the war effort at Aberdeen, but I will not go into detail on that as no tunneling was involved.

Toward the end of the war, I do not know just when, Gurney was assigned to the staff of the Metallurgical Laboratory, a part of the Manhattan District at the University of Chicago. In accordance with the well-known principles of compartmentalization of knowledge, I do not know what he did there, but he worked effectively there until the end of the war and longer.

Then in the summer of 1946 when Brookhaven National Laboratory was being organized on Long Island with Professor P.M. Morse as its first director, Morse sought to have Gurney come to his staff, and such a position was accepted. He resigned from Argonne, gave up his apartment in Chicago and was honored at farewell parties.

Now I come to the sad part of my story, one more chapter in the horrible series of national disgraces which involved many American and foreign scientists in security troubles. Morse found that he could not get Gurney cleared for Brookhaven, so that job fell through, but as proof of the total irrationality of the whole situation, he was still cleared for Argonne and could pick up again his old job which he did for a time.

Then he got a two-year appointment at the Johns Hopkins Laboratory. During this time he had his first stroke which I am sure was aggravated by the nervous worry associated with his Kafka-like situation in which security doubts continued to hang over his head without there being any way at that time they could be brought to a head.

While at Hopkins, Natalie began to do graduate work with Owen Lattimore, the distinguished American scholar of Mongolian politics who was himself under investigation about that time — see his book, "Ordeal by Slander."[10] While in Washington I became well acquainted with Lattimore as I have tried to do with most of the persons that have been attacked in the wave of hysteria which began to develop in 1946 and did not begin to abate until eight years later. I remember how Lattimore told me of a seminar course he gave one of these summers at Hopkins in Mongolian politics which met under the shade trees in the backyard of his home in Baltimore. That is such a narrow academic specialty that all the people in it are closely acquainted, so the three or four students were somewhat mystified at one student who registered for the course, who attended regularly but quite clearly did not have the native intelligence or the prerequisites for graduate work in Asian political science. The others took him out one night and got him drunk and he admitted to them that he was the FBI man assigned to keep an eye on Lattimore.

These anecdotes are not funny because they cause the victim a great deal of worry and expense and often loss of employment. Lattimore was of the Walter Hines Page school of international affairs which Hopkins abolished after Lattimore became notorious under circumstances which strongly suggested that these two things were interrelated. Later Lattimore went to England and has a distinguished professorship there, I think in the University of Manchester. America did many things like that to weaken her academic assets at that time.

After Hopkins, the Gurneys went to the University of Maryland. By this time, Ronald's health was quite poor. He had extremely high blood pressure after his first stroke and had to conserve his health more carefully. During this period, he managed to get some formal charges out of the Air Force on the basis of which he was given a security clearance hearing. His health was so poor that the hearing board allowed Natalie to do most of the testifying for him. The result was a complete clearance for access to classified military work such as he had had at Aberdeen and at Argonne during the war, and the lack of which barred him from going to Brookhaven at a time when it was being boasted that that laboratory was not engaged in secret work. The fact that it really was, being itself kept secret.

At the end of the year or two in Maryland, there happened one of the most awful incidents on the whole hysterical period. The then president of the University of Maryland, a former football coach, abruptly and with a loud public fanfare fired six members of the faculty, including Ronald. All of these faculty members had had loyalty hearings of one kind or another

---

[10] O. Lattimore, *Ordeal by Slander* (Little, Brown, Boston, 1950).

and all had been cleared (the cases were independent and unrelated to each other). But the Maryland president loudly told the press, patriotic fellow that he was, that "I don't want anyone on my faculty about whom there has ever been the slightest doubt!" So these cleared people found themselves out of work at a time when it was impossible to find employment anywhere after having been stigmatized in this way.

The Gurneys took an apartment on Riverside Drive in New York and tried to make a life, living near Columbia University, with him eking out a small income by doing some industrial consulting work. His health continued to decline. Old friends on the Columbia faculty were so frightened by the mood of the times that the Gurneys hardly ever saw any of them. The crowning blow came when Columbia denied Ronald the use of the physics department library for some months until I vehemently intervened. Finally during the night of April 14, 1953, he had another and final stroke and died in his wife's arms.

She called the family physician who made the necessary arrangements that night. Early that morning, however, the doctor was interviewed by two FBI men who wanted to know whether Gurney had talked before he died. He had not.

I have never been able to understand what these crazy fellows suspected Gurney of doing. I feel certain that he was not engaged in any political work or any espionage work, and the same goes for Natalie, but it must be admitted that in the mood of those times a graduate student of Owen Lattimore's was a pretty suspicious person.

I feel very bad about the fact that there was little or nothing that I could do for Natalie after Ronald's death. You see my own clearance had been suspended by the Navy security officers at Buffalo and I was marking time before being given a new hearing under the new and improved procedures that were being worked out by Eisenhower's staff. he had campaigned against Adlai Stevenson on the promise that he would not be soft on communism as he and his vice-presidential candidate repeatedly said or at least inferred that Truman had been. Even so I could have offered Natalie more personal consolation in her sorrow than I did. I was not afraid but I was too terribly distracted at the time to recognize even the simple duties of civilized friendship. Natalie tried in various fruitless ways to clear Ronald's name after his death and was told by rude and heartless officials that there was no procedure for clearing dead men since they had no "need to know."

After a while she gave up America and returned to London where she has made her home ever since these horrible happenings. She was raised in the tradition of civilized British liberalism and I do not believe she ever realized the depths of degradation and fear into which American academic people had sunk in that period.

In the meantime she has remarried. Her husband, Peter Taylor is a psychiatrist attached to one of the major London hospitals. My wife and

I keep up a correspondence with them and have visited them in London. Gradually the wounds heal a little. But America still has some of her best scientists living abroad as scientists-in-exile as reminders of that shameful period. I think particularly of Dave Bohm and Bernard Peters who are professors in London and Copenhagen, respectively, who are not allowed to come to this country to see their aging parents.

Perhaps it is not proper of me to present such a serious topic as an after-dinner talk at what has otherwise been such a pleasant occasion. But with the passage of time, I see the name of Ronald Gurney being more and more forgotten. I have even met young people who always refer to Condon and Gurney, rather than the other way round, and suppose that Gurney was merely some transient student of Condon's who did not turn out well, when as a matter of fact it was he and not me who thought of barrier leakage as the model to explain α radioactivity.

Finally I tell you all of these dismal things not to wallow in despair, but because I still hope, sixteen long years after the peak of the fever was reached in 1954, that the American government may not only cease such persecutions in the future, which it pretty well seems to have done, but may even go a little farther and make amends to those which it so wrongly persecuted for so many years.

Thank you for your careful attention. Possibly governments are not human enough to be able to admit that they ever make errors. But some of those who hold prominent positions in the present government had earlier had prominent roles on the shameful events of the years 1046–54, and maybe they could, as human individuals, do a little something on behalf of the government of which they are now a part.

Silver Anniversary of Atomic Clocks

Edward U. Condon

That's one of those titles that was assigned to me, and I really don't know what to make of it, because it's so obvious that part of the Bureau's main responsibility is to do the best job of measuring all physical quantities of any importance whatever, so that when an idea comes along to improve the measurement of anything as basic as time, that's all the employee really needs in a fundamental sense. And I don't think at that time we thought very much about all of these applications that have been described that came later and grew out of this, so that Harold Lyons will go more into the details of getting the project started and doing it, because I think my main claim to fame in this is that I didn't impede his work too much, which is something that every director needs to learn. So, in lieu of talkig that way very much, I thought I'd tell an antecdote or two that might amuse you out of those days.

The people that work out in the labs of an agency like the Bureau don't realize some of the things that the Bureau Chief is up against as Director, especially in dealing with Congress. Congress is probably one of the most backward institutions in America. For example, in the area of time we all take it for granted that about the setting of clocks just uniformly with hour time zone differences, Mountain Central and Pacific and I think there's an Eastern time. Before that was done, every little town had to set its clock by the sun, by the church clock tower, whcih was regulated by the sun by some local father who set it. And that was awkward for the railroads when the railroads came in. Earlier for carriages and wagons and things it didn't matter much because they took so long to get there. The point I'm trying to make was that the railroad tried for decades, several decades, in the early 19th century to get Congress to establish standard time, which they themselves had invented, and Congress was just as vigorous and active about that as it has been in encouraging the metric system a century later — more than a century later. And so finally the railroads just plain sat down and adapted this by convention, and of course the country went along with it. I think later Congress legalized it, but only later.

If you would think back to the year 1949, there were a lot of things that happened that year. I've just refreshed my mind by pumping Harold Lyons and probably stealing some of the things that he was going to say. The picture that's been focused on here, the clock itself, was taken in January of '49. That meant that this job that he had worked on for a year or two had been completed, and the clock was running at that time, but let me recall some other things about '49. '49 was the year that I first came out here and fell in love with the place, and we decided to put the Bureau out here. And then, but '49 in September was the year when the Russians surprised all of the ignorant fellows in the Defense Department, like General Groves, who thought that the Russians would take centuries to get an Atom Bomb. Instead they had an Atom Bomb within four years after we had an Atom Bomb. So it became very fashionable to assume that, since they're obviously yokels, that that must have been entirely by espionage. I think one of the great untold stories of history is how much was accomplished by espionage. There is no doubt but what they reside on it to some extent, somewhat as all countries do, but whether they could have made it pretty much on their own or not I don't know.

Looking back through my files I found I've always been a collector of stories about appropriations hearings. I am going to tell one that relates to the computation business, because the automatic computers were just getting started about that time, and various agencies of the Government were just using things in that field that are as primitive in that field as the amonia clock is that we honor today. I found this gem from the appropriation hearings of the Defense Department on page 1177, and I quote: Senator A, "How many equations have to be developed in the Army?" Colonel X, "At the present moment it is estimated we will have to develop 135. That, however, is strictly an estimate. The Air Force has had a requirement for around 80 because they have a bit more simplified structure." What the hell either side is talking about I don't know. Then, later on, Senator A says, "Now is there any relationship between the number of equations that have to be developed and the time the machine (that's some computer that they're talking about) is in operation." And this same Colonel replies, "Electricity travels a 186,000 miles per second, Sir, so it is a infinitesimal difference." That was also the same when government economy reached a new high and the appropriates committee cut out of the AEC's budget provision for a dog and cat hospital down in Los Alamos. I don't remember quite how much that was.

My favorite story, that relates to the radio business, I'm going to tell, and good old Newburn Smith who is head of the division in those days is with us today. Newburn was an honest radio engineer, so far as I could ever make out. We took these fellows along, various ones, to help with the appropriations hearings, and our old friend John Rooney was conducting the hearings, but there was another Congressman, Daniel Flood, and I checked and found that he's still in the Congress but not on our appropriations committee anymore, so I guess it'll do no harm to tell this story. Anyway, Flood was one of those fellows with a wax mustache, very

"Dapper Dan" type, and fancied himself to be quite a scientist, and his worst trait was you couldn't keep him on the subject you were trying to expound because he had ideas of his own and wanted to get at them. I'll never forget one day when we were having such a hearing, Flood's turn came to ask questions of Newburn, and he said, "Dr. Smith, I understand among the scientists there are two theories. Some think that space is infinite. Some think it's finite. Where do you stand?" Well, Newburn was a good loyal bureau man. Perspiration broke out on his forehead. He looked at me for some kind of a cue as to whether I knew. He was willing to make it infinite or finite, but I had to give him a little, imperceptible gesture to indicate I didn't know. So, finally, Newburn choked two or three times and says, "I think it's infinite." So you'll find the bureau's on record on that. Whereupon Flood, who didn't give a damn in the first place, was just showing off, said "Thank you very much, Dr. Smith, that's all I wanted to know."

So that's what it's like working for the bureau at the director's level. I dont's know whether it's more fun than working in the lab, but both of them have their great advantages. Thank you very much.